INTELLIGENT ADAPTIVE SYSTEMS

An Interaction-Centered Design Perspective

INTELLIGENT ADAPTIVE SYSTEMS

An Interaction-Centered Design Perspective

Ming Hou ◆ Simon Banbury ◆ Catherine Burns

CRC Press
Taylor & Francis Group
Boca Raton London New York

CRC Press is an imprint of the
Taylor & Francis Group, an **informa** business

CRC Press
Taylor & Francis Group
6000 Broken Sound Parkway NW, Suite 300
Boca Raton, FL 33487-2742

First issued in paperback 2017

ISBN-13: 978-1-4665-1724-0 (hbk)
ISBN-13: 978-1-138-74778-4 (pbk)

Library of Congress Cataloging-in-Publication Data

Hou, Ming (Computer scientist)
    Intelligent adaptive systems : an interaction-centered design perspective / authors, Ming Hou, Simon Banbury, Catherine Burns.
        pages cm
    Includes bibliographical references and index.
    ISBN 978-1-4665-1724-0 (hardback)
    1. Adaptive computing systems. 2. Human-machine systems. I. Banbury, Simon. II. Burns, Catherine M., 1968- III. Title.

QA76.9.A3H68 2014
004--dc23   2014039346

Visit the Taylor & Francis Web site at
http://www.taylorandfrancis.com

and the CRC Press Web site at
http://www.crcpress.com

Dedication

Ming dedicates this book to the late Bo Wen and Jing Zhen; Simon dedicates this book to Jennifer; and Catherine dedicates this book to Gary.

Contents

SECTION I   Theoretical Approaches

SECTION II   Analysis and Design of Intelligent Adaptive Systems

SECTION III   Practical Applications

List of Figures

List of Tables

Abbreviations

ACTA	applied cognitive task analysis
ANN	artificial neural network
API	application programming interface
ATC	air traffic control
ATM	automated teller machine
AugCog	augmented cognition
BDI	belief-desire-intention
CAF	Canadian Armed Forces
COGMON	cognition monitor
CommonKADS	common knowledge acquisition and documentation structuring
CONOPS	Concept of Operations
CTA	cognitive task analysis
CWA	cognitive work analysis
DERA	Defence Evaluation and Research Agency (the United Kingdom)
DNDAF	Department of National Defence Architecture Framework (Canada)
DoDAF	Department of Defense Architecture Framework (the United States)
DRDC	Defence Research and Development Canada
ECG	electrocardiogram
EDR	electrodermal response
EEG	electroencephalogram
EFS	electronic flight strips
EID	ecological interface design
EMD	explicit models design
EMG	electromyography
EOG	electrooculogram
ERP	event-related potential
GCS	ground control station
GDTA	goal-directed task analysis
GIFT	generalized intelligent framework for tutoring
GPS	global positioning system
GSR	galvanic skin response
GUI	graphical user interface
HCI	human–computer interaction

HF	human factors
HGA	hierarchical goal analysis
HMI	human–machine interface
HMS	human–machine system
HRV	heart rate variability
HTA	hierarchical task analysis
HV	human view
IAA	intelligent adaptive automation
IAI	intelligent adaptive interface
IAS	intelligent adaptive system
IDEF	integrated computer-aided manufacturing DEFinition
IED	improvised explosive device
IEDD	IED disposal
IPME	integrated performance modeling environment
JAD	joint application design
MAS-CommonKADS	multiagent system extension of the CommonKADS methodology
MFTA	mission function task analysis
MMI	man–machine interaction
NDM	naturalistic decision making
NIRS	near-infrared spectroscopy
OMI	operator–machine interface
OSD	operational sequence diagram
OV	operational view
PA	Pilot's Associate program
PACT	Pilot Authorisation and Control of Tasks
PCS	pre-collision system
PCT	perceptual control theory
RPA	Rotorcraft Pilot's Associate program
SA	situation awareness
SASS	situation assessor
SME	subject matter expert
SOI	statement of operating intent
SV	systems view
TACPLOT	tactical plot
Team CTA	team cognitive task analysis
TIM	tasking interface manager
TV	technical standards view
UAV	uninhabited aerial vehicle
UML	unified modeling language
WCSS	work-centered support system

Preface

When a project authority writes a request for a proposal to scope out systems design activities for achieving envisioned system goals, what systems design approaches should be requested in the statement of work? When systems designers respond to this request, what analytical techniques and design methodologies should they propose to identify the detailed design requirements and associated constraints that will allow them to further the design? When the required information is available for the design, what implementation and evaluation methods should design engineers follow?

These are common, critical questions that many project authorities, systems designers, and design engineers must address in their day-to-day jobs. Answering these questions is not easy, as many different systems design theories, analytical techniques, and design methods are available in various application domains. It is especially difficult when considering the advance of modern technologies, such as robotics, artificial intelligence, cloud-based computing, and social media technologies.

As more and more intelligent and adaptive systems surround us in our daily lives, it is increasingly important that systems designers incorporate knowledge of human capabilities and limitations into new designs. For example, in today's digital age, the amount of accessible information continues to increase dramatically. Intelligent adaptive systems (IASs) offer opportunities to tailor and manage the deluge of information in an interaction-centered way. Responsive and context-sensitive design is the most promising solution to the overwhelming amounts of data we face in our daily working and personal lives.

Through environmental sensing, operator sensing, work process awareness, and data mining, it is becoming increasingly possible to design systems that are both adaptive and, even more importantly, intelligent. Intelligent systems have the ability to keep their human partners safe in dangerous situations, and the best of these systems work fluidly and efficiently with their human partners, allowing operators to experience a more productive and rewarding life. Systems designers need to design IASs so that humans can work with them not only efficiently, but also effectively and safely.

To design a safe and effective IAS, methodologies must be properly defined and meticulously followed. This book walks through the process of designing an IAS on a conceptual level from an interaction-centered perspective, and provides a generic road map that can be used for any IAS project. It is not intended to offer advice on software design, but, rather, to alert designers to methodologies for understanding human work in complex systems, and to offer an understanding about why and how optimizing human–machine interaction (i.e., communication and collaboration between human intelligence and artificial intelligence) should and can be central to the design of adaptive systems for safety-critical domain applications. To this end, the individual chapters discuss the IAS concept, conceptual architecture, analytical techniques, design methods, operator state monitoring approaches, and implementation guidelines, and provide worked examples of real-world IAS projects.

IAS design is about creating systems that sense the operator and the environment so responsively that the machine presents information and interaction in direct response to the evolving situation. This can be difficult for systems designers and design engineers, as there is currently no coherent body of theoretical frameworks and associated design methodologies to guide them in answering critical questions. As a result, this book targets three primary audiences: systems developers, design engineers, and researchers (e.g., senior students). Systems developers will find the past experiences and lessons learned a valuable resource; design engineers will find the design guidelines and case studies useful design tools; and students will find this text a solid resource to help decide future directions within IAS design.

This book is divided into eight chapters, which are loosely organized into an introduction and three additional sections. Section I discusses theoretical approaches for designing IASs. Here, we review the evolution of two essential IAS components: interface and automation. Then, a conceptual architecture is provided to illustrate how an IAS functions as a whole. Section II discusses IAS analytical techniques, agent-based design methodologies, and operator state monitoring approaches. It culminates in a review of the previous chapters as a road map, and offers guidance on adaptation implementation. Section III examines two case studies in detail, and ties together the content of the first two sections to illustrate the use of IAS design methodologies in the real world.

Each chapter emphasizes the resolution of common design issues raised in the introductory chapter, and considerable interlinkage within the chapters has been provided to assist designers in developing a coherent body of knowledge. The most up-to-date information about various design methodologies is provided, including analyses of their strengths and limitations for IAS design. We hope that those involved in interface and automation design from the human factors, human–computer interaction, and education domains can appreciate the importance of our generic design guidance.

Acknowledgments

This book was a long time in the making and was made possible by the support of Defence Research and Development Canada (DRDC). We express our heartfelt thanks to friends and colleagues at DRDC who served as sounding boards, stimulated our thinking, and offered invaluable feedback: Robert Arrabito and Justin Hollands for their involvement in the planning of this project in its early phase; Stewart Harrison for tracking down hundreds of references; and Syed Naqvi for his invaluable assistance in cutting through red tape. Brad Cain, Cali Fidopiastis, and Robert Kobierski provided much-needed feedback on analytical techniques, operator state monitoring approaches, and case study 1, respectively. Serge Pelletier created artistic drawings that helped make this work exceptional. Plinio Morita, Yeti Li, and Leila Rezai, graduate students at the University of Waterloo, read through the book from the student's perspective and provided many useful comments. We also owe special thanks to our dedicated assistants—Collin Lui, Andrew Sun, Paul Hillier, and Stacey Plowright, who worked with us on reference gathering, literature review, drafting, editing, and other vital tasks. Paul and Stacey not only helped improve the organization and clarity of our hard work, but also provided fresh perspectives that resulted in major improvements in several chapters. This book would not have been made a reality without their skills and commitment. Working with our editors LiMing Leong, Cindy Carelli, and Laurie Schlags has also been deeply rewarding.

Ming is grateful for the contributions of several special individuals. His thanks go out to his good friend Paul Milgram, who introduced him to human factors and emphasized the importance of balance in human–machine partnerships, and to his esteemed colleague Keith Hendy, who stimulated the idea for this book through his emphasis on the need to bridge the gap between science and engineering and develop tools for designers. Deepest gratitude to his lovely wife Kaiyan, who encouraged and supported him during the years spent on the project; and special thanks to his beautiful daughter Katelyn, for her enthusiastic help and artistic touch on the book cover design, and to his adorable son Benjamin, for his ability to make Daddy laugh every day, which made writing this book a joyful journey filled with many fond memories.

Authors

Dr. Ming Hou is a senior defence scientist at Defence Research and Development Canada (DRDC)-Toronto, where he is responsible for providing science-based advice to the Canadian Armed Forces about the investment in and application of advanced technologies for human–machine systems requirements. His research interests include applied cognition, intelligent adaptive interface and systems design, human-technology/automation interaction, intelligent tutoring, and stereoscopic virtual and mixed reality displays. Dr. Hou is the Canadian National Leader of the Human Systems Performance Technical Panel for the Air in The Technical Cooperation Program (TTCP). He also serves several NATO working groups. Dr. Hou is a senior member of the Institute of Electrical and Electronics Engineers (IEEE), a member of the Human Factors and Ergonomics Society, and a member of the Association of Computing Machinery.

Dr. Simon Banbury is the owner and president of Looking Glass HF Inc., an independent Canadian-based human factors consultancy specializing in optimizing how people interact with technology. He is also a Professeur Associé of the School of Psychology at Université Laval (Canada), where he supervises PhD students and supports teamwork and medical decision-making research. Dr. Banbury has almost 20 years of human factors consultancy and applied research experience in the defense, industrial, and academic domains; he has worked as an human factors consultant in the defense and industrial sectors, as a lecturer in psychology at Cardiff University, and as a defense scientist for the United Kingdom's Defence Evaluation and Research Agency (DERA).

Dr. Catherine Burns is a systems design engineering professor and the founding director of the Centre for Bioengineering and Biotechnology at the University of Waterloo (Canada). At Waterloo she also directs the Advanced Interface Design Lab. Her research examines user interface design, visualization, and cognitive work analysis, and her work has been applied in military, healthcare, power plant control, and oil and gas refining domains. She regularly consults with companies in the areas of human performance in complex systems, interface design, and traditional human factors engineering.

Dr. Burns has authored over 200 publications, coauthored a book on ecological interface design, and coedited *Applications of Cognitive Work Analysis*. She has been the program chair for the Cognitive Engineering and Decision Making Technical Group of the Human Factors and Ergonomics Society and has been awarded both teaching and research excellence awards at the University of Waterloo.

Introduction

CHAPTER 1: UNDERSTANDING THE HUMAN–MACHINE SYSTEM

Chapter 1 starts by defining the concepts of human–machine system (HMS) and Intelligent Adaptive System (IAS) with examples of passive and active systems to illustrate the range of human–machine interactivity. Instances of good and poor systems design are discussed and the issues frequently encountered by operators interacting with machines are addressed, notably technological issues, human performance issues, and communication issues. We then explain the scope of human–machine interaction and human–automation interaction, compare overlapping automation taxonomies from the human factors (HF) and human–computer interaction (HCI) domains, and introduce the potential that results when technology becomes an active and effective partner for human work. This is in many ways the overall goal of IAS design—to develop an ideal partnership between people and technology by optimizing the interaction between human intelligence and artificial intelligence. When this partnership is optimized, humans work effectively and efficiently and can handle even difficult challenges safely. We also emphasize the importance of applying unambiguous and coherent design methodologies to ensure a safe and effective IAS for appropriate application domains.

CHAPTER 2: AN OVERVIEW OF INTELLIGENT ADAPTIVE SYSTEMS

Chapter 2 illustrates the origins of the interface and automation systems that were integral to the development of IASs. We review the history and evolution of interface technology over the years and provide examples of interface, intelligent interface, adaptive interface, and intelligent adaptive interface. Then we review automation concepts and technologies, and provide examples for static automation, flexible automation, adaptive automation, adaptable automation, and intelligent adaptive automation. The design of these two technologies shifts quickly from a technology-centered approach to a user-centered approach to an interaction-centered approach. Modern IAS design is a direct development from previous work in human–machine interaction. However, modern capabilities in computation, sensing, and advanced information display allow designers to take their systems designs to the next level. There are many different levels of adaptation and opportunities available for interactions, all of which must be considered carefully to match the technology to the intentions of the workspace. The many examples in this chapter clearly show that automated technologies are quickly becoming a part of our daily lives. In the next few years, IASs will likely be very common and the design of such systems should follow an interaction-centered approach to facilitate optimal human–machine partnerships. This chapter clearly identifies the various levels and roles we can expect machine intelligence to play. In this way, we provide guidance on the high-level scoping for IAS design projects.

CHAPTER 3: A CONCEPTUAL ARCHITECTURE FOR INTELLIGENT ADAPTIVE SYSTEMS

Chapter 3 presents the conceptual architecture for IASs that guides the IAS design process. In many ways the architecture is very similar to other software architectures that guide design processes, and in many ways this design process is very similar to other software design processes. The key differences unique to IAS architecture lie in the closed-loop feed-forward feature. IASs have more than just a feedback property, so that well-designed systems can act as active partners to truly assist their human counterparts. To achieve the goal of collaboration and allow automation to understand its human partner's limitations and capacities in a specific situation at a particular moment, a generic IAS conceptual architecture is needed to help develop both the situation assessment module and the operator state assessment module, and to adapt the information sent to the operator–machine interface (OMI). There are several ways to design the conceptual architecture of an IAS, depending on the contextual features of the work being supported. A number of case studies are examined to provide an overview of how IASs have been implemented in the past and to explain their contextual features. The goal is to define and describe the core components of these systems to provide a unified conceptual architecture that, ideally, should underpin the development of all IASs.

CHAPTER 4: ANALYTICAL TECHNIQUES FOR IAS DESIGN

Chapter 4 reviews analytical techniques and design techniques that can be useful in several module design steps. The analytical techniques include mission function task analysis, hierarchical task analysis, hierarchical goal analysis, goal-directed task analysis, cognitive task analysis, and cognitive work analysis. The design techniques include joint application design, the Department of Defense architecture framework, and ecological interface design. These methods are drawn from the field of HF engineering and focus on identifying the components of cognitive work and decision making that must occur. All these techniques seek to better understand the work of the operator, the information required by the operator, and the decisions and actions that must take place to complete tasks and achieve system goals. This information is useful in multiple ways—it can be used to select the functions that are the best candidates for automation, it can inform the OMI design, and it can help develop simulations and test environments to evaluate the IAS. Each approach is discussed in a stand-alone format, making this chapter a useful reference for future work. Although the techniques may differ in their details and origins, they share many of the same goals. Finally, the use of multiple techniques to support an IAS design for the control of Uninhabited Aerial Vehicles (UAVs) is examined to show the benefits of using multiple analytical techniques, as well as the ways an analytical technique can contribute to the design process. When designing a system, choose the techniques that fit best with your contextual design problem and the type of solution that you are seeking to design. Using more than one method can increase the amount of information that you collect and your understanding of the problem domain, and improve your overall design.

CHAPTER 5: AGENT-BASED, INTERACTION-CENTERED IAS DESIGN

Chapter 5 addresses issues relevant to the understanding and interaction between human and automation agents in IAS design, such as organization and teamwork. Agent design, agent-based design methods, and agent architecture are discussed here. The goal is to provide a general principle for optimizing human–agent interaction in IAS design. In essence, agents provide intelligent support for roles and capabilities traditionally performed by human operators. Designed well, operator–agent interactions allow the human operator to function at a higher level as an overall controller of an artificially intelligent team that manages task details. The particular design methods we explore are common knowledge acquisition and documentation structuring, integrated computer-aided manufacturing definition, explicit models design, ecological interface design, belief-desire-intention, and blackboard. These methods target different parts of the IAS design processes, from the design of the situation assessment module to the design of the OMI. A conceptual operator–agent interaction model and an associated interaction-centered hierarchical architecture provide design guidance for IAS and associated OMI and other system components.

CHAPTER 6: OPERATOR STATE MONITORING APPROACHES

Chapter 6 discusses operator state monitoring approaches and offers guidance on determining appropriate options. Monitoring the operator and continually updating agents on operator states is a key driver for optimal human–machine interaction and intelligent adaption in IASs. These monitoring approaches include behavioral-based, psychophysiological-based, contextual-based, and subjective-based monitoring technologies. We introduce electroencephalogram (EEG), near-infrared spectroscopy (NIRS), electrodermal measurements, electrocardiogram (ECG), heart rate variability (HRV), eye tracking, respiration measurements, skin temperature measurements, and electromyography (EMG) technologies with associated benefits and limitations. To use these technologies effectively, designers need to understand what they measure and its impact on performance—key issues discussed in this chapter. As well, some of the risks associated with interpreting this data are discussed. To moderate these risks, many people have taken combination-based approaches, in which they use more than one method combined algorithmically to paint a clearer picture of operator state. By creating the most accurate operator picture possible, the partnership between operator and technology can begin. We also review the current possibilities in biofeedback sensing technology and their fit to various implementation challenges, and discuss the cognition monitor as an example of how sensing technologies were applied in a real-world IAS.

CHAPTER 7: KEY CONSIDERATIONS FOR IAS DESIGN

This book is intended as a resource for systems designers and developers to assist in IAS design from an interaction-centered point of view using coherent systems design guidance and methodologies. Chapter 7 provides guidance, based on our own practical experience, for complying with established guidelines for the design and development of IASs. IASs must be built and used in the real world, where there are often constraints and complications of a diverse nature. We expose you to some of these considerations and present a road map and strategies to make your IAS development process more successful. Designs that are useful, meet real-world application needs, and are developed effectively and within resource constraints are more likely to be acceptable. We then discuss issues related to the ethical, legal, social, and cultural challenges of IASs and present real examples of some of the challenges and perceptions of operators of these systems. Finally, strategies to increase technology acceptance are given to show how designs can be introduced and adopted in the most effective way. The adaptation taxonomy gives an overall view of how to position a project and technology appropriately for success. This chapter also provides guidance on how to trigger or initiate an adaptation. Once the situation assessment module, the operator monitoring system, and the OMI are completed, adaptation must be triggered at the most effective times and in ways that are consistent with a smooth user experience.

CHAPTER 8: CASE STUDIES

Chapter 8 examines the application of IAS design methodologies in the context of interface design for the control of multiple UAVs and the development of an intelligent tutoring system. Detailed descriptions of two IASs provide worked examples to illustrate what, where, when, why, and how analytical techniques, agent-based design methods, operator monitoring approaches, and design frameworks were considered and applied to facilitate operator–agent interactions and improve overall system performance. Although two IASs were designed for different applications, they share commonalities in the application of interaction-centered IAS design methodologies. From conducting stakeholder and mission analyses to identifying operational priorities; from generating scenarios to deciding operational flow and functional requirements; from gathering information requirements to understanding operator intercommunications; from performing cognitive processing analyses to allocating tasks and functions between operators and automation agents; from developing task network models to determining adaptation mechanisms; from designing adaptation modules to developing system software architecture; from beginning the original performance modeling to completing the IAS prototype design, implementation, and evaluation—two entire systems have been described in detail to walk systems designers through the process of applying the IAS design methodologies introduced in this book.

Across all chapters, this book takes a strongly interaction-centered approach to IAS design. This differs from, but should be used in conjunction with, software engineering approaches. Maintaining an interaction-centered approach is critical. Ultimately, every IAS interacts with people. The nature of this interaction between human intelligence and artificial intelligence, whether it is functional and helps human performance or is awkward and degrades human performance, influences the success of the IAS. An ideal IAS supports human work as a responsive, interactive, and tactful partner, and understands its human partner's objectives, work processes, contextual situation, and mental state.

IASs are technologies destined to become more pervasive in the near future, and acceptance of these new technologies is often more challenging than developing them. The underlying message of this book is clear—technology must be designed from an interaction-centered perspective. This means understanding the work that must be performed, planning automation carefully to relieve human tasks, choosing wisely the ways in which automation interacts with and informs the operator, and monitoring the operator so that automation can be employed intelligently at the right time and in the right way. These are all challenges, but not impossibilities. The methods presented in this book provide guidance on how to achieve these objectives. The increased ability to fine-tune technology to meet our needs offers humans and their technological partners a very bright future indeed.

1 Understanding the Human–Machine System*

Good design means not only designing for normal human use, but also designing against misuse, unintended uses, and abuses—Chapanis (1996).

1.1 OBJECTIVES

- Introduce the concepts of human–machine system (HMS) and intelligent adaptive system (IAS)
- Discuss human–machine interaction and why it applies to HMS design
- Discuss the scope of human–machine interaction and the importance of human–automation interaction taxonomies
- Discuss common technological, human performance, and communication issues that result from poor human–machine interaction
- Discuss the need for consistent human–machine systems design methodologies

Society takes for granted the good design that surrounds it. The car you drove to work today had to be ergonomically conceptualized, designed, and tested considering human physical dimensions and psychological characteristics. The traffic signals that control the chaotic morning flow of traffic had to be conceptualized, designed, and tested while considering human reaction to external stimuli and human vehicle control ability.

Good systems design is ubiquitous. It is transparent. And yet, despite its invisibility, systems design is one of the most vital and fundamental processes that affects our daily lives. For design to appear seamless, it must take into account human abilities and limitations. The unspoken acknowledgment of human capabilities is a fundamental objective for the field of ergonomics, also known as human factors (HF).

The modern understanding of ergonomic issues stems from World War II, when new human-centered challenges were being discovered. For example, pilots exposed to higher altitude required supplemental oxygen to breathe in the thinner air (Chapanis, 1996) and increased aircraft speed exposed pilots to g-forces that forced blood down from the brain toward the legs, often causing impaired vision and unconsciousness. Awareness of these issues spawned a need to understand the physical and psychological limits of the human body when airborne. Physiologists, psychologists, and physicians collaborated in the study of pilots at work and explored

---

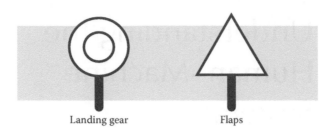

Landing gear   Flaps

FIGURE 1.1   An example of shape-coded aircraft levers. To prevent landing accidents, human factors pioneer Alphonse Chapanis proposed shape coding, such as a circular knob for the landing gear lever and a triangle-shaped knob for the flap control lever.

various options for aviation design. As a result, they published a book in 1946 and used "human factors" in a book title for the first time (Chapanis, 1996).

Alphonse Chapanis was one of the earliest pioneers in the HF field. He began his career in 1943 investigating numerous runway crashes of U.S. Army B-17 Flying Fortress bomber aircraft (Lavietes, 2002). Chapanis discovered a critical flaw in the B-17 cockpit design: the controls used to operate the landing gear and flaps were identical. As a result, pilots often confused the two levers and retracted the landing gear instead of deploying the flaps. Figure 1.1 illustrates Chapanis' proposed solution: a wheel-shaped knob attached to cap the landing gear lever, and a triangle-shaped knob for the flap lever. This well-known cockpit design is referred to as shape coding (Woodson and Conover, 1964). From an HF perspective, shape coding is an example of a passive system. Passive systems accept the operator's manual control without resistance or active response.

As World War II ended, technology rapidly progressed, ushering in the space age and then the computer age, also known as the information age. Raw muscle power was no longer the focus of systems design. The focus became improving technology in the wake of the computer revolution. With today's advanced computing technologies, systems are more complex and less passive, acting as an active partner to operators. This chapter begins by defining the concepts of HMS and IAS, both of which are central to the topics discussed throughout this book. It then presents two events that illustrate success and failure in HMS design. The need for design guidance in understanding the scope and nature of human–machine interaction by comparing and contrasting various human–automation interaction taxonomies is subsequently addressed. Technological, human performance, and communication issues regarding human–automation interaction that should be proactively considered to guide the design of a safe and effective HMS are then examined. Finally, there is a discussion of how the HF and human–computer interaction (HCI) research literatures view HMS design, and why HMS designers need to follow appropriate and coherent design methodologies rooted in HF and HCI, regardless of their target application domains.

1.2   DEFINING HUMAN–MACHINE SYSTEM AND INTELLIGENT ADAPTIVE SYSTEM

There is incredible diversity in both the scope and the capability of technology today. Machines are increasingly able to perform tasks traditionally done by humans, and in certain cases they are even able to perform actions that humans cannot. Systems designers need to be aware of human and machine limitations and capabilities in order to exploit these technologies appropriately. To do this, they need to understand the interactions that occur between human and machine.

1.2.1 THE HUMAN–MACHINE SYSTEM

Formally defined, an HMS is a system in which the functions of an operator (or a group of operators) and a machine are integrated (International Federation of Automatic Control, 2009). An HMS has three basic components: a human, a machine, and an interface. The machine is the device or mechanism that performs, or assists in performing, the task or tasks for the human. It can be almost anything a human operator interacts with, including a vehicle, a smartphone, a robot, or an aircraft. The interface is the means through which the human interacts with the machine. The human and machine integrate, working together to perform tasks as a cohesive entity. A task is an operation or interaction between the human and the machine intended to achieve a goal. Task models are discussed in Chapter 2.

Figure 1.2 illustrates the HMS concept using a pilot and an aircraft. The human operator in this scenario is the pilot, and the machine is the aircraft. The operator is able to use the interface, in this case the cockpit, to communicate with the machine. The HMS is the synergy of human and machine (i.e., the combination of both pilot and aircraft).

With the advance of computing technologies, more functions and capabilities can be built into machines, dramatically altering HMS design. From a systems design perspective, the early B-17 bomber aircraft was designed as a passive receiver, and could only provide basic feedback to its human partner, using technology such as the wheel-shaped knob for the landing gear and the triangle-shaped knob for the flaps. Today's machines are not always passive receivers of human operator control—they can be designed to actively interact with their human partners. Commercial aircraft are now capable of

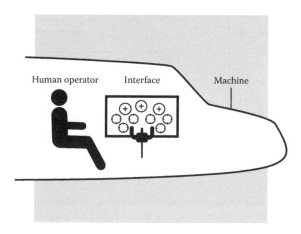

FIGURE 1.2 A human–machine system, as represented by a human operator (i.e., pilot), a machine (i.e., aircraft), and the interface that allows them to communicate.

flying themselves, and computers, software, and associated control algorithms are built into aircraft to update pilots automatically about critical information, such as weather conditions, altitude, current headings, and the status of aircraft components.

When computing and artificial intelligence technologies allow the design of the machine to follow quantitative performance models of human information processing, signal detection, and control, HMSs can be designed to provide humans with both feedback and assistance. For example, an in-car navigation system, or global positioning system (GPS), is an intelligent machine partner. When instructed by a driver, a GPS can act as a navigator to assist in traveling from point A to point B. It visually displays the current position of the vehicle and states the distance remaining before the next turn, helping the driver maintain situation awareness (SA). SA is defined as "the perception of the elements in the environment within a volume of space and time, the comprehension of their meaning, the projection of their status into the near future, and the prediction of how various actions will affect the fulfillment of one's goals" (Endsley, 1999). Good SA allows the driver to anticipate the next set of instructions and make informed decisions, reducing the driver's cognitive workload in busy traffic.

In the GPS example, human interaction with the machine partner is active and broad, and may involve different interface technologies, such as a trackball, buttons, or a touch screen. These machine partners are normally not active enough to provide additional assistance to the operator based on changes in either the human or the environment. More sophisticated navigation systems, however, can change their instructions based on traffic conditions, changes in weather, or driver preference.

Figure 1.3 illustrates passive human–machine interaction and active human–machine interaction in the examples of passive and active HMSs. Passive systems limit human interaction with the machine partner and typically involve conventional interfaces, such as mechanical knobs and dials. The design of passive HMSs follows traditional biomechanical ergonomic approaches. Active HMSs are designed

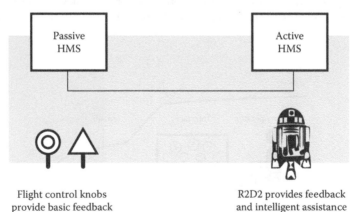

Flight control knobs
provide basic feedback

R2D2 provides feedback
and intelligent assistance

FIGURE 1.3 Passive interaction and active interaction, as illustrated by early interface technology and futuristic machine technology. Passive systems limit human interaction with the machine partner and typically involve conventional interfaces, such as mechanical knobs and dials. Active systems allow machines to offer intelligent and adaptive assistance to their human partners at a variety of levels.

primarily following HF and HCI principles. In an active HMS, the machine component is designed to provide its human partner with intelligent and adaptive assistance, and human interaction with the machine varies in both level and scope. For example, the well-known robot R2D2, one of the main characters in the Star Wars universe, is an active machine partner. Although R2D2 was only able to communicate through beeps and whistles, the machine was able to help its human partners save the galaxy time and time again. Like a human copilot in a fast jet aircraft, R2D2 was able to assist Luke Skywalker during the attack on the dreaded Death Star by repairing damaged systems and alerting Luke to imminent danger. The machine was also able to operate autonomously and take initiative, such as during the rescue of Luke Skywalker, Han Solo, and Princess Leia from the clutches of Jabba the Hutt.

The intelligent and adaptive behavior that R2D2 exhibits makes the interaction between human and machine to achieve a common goal similar to the interaction that occurs between two humans and demonstrates an essential characteristic of active HMSs, also called IASs. Systems that fall between the two extremes of passive and active HMS, such as GPSs, are also considered IASs, have some degree of both passive and active features, and may exhibit some intelligent and adaptive behavior.

1.2.2 THE INTELLIGENT ADAPTIVE SYSTEM

HMSs require human operators because humans are good at reasoning, interpretation, and problem solving, especially for unstructured problems for which rules do not currently exist: for example, radiologists diagnosing diseases, lawyers writing persuasive arguments, or web designers creating new online applications. However, the human mind has well-known limits with regard to information processing. For example, humans have a limited capacity to remember things—around seven items (±two) (Miller, 1956; Wang et al., 2009; Lewis et al., 2010) or even fewer (Wickens et al., 2013). Given limited memory and attention capabilities, humans are easily distracted by background noise and find it hard to resume tasks that have been interrupted. Humans can also have difficulty with detecting subtle changes in their environment due to limited signal detection and vigilance capabilities (Wickens et al., 2013), and this is exacerbated by the abundance of information that human populations are currently subjected to. In today's information age, the volume of information has increased, knowledge spectrums have expanded, and the pace of information distribution has accelerated. Humanity is overwhelmed by a vast amount of information through various media, including newspapers, magazines, web sites, and social media, leaving us vulnerable to information overload. Information overload can compromise attention, making it difficult for human or machine to perceive, interpret, and sort relevant information, and to choose the proper response within the time constraints of the situation.

To mitigate the limitations of the human brain in the age of information technology, more and more technologies, such as automation, are designed and built into the machine component of an HMS to assist the human partner. In the HF domain, automation is defined by Wickens et al. (1997) as "a device or system that accomplishes (partially or fully) a function that was previously carried out (partially or fully) by a human operator."

With the growth of automation technology in HMS design, automation can now not only replace functions of machine mechanical action, but also perform machine data gathering, processing, and decision making, as well as machine information action and communication (Sheridan and Parasuraman, 2006). Automation no longer simply performs tasks for humans—it changes the nature of the tasks that humans perform (Bratman et al., 1988). As a result, it has become increasingly possible for operators to stay on-the-loop (i.e., supervisory control) rather than in-the-loop (i.e., active control) (Chen and Barnes, 2014).

As machines become more automated, they also become more sophisticated. There is a risk of unintended or unexpected human and system performance issues associated with unintelligent automation, such as tunnel vision, degraded situation awareness, and the abuse, misuse, and disuse of automation (Parasuraman and Riley, 1997; Lee, 2008; Chen et al., 2011). The evolution of automation in HMS design and additional human–automation interaction issues are discussed in Chapter 2. There is an obvious need to develop intelligent automation technologies to preserve the pros and mitigate the cons of automation technology, which is the goal of active HMSs and, more specifically, IASs.

As noted earlier, R2D2 from the Star Wars universe demonstrates essential IAS abilities. First, the machine is intelligent enough to assess the situation. It knows exactly what is happening during the attack on the Death Star and what help Luke Skywalker needs, including the reparation of damaged systems and alerts about imminent danger. Second, the machine intelligently adapts to situations and acts autonomously, such as in the rescue of Luke, Han, and Leia from Jabba the Hutt.

This ability to feed forward—to predict what will happen next and autonomously take action to assist their human companions—is a key feature of IASs. It often mitigates human limitations and can play a critical role in keeping humans safe in dangerous situations where machine reaction time is quicker than human reaction time.

A real-world example of an IAS is Toyota's pre-collision system (PCS), implemented to enhance the safety of operators (i.e., drivers) and their passengers. The PCS was the first forward-warning collision system available in the North American automobile market. It uses radar, infrared, stereo and charge-coupled device (CCD) cameras, and other monitoring technologies as part of the vehicle's HMS to protect both driver and passengers in the event of an unavoidable collision. The system is intelligent enough to determine whether or not a collision can be avoided and uses these monitoring technologies to maintain awareness of the vehicle's environment. More recent implementations of the PCS have added a driver monitoring system that can determine where the driver is looking and alert the driver if an obstacle is detected while the driver's gaze is diverted elsewhere. The PCS is able to intelligently adapt to changes in both the environment, such as obstacles, and the operator, such as where the operator is gazing, and take the initiative autonomously. Adaptive measures undertaken by the PCS to protect the driver and passengers include tightening the seat belts, precharging the brakes to provide maximum stopping power once the operator presses the brake pedal, and modifying the suspension, gear ratios, and torque assist to aid evasive steering measures.

As illustrated by the examples of R2D2 and the Toyota PCS, an IAS can be defined as an active HMS that has the ability to intelligently adapt to the changes of the operator, the working environment, and the machine. In an IAS, the machine component not only provides feedback to its human partner but also provides feedforward intelligent assistance based on the needs of the operator in a given situation.

The goal of IASs is to intelligently help operators achieve the overall HMS objectives. Technological advances allow machines to become more intelligent, more sophisticated, and more capable of helping their human partners, but the human brain does not advance at the same rate, which causes a human–technology capability gap. This gap is a large issue for IAS designers—to design an effective IAS, it is critical to understand the nature of human–machine interaction and the implications of poor human–machine interaction design.

1.3 SYSTEMS DESIGN AND HUMAN–MACHINE INTERACTION

Since World War II, there has been steady progress in breaking technological boundaries, such as space travel and high-performance computing. A focus on solving problems related to environmental forces has evolved into a focus on solving information-processing problems. The established principle of designing an active HMS around human capabilities still holds true, but as technology advances more design guidelines are needed. To help predict and prevent future disasters, humanity often recalls important lessons learned from historical events.

1.3.1 THE IMPORTANCE OF GOOD SYSTEMS DESIGN

At 3:26 p.m. on January 15, 2009, US Airways Flight 1549 departed LaGuardia Airport in New York City, en route to Charlotte, North Carolina. A minute into the flight, at approximately 3000 feet, the aircraft lost engine power, and 3 min into the flight it hit the water (McFadden, 2009). The Airbus A320 had struck migratory Canadian geese and damaged its engines (Smithsonian, 2009). The pilot made a quick decision to turn the aircraft south and make an emergency landing in the Hudson River. The aircraft splashed into the frigid water, fully intact when it landed. Figure 1.4 illustrates the A320 after landing on the Hudson River; all 150 passengers and 5 crew members survived. Three days later, the aircraft's black box data was

FIGURE 1.4 Airline passengers wait to be rescued on the wings of US Airways Flight 1549 after landing in the Hudson River on January 15, 2009. (Photo by Day, S., with permission from Associated Press, 2009). Good systems design allowed the pilot to take control of the aircraft in a critical situation.

FIGURE 1.5 Recife—The frigate Constituição arrives at the Port of Recife, transporting wreckage of the Air France Airbus A330 that was involved in an accident on June 1, 2009. (Roberto Maltchik, Photo courtesy of TV Brasil). Bad systems design contributed to the accident in multiple ways, such as in the stopping and starting of a stall warning alarm, which contradicted the actual aircraft status.

made public, confirming that bird strikes had caused the engine failure (Elsworth and Allen, 2009).

On June 1, 2009, Air France Flight 447 from Rio de Janeiro, Brazil to Paris, France crashed into the Atlantic Ocean, killing all 216 passengers and 12 crew members. Figure 1.5 illustrates the plane's vertical stabilizer. Most of the rest of the plane was never recovered, including 74 of its passengers (Vandoorne, 2011). The accident was the deadliest in the history of Air France, and the second and most deadly accident in the history of the Airbus A330. The investigation into the disaster concluded that the aircraft crashed after temporary inconsistencies in airspeed measurements caused the autopilot to disconnect. In particular, the misleading stopping and starting of a stall warning alarm, which contradicted the actual aircraft status, contributed to the crew's difficulty in analyzing the situation (Hosford et al., 2012). As a result, and despite the fact that they were aware that their altitude was declining rapidly, the pilots were unable to determine which instruments to trust.

The Hudson River incident demonstrates the importance of maintaining SA by keeping operators in-the-loop when interacting with airplane cockpit systems. Good design decisions for human–machine interaction within the pilot–aircraft system facilitated the pilot knowing when to take control of the aircraft and allowed him to override the automated functions of the plane. Consequently, 155 lives were saved. If the pilot had not been made aware of the situation, or had not been able to override the machine's decision and had been forced to allow the autopilot to decide where to land, a disaster might have occurred. By being aware of what was happening, the pilot was able to make an informed decision and effectively interact with his machine partner. In this case, the operator was in control of the machine.

Alternatively, the Air France disaster illustrates the consequences of poor design decisions for human–machine interaction within the pilot–aircraft system.

Inconsistent airspeed readings were sent to the autopilot, resulting in its disconnection and the crew's confusion about aircraft speed and altitude. This led to the crew members losing SA about the aircraft status and no longer trusting the machine. Additionally, when the crew made the decision to act, they could not take control of the aircraft. The human–machine interaction design of this system did not allow the crew to intervene, even though they fought with the autopilot to do so. In this case, the machine was in control of its human partners—which resulted in a deadly crash in the Atlantic Ocean.

The markedly different conclusions to the two events lead systems designers to consider how much humans should depend on their machine partners and how much authority machines should be given so that they may complete tasks while operators remain in control. Systems designers must then consider the best ways of designing an IAS to provide operators with the right information to ensure that they maintain SA. Ideally, IASs should be more human like and resemble human behavior to facilitate effective interaction for both the human and machine components of the system. Human–machine interaction should adopt human–human interaction behaviors, including physiological attributes, such as eyes, head, and hands; cognitive characteristics, such as capacity, recognition, learning, and decision making; knowledge bases, such as knowledge of environment, machine, task, and operator; and human psychological states, such as varying levels of attention, vigilance, fatigue, and patience (Hou et al., 2011). Human like attributes related to IAS design are further discussed in Chapter 5.

1.3.2 DETERMINING THE SCOPE OF HUMAN–MACHINE INTERACTION

Design is a key factor in enabling an HMS in which the human is in control of the machine or an HMS in which the machine is in control of the human. As such, systems designers need to understand the scope of how an operator can interact with an intelligent machine partner when designing a particular IAS.

With advances in automation technology, automation can replace more and more system functions that humans once performed, such as information analysis, communication, and decision making. Automation is the key HMS component provided by the machine. Hence, to a certain extent, human–machine interaction in HMS design is human–automation interaction. This is especially true in IAS design. Thus, it is critical to understand how operators interact with automation before starting design activities.

Taxonomies provide good guidance to understand the nature of human–automation interaction, and it is essential to review the best-known automation taxonomies. The precise amount of human–machine interaction or human control over the machine can be described on a scale of levels of automation. Researchers in the HF field use a variety of levels of automation to describe how humans interact with automation in the design of HMSs. For instance, Sheridan and Verplank (1978), Endsley and Kaber (1999), and Parasuraman et al. (2000) have all proposed different taxonomies for the use of automation in HMS design. Each taxonomy addresses similar issues, and divides automation into levels, but considers slightly different automation factors. For example, Sheridan and Verplank's taxonomy and its modified version (Parasuraman et al., 2000) address automation in 10 levels, with no mention of who monitors the automation; Endsley and Kaber's taxonomy discusses who initiates the automation and who monitors the automation. In addition to these three taxonomies, there are domain-specific levels of automation that illustrate the scope of human–automation interaction, notably the Pilot Authorisation and Control of Tasks (PACT) automation taxonomy developed by the United Kingdom's Defence Evaluation and Research Agency (DERA).

1.3.2.1   The Pilot Authorisation and Control of Tasks Taxonomy

It is unrealistic to expect that all pilots will require the same level of automation from an HMS at all times. DERA developed PACT as part of the overall Cognitive Cockpit project (further discussed in Chapters 3 and 6), which was tasked with improving pilot aiding to increase mission effectiveness and safety (Taylor et al., 2001). The PACT taxonomy was developed to support the system requirements of military pilots, where circumstances of a particular mission and pilot preferences dictate the appropriate level of automation (Taylor, 2001a; Edwards, 2004). PACT is based on the notion of contractual autonomy, in which the operator and the system establish an agreement about the system's responsibilities. Figure 1.6 illustrates the contracts that are made using a six-level system from 0 (no automation) to 6 (automatic system control). PACT allows operator intervention even at the highest level, reflecting the Cognitive Cockpit project's overall goal of developing an integrated system using a human-centered approach that keeps the operator in charge (Taylor et al., 2001).

1.3.2.2   Sheridan and Verplank's Levels of Automation Taxonomy

Sheridan and Verplank's taxonomy was originally conceived for tele-operation activities in the HF domain as an alternative form of supervisory control interactions between

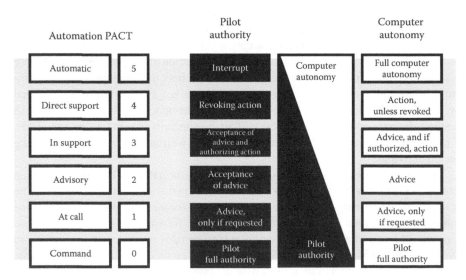

FIGURE 1.6 Pilot Authorisation and Control of Tasks (PACT) automation taxonomy. (Redrawn from Taylor, R.M., *Technologies for Supporting Human Cognitive Control*, Defence Science and Technology Laboratory, Farnborough, UK, 2001b.)

the operator and the machine. Tele-operation is related to controlling machines, such as robots, uninhabited aerial vehicles (UAVs), and other surveillance devices, from a distance. Sheridan and Verplank's taxonomy remained essentially identical for more than two decades due to its extreme simplicity and intuitivism. The taxonomy was once thought of as relatively binary—the authority rests with either the human or the machine. However, this is not applicable to many systems, such as military systems, which are complex enough to require automation working on different tasks and task types concurrently. Additionally, many tasks can be performed by various mixes of both human and automation. Automation levels and automated tasks cannot be averaged.

Another issue with Sheridan and Verplank's taxonomy is that it is ambiguous about the precise application domain of the human–automation relationships. It combines several potential behaviors that automation can perform: making decisions, providing suggestions, and executing actions. However, it focuses very little on the roles or functions that can be performed by human or automation or both and was later modified by Parasuraman et al. (2000). They added another dimension: the function or the task over which the human–automation relationship is defined. The taxonomy has been reapplied to a more general model for human–automation interaction, and they emphasize that different process stages of complex systems are automated appropriately to different degrees.

Parasuraman et al. (2000) implemented a four-stage model of information processing into the original levels of automation taxonomy:

1. *Information acquisition*: Obtaining information from different sources and presenting the information to the operator
2. *Information analysis*: Providing filtering, distribution, or transformation of data; providing confidence estimates and integrity checks; enabling operator requests; and managing how the information is presented to the operator
3. *Decision making*: Providing support to the operator's decision-making processes, either unsolicited or by operator request, by narrowing the decision alternatives or by suggesting a preferred decision based on available data
4. *Action*: Acting, responding to selections, and implementing, such as executing actions or controlling tasks with some degree of autonomy

Table 1.1 shows the updated 10 levels of automation taxonomy originally conceived by Sheridan and Verplank, which includes a clear division of authority. It addresses the question of who initiates, what needs to be done, and by whom. At lower levels of automation, the computer offers minimal or no assistance; at higher levels of automation, the computer offers partial or full assistance. The human controls automation from levels 1 to 5, and the computer controls automation from levels 6 to 10.

For example, in level 9 automation the computer is capable of independently deciding whether a task needs completion and taking the appropriate action, and only informs the human after the task has been completed. Imagine driving down an icy, pitch-black road. It begins to snow. A soothing female voice says, "Your vehicle detects ice and snow conditions ahead. Activating stability control and activating high beams." The traction control engages, the heater turns on, the leather seat warms up, and the brighter lights illuminate the path to safety.

TABLE 1.1

Ten Levels of Automation Taxonomy

Level	Decision or Action Performed by Computer
10	Decides everything and acts autonomously, ignoring the user
9	Informs the user after execution, only if it decides to
8	Informs the user after execution, only if asked
7	Executes a decision or action automatically, then informs the user
6	Allows the user a restricted time to veto before automated execution
5	Executes the suggested alternative, if the user approves
4	Suggests one alternative
3	Narrows decision/action alternatives to a few
2	Offers a complete set of decision/action alternatives
1	Offers no assistance; the user makes all decisions and takes all actions

Source: Data from Parasuraman, R., Sheridan, T.B., and Wickens, C.D., *IEEE Transactions on Systems, Man, and Cybernetics—Part A: Systems and Humans*, 30, 286–297, 2000.

Levels of automation can also be compared in the context of aircraft control. An autopilot can fly an aircraft on its own with minimal human supervision, but human supervision is required to alter course if there is inclement weather. This is an example of level 5 automation, where the automation system executes the alternative only if the operator approves.

In general, the first five levels of automation indicate that the operator is in control of automated tasks. For levels 6–10, automation is controlled by the computer and is mostly out of the operator's reach. The modified taxonomy of Parasuraman et al. (2000) can be used to categorize many complex HMSs into a mix of levels of automation across the four information-processing stages.

1.3.2.3 Endsley and Kaber's Automation Taxonomy

Similarly to the modified functions model by Parasuraman et al. (2000), Endsley and Kaber (1999) developed an automation taxonomy that addresses not only the question of authority but also the monitoring responsibilities between human and automation, which is another dimension of human–automation relationships. Table 1.2 shows how the taxonomy implemented four generic functions intrinsic to numerous domains, including air traffic control, piloting, advanced manufacturing, and tele-operations:

1. *Monitoring*: Scanning displays and retrieving data to determine system status
2. *Generating*: Formulating options or strategies for achieving goals
3. *Selecting*: Deciding on a particular option or strategy
4. *Implementing*: Carrying out a chosen option or strategy

TABLE 1.2

Endsley and Kaber's Levels of Automation Taxonomy

Level of Automation	Monitoring Role	Generating Role	Selecting Role	Implementing Role
10	Full automation	Computer	Computer	Computer	Computer
9	Supervisory control	Human/Computer	Computer	Computer	Computer
8	Automated decision making	Human/Computer	Human/Computer	Computer	Computer
7	Rigid system	Human/Computer	Computer	Human	Computer
6	Blended decision making	Human/Computer	Human/Computer	Human/ Computer	Computer
5	Decision support	Human/Computer	Human/Computer	Human	Computer
4	Shared control	Human/Computer	Human/Computer	Human	Human/ Computer
3	Batch processing	Human/Computer	Human	Human	Computer
2	Action support	Human/Computer	Human	Human	Human/ Computer
1	Manual control	Human	Human	Human	Human

Source: Data from Endsley, M.R. and Kaber, D.B., *Ergonomics*, 42, 462–492, 1999.

Similarly to the taxonomy of Parasuraman et al., these functions in Endsley and Kaber's taxonomy are also related to 10 levels of automation, which range from manual control, where the human performs all roles, to full automation, where the system carries out all actions.

Consider Endsley and Kaber's taxonomy in action: an aircraft has automation that aids the autopilot. The autopilot must acquire information about the external environment, such as wind speed, altitude, and so on. Then, the autopilot must analyze the gathered information, and decide how much thrust to give the engines to keep the aircraft flying smoothly in the correct direction. Based on the retrieved information, the automation must decide whether or not to perform the action.

Endsley and Kaber's automation taxonomy reflects the booming information age of the late 1990s, as increased computing power is vital to automation. The monitoring duties shared between human and computer in levels 1–9 illustrate the importance of keeping the operator in-the-loop and the importance of a human-centered approach for automation level design models. Keeping the human in-the-loop by monitoring and selecting automation actions is essential to maintain operator SA and keep the machine within the human's control. In-the-loop monitoring is one of the reasons why successful outcomes in crisis situations, like the miracle in the Hudson River, occur.

1.3.2.4 The Human–Human Interaction Taxonomy

Despite conceptual similarities among the human–automation interaction taxonomies discussed above, these theories are not always acknowledged or cited by each other. More importantly, it should be considered how useful these taxonomies are for designing an HMS from a practical perspective. Although the taxonomies proposed by Parasuraman et al. and Endsley and Kaber provide a better understanding of authority, role, and human–automation relationships at the conceptual level, and represent a major advance over the earlier binary taxonomy proposed by Sheridan and Verplank, they arguably do not go far enough. An implication of these two latter taxonomies and the PACT taxonomy is that a task can be decomposed into subtasks so that a single automation level can be appropriately assigned. However, the decomposition of a parent task into four information-processing stages or action-monitoring and selection functions represents only a single level of subdivision into abstract task categories. Practically, a parent task is not accomplished by abstract functions but by many levels of subtasks, which are hierarchically decomposable sequences of specific activities. Thus, the relationships between automation level and task decomposition are still more complex, though there are many analytical techniques in HF to perform task decompositions in a hierarchical fashion.

Additionally, an automation taxonomy needs to address the question of who performs which functions when tasks are assigned to either human or machine. Ideally, for tasks that can be done by both human and machine, the tasks and levels of automation should be assigned to the human–machine partnership. In these cases, the levels of automation cannot be fixed.

Notably, a recent U.S. Defense Science Board report cautioned the use of the 10 levels of automation in the design of HMSs. The rationale for this was that the 10 levels of automation models suggest that discrete or fixed levels of autonomy and intelligence focus too much on the machine (rather than on human–machine collaboration/interaction), whereas, in reality, various combinations of human–automation input at different levels of automation may occur throughout the mission's duration (Defense Science Board, 2012). This suggests that automation design architecture should be compatible with, or perhaps even mimic, human cognition. In other words, for effective human–machine collaboration, automation should be designed to resemble human–human interaction behavior. HMS design should focus not solely on human or machine, but on human–machine interaction as a partnership.

For example, the belief-desire-intention (BDI) agent taxonomy, used primarily for software agent design, uses a mentalist metaphor that emulates human cognition, motivation, and adaptation processes (Bratman et al., 1988; Georgeff et al., 1999; Jarvis et al., 2005; Sudiekat et al., 2007). BDI suggests that systems designers should consider not only the relative technical merits of software agent and human capabilities, but also the mutual intelligibility between the two parties. The BDI taxonomy has been used to develop IASs in various contexts where an agent's understanding (i.e., beliefs) is updated based on its perception of the environment, its communication with other agents and humans, and its inference mechanisms (Ferguson and Allen, 2007; Briggs and Scheutz, 2012). Once the agent's beliefs are updated and new tasks are identified, the agent proceeds to select goals (i.e., desires), and then plan or execute goals or tasks (i.e., intentions). In order for the agent to be optimally transparent to the operator, its BDI needs to be effectively conveyed to the human partner (Chen and Barnes, 2014). BDI is further discussed in Chapter 5.

TABLE 1.3

Behavior and Corresponding Components of Human–Human Interaction

Behavior	Component
Observation/Perception	Aural, taste, vision, touch, olfactory
Communication	Mouth, eyes, body, scent
Cognitive process	Learning, understanding, reasoning, reference, trust
Adaptation	Belief influence, behavior change
Collaboration	Reduce workload, improve situation awareness, improve operational effectiveness

The BDI agent model only resembles the cognition, motivation, and adaptation processes of human behavior—there are other human–human interaction behaviors that should be imitated when designing automation, including physiological attributes, cognitive characteristics, knowledge bases, and psychological states. HMS designs that incorporate these behaviors promise better human–machine interaction with greater benefits:

- *More effective interaction*: Doing the right thing at the right time, and tailoring the content and form of the interaction to the context of operators, tasks, system, and communications
- *More efficient interaction*: Enabling increased rapid task completion with less work
- *More natural interaction*: Supporting spoken, written, and gestural interaction, ideally as if interacting with a human interlocutor

From a practical perspective, an HMS design should emulate human–human interaction behavior in perception, communication, cognition, adaptation, and collaboration. Incorporating proactive and personalized automation so that the machine partner can take on responsibilities delegated by humans optimizes human–machine interaction in an intuitive fashion, resulting in maximized overall system performance. Table 1.3 shows the components of these interaction behaviors. The following section examines technological, human performance, and communication issues of human–automation interaction that should be proactively considered in the design of safe and effective HMSs.

1.4 COMMON CAUSES OF POOR HUMAN–MACHINE INTERACTION

The miracle in the Hudson River and the disaster in the Atlantic Ocean raise many questions. For example, in the realm of aircraft design, how do engineers design an aircraft cockpit to avoid future accidents? In the realm of systems design, how do engineers develop a system that is able to avoid catastrophic failure? The pilot of the Airbus A320 involved in the Hudson River incident was able to safely land the plane, in part, because of adequate SA and the ability to intervene and take control of the aircraft. In this case, the human was in control of its machine partner.

The pilots of the Airbus A330 involved in the Atlantic Ocean disaster had inadequate SA because their autopilot provided inconsistent information and, as a consequence, they did not trust the automation. They were also unable to intervene and take control of the aircraft when they needed to. In this case, the machine was in control of its human partners.

In both the Airbus A320 and Airbus A330 events, SA, human error, trust, and control authority were critical elements. If an HMS does not allow pilots to maintain SA, reduce human error, and keep a balance of trust and authority between human and machine, the system is neither safe nor effective. Key design principles are needed to address these issues and guide the design of safe and effective HMSs.

The general intention of automation is to increase productivity, safety, and efficiency by moderating human workload. As automated systems become more intelligent and sophisticated due to advances in automation technology, machines are able to perform an increasing number of tasks once handled by humans, resulting in continuously changing roles of human function in these environments. For example, automation appears in almost every aspect of daily life, from automated teller machines (ATMs) (a basic level) to unmanned drone aircraft and self-driving cars (a complex, safety-critical level). Human roles are increasingly supervisory in nature; humans make contextual decisions, which are then applied to automated systems (Parasuraman and Manzey, 2010). Unsurprisingly, this change in roles has the potential to create issues for humans interacting with automation. Technological issues, human performance issues, and communication issues are the three primary types of design issues inherent in human–automation interaction.

1.4.1 Technological Issues

The key technological issue faced in human–automation interaction can be more accurately defined as increased system complexity (Wickens et al., 2013); as technology advances, more functions are considered for automation. Generally, if automation can surpass human performance, the function should be allocated to an automated system. This particular type of function allocation is called static automation, and is further discussed in Chapter 2. The addition of automated functions, however, tends to increase system complexity—not only is the initial system present, but the new system then automates a function, which results in more components for the operator to monitor and more subsystems to understand.

There is an increased probability of system failure associated with an increased number of subsystems, adding to the complexity of the operator's role. Increased system complexity also creates the problem of observability, wherein the operator is unable to understand why the automation performs an action because the algorithms behind the action are both numerous and complex (Wickens et al., 2013). The greater the number of automated functions an HMS has (whether these functions are physical or internal processes), the more complex an HMS becomes and the greater the opportunity for failure. There may also be little recourse when failure occurs. For example, systems that rely on outdated computer code may fail and there may be no technicians knowledgeable enough to resolve even minor issues (Landauer, 1995).

1.4.2 HUMAN PERFORMANCE ISSUES

Human performance issues occur because no two humans are alike. Humans are conscious beings comfortable with human-to-human interaction and subject to feelings, instincts, and behaviors. Individuality has benefits, but it also means that inconsistencies in function are inevitable. Humans have different skill sets and different capabilities, and they react to unexpected situations in different ways (Ezer et al., 2008; Merritt and Ilgen, 2008; Warm et al., 2008; Finomore et al., 2009; Szalma and Taylor, 2011; Merritt, 2011; Reichenbach et al., 2011; Neubauer et al., 2012). As previously noted, automation is introduced to mitigate the effects of limited human attention span, which is a result of limited working memory capacity. In some cases, however, removing operators completely from decision-making tasks results in efficiency issues—in at least one instance, adding human input actually increased efficiency by 12% (Clare et al., 2012). HMS design issues related to human performance include loss of situation awareness, loss of skills, overtrust, and undertrust.

Loss of SA can occur because humans have limited working memory and are neither efficient information processors nor effective monitors. As more and more functions are allocated to automation, operators have fewer functions to execute but more automated tasks to monitor and are not always able to process the information presented in the time available to them. As a consequence of increased monitoring load, operator SA of automated tasks is often degraded. Numerous studies have shown that the implementation of automation may make humans slower and less accurate at failure detection when they become passive decision makers (Endsley and Kiris, 1995; Moray et al., 2000; Chen and Barnes, 2012; Onnasch et al., 2013). Loss of situation awareness under these circumstances is also referred to as being out-of-the-loop.

Operators who are out-of-the-loop may lose, or even unlearn, their skills due to lack of use (Lee and Moray, 1994; Ferris et al., 2010; Geiselman et al., 2013). This renders operators less able to perform vital functions if they are required to take manual control of systems. Skills loss, often a result of increased supervisory tasks and reduced manual control capability, can also lead to increased human error and negative impact on task performance. Skills loss may also occur due to generation effect, which happens because humans find it easier to remember actions they have performed themselves than actions they have watched someone else (either human or machine) perform (Slamecka and Graf, 1978; Farrell and Lewandowsky, 2000). One of the common myths about the impact of automation on human performance is that, as automation investment increases, investment in human expertise can decrease. In fact, increased automation creates new knowledge and skill requirements, which require greater investment in operators. Today, this issue is increasingly relevant as technology is more capable and there is an increased potential for automation and support at higher levels.

Overtrust and undertrust are two variations on the same theme: automation trust. Automation trust is a cognitive state that reflects the amount of faith humans have in their machine partners (Parasuraman et al., 1993; Jian et al., 2000). Automation dependency is an objective behavior that fosters overtrust (Lee and Moray, 1992). If humans trust automation too much, or too little, the HMS will not achieve optimal performance levels.

Overtrust, sometimes referred to as complacency, can occur as a result of expectation. Operators who expect machines to function a certain way and have not witnessed irregularities in function are more likely to trust that the machine will continue to function without error (Mosier and Skitka, 1996). Complacency may lead operators to overlook visible irregularities and exhibit a lack of proper monitoring practices; ignorance of minor issues increases when automated systems closely resemble real systems (Galster and Parasuraman, 2003). Additionally, human reliance on technology increases as personal risk increases (Lyons and Stokes, 2012).

Complacency can also become an issue in dual task situations where operator attention is split. Operators often focus more attention on manual tasks, leaving automated tasks to be completed without supervision (Wickens and Dixon, 2007). When operators rely heavily on automation, automation bias may cause issues of misuse, or even abuse, which can lead to missed automation failures and ignorance of the human role in system processes and protocols (Lee, 2008). For example, in the case of the Solar and Heliospheric Observatory (SOHO) spacecraft used by the European Space Agency (ESA) and National Aeronautics and Space Administration (NASA) to study the Sun and discover comets, dependency led to complacency and overconfidence in the system, which led to inadequate testing and change reviews (Leveson, 2009).

Undertrust, or skepticism, occurs when human operators aware of potential automation failure do not trust automated systems, or operators confident in their own ability to perform particular tasks are less trustful of automation than they should be (Lewandowsky et al., 2000; Lee and Moray, 1994; de Vries et al., 2003; Chen and Terrence, 2009). The first time operators encounter a system failure they may turn from overtrust to distrust of the system, and may potentially abandon the automation completely (Parasuraman and Riley, 1997). The actual extent of undertrust is often a result of operator perception of how bad an automation error is (Merritt and Ilgen, 2008; Krueger et al., 2012). System abandonment when automation might be beneficial is a disuse issue (Lee, 2008), and can create increased workload for human operators (Lee and See, 2004).

Automation complexity is another factor that can contribute to undertrust. If an automated system is difficult to interact with, or perceived as difficult to work with because of poor feedback, operators may abandon the interaction (Parasuraman and Riley, 1997). If complex automated interactions exist outside human understanding and are completed without operator input or observation, such as the processes completed by a complicated algorithm, operators may not trust the results. If humans act despite not understanding how the automation came to a particular conclusion, fatalities may result (Degani, 2004).

Ideally, operators should understand automation limits and benefits, and have an appropriate trust relationship with their automated partners (Hoffman et al., 2013; Chen and Barnes, 2014).

1.4.3 COMMUNICATION ISSUES

Communication issues are common in human–automation interaction. Operators require feedback from automated systems so that they can understand what the automation will do (Olson and Sarter, 2000), but automated systems do not always provide the types of feedback that operators would like. Automated systems perform better if they are told about what humans intend to achieve, and not about the working environment (Harbers et al., 2012). Thus, a common issue demonstrated by automated systems is conversational inflexibility, which occurs when automated systems and agents are unable to react to the information of queries being offered by human participants. For instance, automated phone systems often require callers to listen to all options before they are able to make a decision.

Automation is becoming increasingly prevalent as humanity continues to traverse the information age. As machines replace humans in a variety of tasks and slowly turn into independent entities, issues regarding human–machine interaction come to the forefront. Sheridan (2002) comments that, as the frontiers between automation and operators blur, it becomes "increasingly critical" that automation designers realize that they are building not only technology, but relationships. To facilitate the development of these relationships, it is essential for HMS designers to follow coherent design methodologies rooted in their application domain (e.g., aviation, medical, military, etc.).

1.5 THE NEED FOR CONSISTENT HUMAN–MACHINE SYSTEMS DESIGN METHODOLOGIES

Systems design methodologies differ from domain to domain. A domain is the area, or problem space, in which users work. A domain can also be described as a set of constraints, both physical and abstract (Vicente, 1990). Issues can arise because design methodologies are often loosely used. If systems design engineers use unclear design methodologies, it can lead to poor design choices and potential accidents caused by the system. To avoid this, researchers develop different design methodologies and principles for different contexts, and these are often transferred ambiguously and carelessly from one domain to another. For example, design methodology overlaps exist in the HF and HCI domains, even though both domains are considered exclusive. However, the use of HCI design methodologies for HF applications may lead to serious implications, as HCI focuses primarily on users and their interaction with computers, rather than other kinds of machines or design artifacts for safety-critical systems.

1.5.1 DOMAIN DIFFERENCES BETWEEN HUMAN FACTORS AND HUMAN–COMPUTER INTERACTION

HF is concerned with understanding the capabilities and limitations of the human mind and body, and how to design, develop, and deploy systems for human use. It is also concerned with variables that impact humans working with machines, particularly external variables such as workstation and display design, or input and output devices. The overall goal of HF is to improve the performance and safety of persons using machines. HF practitioners include psychologists, physiologists, interaction designers, and industrial designers. Note that the term *human factors* also refers to the physical or cognitive requirements of a person using a machine.

HCI is concerned with the design, evaluation, implementation, and study of major phenomena surrounding interactive computing systems for human use (ACM SIGCHI, 1992). HCI studies the interaction of humans and computers. It draws support for its theories from two sides: human and machine. The human-centered side of HCI focuses on designing effective communication methods: cognitive theory, communication theory, and industrial design are a few of the many disciplines involved. The machine-centered side looks at computer-related areas such as software, programming languages, and computer graphics. Note that HCI is also known as man–machine interaction (MMI) and computer–human interaction.

Without consistent and coherent frameworks of theory and design methodology, confusion arises even when design methods are used within the same domain. The lack of integration between the HF and HCI research domains causes methodology to become ambiguous and misleading when applied globally. Consequently, there is a pressing need to develop a consistent and coherent body of knowledge that integrates the conceptual methodologies of HF and HCI to enable systems designers to effectively represent both research domains.

Advances in one domain can benefit the other more greatly if researchers and systems designers work together to address the lack of established guidance for HMS design across domains. Systems designers must be aware of the application context when applying HF and HCI methods to specific HMS designs, as selecting an incorrect design methodology may lead to disastrous consequences. When designing an HMS, HCI methods are used largely for systems that are static, and not complex or safety-critical. HF methods should then be applied to systems that are dynamic, complex, and safety-critical. Understanding the domain differences between HF and HCI and applications is important, although there are always instances and intermediate cases where both HF and HCI are needed. HCI system applications mostly fall into the coherence domain, while HF applications mostly fall into the correspondence domain.

1.5.2 CHOOSING THE COHERENCE OR CORRESPONDENCE APPLICATION DOMAIN

It is possible for systems designers to be unaware of application domain differences when choosing design methods. They may have different educational backgrounds and possess different domain experience, and may understand HMSs from different perspectives. If systems designers are unaware of domain differences, inappropriate design methods may be used. For example, many people misunderstand the differences between the coherence and correspondence domain applications (Vicente, 1990).

When designing an HMS, systems designers need to consider three context constraints: (a) technological constraints; (b) the operator's psychophysiological constraints (i.e., human capability and limitations); and (c) the functionalities the particular HMS should achieve. According to Vicente (1990), for coherence domain applications, the context constraints of the work domain are static and can be factored in as initial conditions of the design process. Coherence design does not need to correspond to the objective state of the actual world situation, but should maintain coherence with the predefined game plan. In contrast, correspondence domain application constraints are dynamic, complex, and mostly safety-critical, and the design process is mostly goal relevant. Accordingly, correspondence design should reflect objective reality.

To better understand the differences between these two domain concepts, Miller (2000) extended Vicente's (1990, 1999) concepts of coherence and correspondence domains and further explained the different consequences of design flaws within these two domains. In an example of a computer word processor program in the coherence domain, he indicates that the objective reality in this case is how software presents itself to a user. If the word processing program crashes, there will not be immediate, global consequences. Design flaws of systems in the coherence domain do not have severe or expensive consequences.

However, in the correspondence domain, the objective reality is dynamic, complex, safety-critical, and built of constraints that the system must respect to succeed. For example, in the aviation, process control, and medical fields, as well as in warfare, a design flaw in a medical instrument or a weapon can have lethal, catastrophic, and expensive consequences.

Figure 1.7 illustrates the common setup of a nuclear power station, such as the one at Three Mile Island in Harrisburg, Pennsylvania, where a meltdown occurred in 1979. This event was a clear example of catastrophic disaster in the correspondence domain. The catalyst was a clog in the feedwater lines of the primary turbine, which caused the automatic shutdown of the turbine's feedwater pump. Subsequent human error and three critical control panel design flaws led to the meltdown of the nuclear reactor (Wickens, 1992).

Initially, the pipe for the reactor's alternate water supply was blocked off by a maintenance crew. This caused the radioactive core to start overheating, which

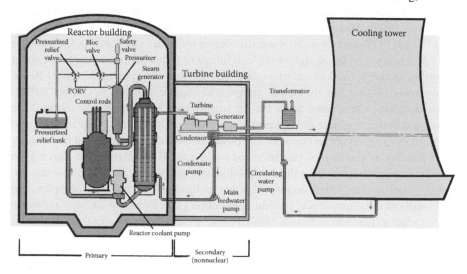

FIGURE 1.7 Schematic of a nuclear reactor. A clog in the feedwater lines, human error, and bad control panel design contributed to the meltdown at Three Mile Island. (Adapted from "Plant diagram" [Drawing]. Retrieved July 16, 2014 from http://www.nrc.gov/reading-rm/doc-collections/fact-sheets/3mile-isle.html#tmiview, 2013.)

subsequently generated extreme pressure. The automatic pressure-relief valve malfunctioned, causing the valves to stay open, and a barrage of alarms, lights, and signals overwhelmed the crew who were attempting to resolve the problem. Adding to the confusion, the displays showed what the valve was commanded to do, rather than displaying what the valve actually did. Readings also showed that there was an overflow of coolant, rather than the perceived shortage of coolant. The crew overtrusted the automated mechanism and subsequently shut down the emergency pump and overheated the core, causing large amounts of radioactive material to be released into the surrounding area.

Many analyses have been conducted to determine why the disaster occurred, and this story has become a classic example of bad HMS design (Wickens, 1992). According to the President's Commission on the accident at Three Mile Island (1979), the major factor that caused the accident was inappropriate operator action. One of the issues contributing to operator actions was deficiency in the design of the control room and process control panel. The coolant flow readings in the control panel contradicted the actual state of the situation, resulting in incorrect SA for the operators, who then shut down the emergency pump, which led to the disaster. The severity of the consequences highlights the importance of distinguishing between the machine's representation of reality and reality itself.

Charles O. Hopkins, technical director of the Human Factors Society, made reference to the control panel involved, stating that "the disregard for human factors in the control rooms was appalling. In some cases, the distribution of displays and controls seemed almost haphazard. It was as if someone had taken a box of dials and switches, turned his back, thrown the whole thing at the board, and attached things wherever they landed. For instance, sometimes 10–15 ft separated controls from the displays that had to be monitored while the controls were being operated" (Chapanis, 1996). According to the President's Commission on the accident at Three Mile Island (1979), there was no standard control room design in the nuclear process control industry, and the design of the control room followed the design concepts of the utility industry. Lack of industry standards allowed the systems designers to use large pistol-grip switches to operate pumps and smaller switches to operate valves. When designing the control room for the pressurized water reactor at Three Mile Island, the design engineer did not perform a requirements analysis but instead read safety analysis reports from other nuclear power plants, including one that had boiling water reactors as a source of guidance and precedent. The design of the Three Mile Island control room, including the control displays and panels, did not follow any HF design guidelines but, rather, an inappropriate method for this dynamic, complex, and safety-critical correspondence domain application (The President's Commission on the Accident at Three Mile Island, 1979). Figure 1.8 illustrates a control panel similar to the one used in the Three Mile Island nuclear reactor control room.

From the perspective of poorly designed human–machine or human–automation interaction in HMS design, three critical errors were made in the design of the control panel. First, the design did not consider human information-processing limitations: the crew was overwhelmed by information from the myriad of alarms, lights, and signals. Second, the automation did not provide the operator with feedback regarding the malfunctioned pressure-relief valve. Third, the displays and system signals

FIGURE 1.8 Nuclear control panel used at the Three Mile Island unit 2 nuclear station. (Adapted from The President's Commission on the Accident at Three Mile Island, *The Need for Change: The Legacy of TMI*, The White House, Washington DC, 1979.)

provided incorrect SA information to the crew. As a result, they misunderstood the situation and took inappropriate action.

The Three Mile Island incident is a reminder of the importance of designing a system around human capacities. It is also a warning about the consequences of using inappropriate design methodologies for different domain applications. The design of the nuclear plant control panel, including both interface and automation, did not follow appropriate guidance for that domain application. Had the systems designers selected the proper design methodologies when designing the control panel, a nuclear meltdown might have been prevented.

More importantly, "overall little attention had been paid to the interaction between human beings and machine under the rapidly changing and confusing circumstances of an accident" (The President's Commission on the Accident at Three Mile Island, 1979). Systems designers should consider not only human capabilities and limitations, technological constraints, and working domain contexts, but also human–machine interaction.

The Three Mile Island catastrophe and the disaster in the Atlantic Ocean both offer lessons for the future. Design flaws in correspondence domain applications, such as aviation, should be closely examined so that patterns are not repeated. While there was no design standard for control room and control panels in the nuclear process control industry 30 years ago, there are many design standards and design guidelines for cockpit design in the aviation industry today. The more pressing issue is whether a coherent design framework exists that contains appropriate theories and methodologies for a systems design team to strictly follow. Systems designers attempting to follow such a framework need to understand the theories and methodologies the framework is based on. More importantly, they need to understand how humans and machines can work together. The more active an HMS becomes, the more priority should be given to understanding the capabilities and functionalities of individual IAS components and how they interact. Specifically, this should be done at the beginning of the design cycle. IAS design should follow an interaction-centered approach rather than a technology-centered, user-centered, or environment-centered approach. General design approaches are further discussed in Chapter 2.

1.6 SUMMARY

This chapter began by introducing and defining basic HF systems design concepts. It looked closely at HMSs and IASs, and demonstrated passive and active HMSs using the B-17 bomber aircraft, in-car GPS, and the well-known R2D2 robot. A real-world IAS example was also discussed in the Toyota PCS, which was designed to enhance the safety of automobile drivers and passengers. The importance of maintaining operator SA was stressed, and the benefits of conscientious systems design were also highlighted.

The concept of good systems design was looked at, and two recent aviation examples—US Airways Flight 1549 and Air France Flight 447—demonstrated the results of well-designed and poorly designed systems in safety-critical situations. Issues frequently encountered by operators interacting with machines, notably technological issues, human performance issues, and communication issues, were also discussed.

Common human–machine interaction taxonomies, specifically the PACT taxonomy, Sheridan and Verplank's levels of automation taxonomy, Endsley and Kaber's automation taxonomy, and the human–human interaction taxonomy, were highlighted to help systems designers understand the scope of automation within particular IASs.

The Three Mile Island nuclear disaster was also recounted, in part to underline the importance of understanding the difference between coherence and correspondence domain applications but primarily to demonstrate that HMS and IAS design should follow a consistent and coherent framework with appropriate theories and methodologies as guidelines. IAS design should not limit itself to technology-centered, user-centered, or environment-centered approaches; it should follow an interaction-centered approach to maximize overall system performance and appropriately address system safety issues. The basic components of IASs are introduced in Chapter 2.

REFERENCES

ACM SIGCHI. (1992). Curricula for human–computer interaction. ACM Special Interest Group on Computer-Human Interaction Curriculum Development Group. Retrieved July 16, 2014 from http://old.sigchi.org/cdg/index.html.

Bratman, M., Israel, D., and Pollack, M. (1988). Plans and resource-bounded practical reasoning. *Computational Intelligence*, 4(3), 349–355.

Briggs, G. and Scheutz, M. (2012). Multi-modal belief updates in multi-robot human-robot dialogue interactions. In *Proceedings of the AISB/IACAP Symposium: Linguistic and Cognitive Approaches to Dialogue Agents*, pp. 67–72. July 2–6, Birmingham, UK.

Chapanis, A. (1996). *Human Factors in System Engineering*. Hoboken, NJ: Wiley.

Chen, J. Y. C. and Barnes, M. J. (2012). Supervisory control of multiple robots in dynamic tasking environments. *Ergonomics*, 55(9), 1043–1058.

Chen, J. Y. C. and Barnes, M. J. (2014). Human-agent teaming for multi-robot control: A review of human factors issues. *IEEE Transactions on Human-Machine Systems*, 44(1), 13–29.

Chen, J. Y. C., Barnes, M. J., and Harper-Sciarini, M. (2011). Supervisory control of multiple robots: Human-performance issues and user-interface design. *IEEE Transactions on Systems, Man, and Cybernetics, Part C: Applications and Reviews*, 41(4), 435–454.

Chen, J. Y. C. and Terrence, P. I. (2009). Effects of imperfect automation and individual differences on concurrent performance of military and robotics tasks in a simulated multitasking environment. *Ergonomics, 52*(8), 907–920.

Clare, A. S., Cummings, M. L., How, J., Whitten, A., and Toupet, O. (2012). Operator objective function guidance for a real-time unmanned vehicle scheduling algorithm. *AIAA Journal of Aerospace Computing, Information and Communication, 9*(4), 161–173.

Defense Science Board. (2012). *Task Force Report: The Role of Autonomy in DoD Systems.* Washington, DC: Office of the Secretary of Defense. Retrieved July 16, 2014 from http://www.fas.org/irp/agency/dod/dsb/autonomy.pdf.

Degani, A. (2004). *Taming HAL: Designing Interfaces beyond 2001.* New York, NY: Palgrave Macmillan.

de Vries, P., Midden, C., and Bouwhuis, D. (2003). The effects of errors on system trust, self-confidence, and the allocation of control in route planning. *International Journal of Human-Computer Studies, 58*(6), 719–735.

Edwards, J. L. (2004). A generic agent-based framework for the design and development of UAV/UCAV control systems (Report No. CR 2004-062). Toronto, Canada: Defence Research and Development.

Elsworth, C. and Allen, N. (2009). New York plane crash: Black box shows both engines failed. *The Telegraph.* Retrieved July 15, 2014 from http://www.telegraph.co.uk.

Endsley, M. R. (1999). Situation awareness in aviation systems. In D. J. Garland, J. A. Wise, and V. D. Hopkin (eds), *Handbook of Aviation Human Factors*, pp. 257–276. Mahwah, NJ: Lawrence Erlbaum Associates.

Endsley, M. R. and Kaber, D. B. (1999). Level of automation effects on performance, situation awareness and workload in a dynamic control task. *Ergonomics, 42*(3), 462–492.

Endsley, M. R. and Kiris, E. O. (1995). The out-of-the-loop performance problem and level of control in automation. *Human Factors: The Journal of the Human Factors and Ergonomics Society, 37*(2), 381–394.

Ezer, N., Fisk, A. D., and Rogers, W. A. (2008). Age-related differences in reliance behavior attributable to costs within a human-decision aid system. *Human Factors: The Journal of the Human Factors and Ergonomics Society, 50*(6), 853–863.

Farrell, S. and Lewandowsky, S. (2000). A connectionist model of complacency and adaptive recovery under automation. *Journal of Experimental Psychology: Learning, Memory, and Cognition, 26*(2), 395–410.

Ferris, T., Sarter, N., and Wickens, C. D. (2010). Cockpit automation: Still struggling to catch up. In E. Salas and D. Maurino (eds), *Human Factors in Aviation*, pp. 479–503. Waltham, MA: Academic Press.

Ferguson, G. and Allen, J. (2007). Mixed-initiative systems for collaborative problem solving. *AI Magazine, 28*(2), 23–32.

Finomore, V., Matthews, G., Shaw, T., and Warm, J. (2009). Predicting vigilance: A fresh look at an old problem. *Ergonomics, 52*(7), 791–808.

Galster, S. M. and Parasuraman, R. (2003). The application of a qualitative model of human interaction with automation in a complex and dynamic combat flight task. In *Proceedings of the 12th International Symposium on Aviation Psychology*, pp. 411–416. April 14–17, Warsaw, Poland.

Geiselman, E. E., Johnson, C. M., Buck, D. R., and Patrick, T. (2013). Flight deck automation: A call for context-aware logic to improve safety. *Ergonomics in Design: The Quarterly of Human Factors Applications, 21*(4), 13–18.

Georgeff, M., Pell, B., Pollack, M., Tambe, M., and Wooldridge, M. (1999). The belief-desire-intention model of agency. In *Intelligent Agents V: Agents Theories, Architectures, and Languages*, pp. 1–10. Heidelberg, Germany: Springer-Verlag.

Harbers, M., Jonker, C., and van Reimsdijk, B. (2012). Enhancing team performance through effective communications. *Paper presented at The Annual Human-Agent-Robot Teamwork (HART) Workshop*. Boston, MA.

Hoffman, R. R., Johnson, M., Bradshaw, J. M., and Underbrink, A. (2013). Trust in automation. *IEEE Intelligent Systems*, 28(1), 84–88.

Hosford, M., Effron, L., and Battiste, N. (2012). Air France Flight 447 crash "didn't have to happen," experts say. Retrieved July 16, 2014 from http://www.abcnews.go.com.

Hou, M., Zhu, H., Zhou, M., and Arrabito, G. R. (2011). Optimizing operator-agent interaction in intelligent adaptive interface design: A conceptual framework. *IEEE Transactions on Systems, Man, and Cybernetics, Part C: Applications and Reviews*, 41(2), 161–178.

International Federation of Automatic Control. (2009). TC 4.5 Human Machine Systems. Retrieved from http://tc.ifac-control.org/4/5.

Jarvis, B., Corbett, D., and Jain, L. C. (2005). Beyond trust: A belief-desire-intention model of confidence in an agent's intentions. In *Knowledge-Based Intelligent Information and Engineering Systems*, pp. 844–850. Heidelberg, Germany: Springer-Verlag.

Jian, J. Y., Bisantz, A. M., and Drury, C. G. (2000). Foundations for an empirically determined scale of trust in automated systems. *International Journal of Cognitive Ergonomics*, 4(1), 53–71.

Krueger, F., Parasuraman, R., Iyengar, V., Thornburg, M., Weel, J., Lin, M., Clarke, E., McCabe, K., and Lipsky, R. H. (2012). Oxytocin receptor genetic variation promotes human trust behavior. *Frontiers in Human Neuroscience*, 6. Retrieved July 16, 2014 from http://www.ncbi.nlm.nih.gov/pmc/articles/PMC3270329/.

Landauer, T. K. (1995). *The Trouble with Computers: Usefulness, Usability, and Productivity*, vol. 21. Cambridge, MA: MIT Press.

Lavietes, S. (2002) Alphonse Chapanis Dies at 85; Was a Founder of Ergonomics. The New York Times. Retrieved July 16, 2014 from http://www.nytimes.com/2002/10/15/us/alphonse-chapanis-dies-at-85-was-a-founder-of-ergonomics.html.

Lee, J. D. (2008). Review of a pivotal human factors article: "Humans and automation: Use, misuse, disuse, abuse." *Human Factors: The Journal of the Human Factors and Ergonomics Society*, 50(3), 404–410.

Lee, J. D. and Moray, N. (1992). Trust, control strategies and allocation of function in human-machine systems. *Ergonomics*, 35(10), 1243–1270.

Lee, J. D. and Moray, N. (1994). Trust, self-confidence, and operators' adaptation to automation. *International Journal of Human-Computer Studies*, 40(1), 153–184.

Lee, J. D. and See, K. A. (2004). Trust in automation: Designing for appropriate reliance. *Human Factors: The Journal of the Human Factors and Ergonomics Society*, 46(1), 50–80.

Leveson, N. G. (2009). The need for new paradigms in safety engineering. In C. Dale and T. Anderson (eds), Safety-Critical Systems: Problems Process and Practice, pp. 3-20, London England: Springer-Verlag London Limited.

Lewandowsky, S., Mundy, M., and Tan, G. (2000). The dynamics of trust: Comparing humans to automation. *Journal of Experimental Psychology: Applied*, 6(2), 104–123.

Lewis, M., Wang, H., Chien, S., Velagapudi, P., Scerri, P., and Sycara, K. (2010). Choosing autonomy modes for multirobot search. *Human Factors: The Journal of the Human Factors and Ergonomics Society*, 52(2), 225–233.

Lyons, J. B. and Stokes, C. K. (2012). Human–human reliance in the context of automation. *Human Factors: The Journal of the Human Factors and Ergonomics Society*, 54(1), 112–121.

McFadden, R. D. (2009). Pilot is hailed after jetliner's icy plunge. *The New York Times*. Retrieved July16, 2014 from http:///www.nytimes.com.

Merritt, S. M. (2011). Affective processes in human–automation interactions. *Human Factors: The Journal of the Human Factors and Ergonomics Society*, 53(4), 356–370.

Merritt, S. M. and Ilgen, D. R. (2008). Not all trust is created equal: Dispositional and history-based trust in human-automation interactions. *Human Factors: The Journal of the Human Factors and Ergonomics Society*, *50*(2), 194–210.

Miller, C. (2000). Intelligent user interfaces for correspondence domains: Moving IUI's "off the desktop." In *Proceedings of the 5th International Conference on Intelligent User Interfaces*, pp. 181–186. January 9–12, New York, NY: ACM Press.

Miller, G. (1956). The magical number 7 plus or minus two: Some limits on our capacity for processing information. *The Psychological Review*, *63*(2), 81–97.

Moray, N., Inagaki, T., and Itoh, M. (2000). Adaptive automation, trust, and self-confidence in fault management of time-critical tasks. *Journal of Experimental Psychology: Applied*, *6*(1), 44–58.

Mosier, K. L. and Skitka, L. J. (1996). Human decision makers and automated decision aids: Made for each other. In R. Parasuraman and M. E. Mouloua (eds), *Automation and Human Performance: Theory and Applications*, pp. 201–220. Mahwah, NJ: Lawrence Erlbaum Associates.

Neubauer, C., Matthews, G., Langheim, L., and Saxby, D. (2012). Fatigue and voluntary utilization of automation in simulated driving. *Human Factors: The Journal of the Human Factors and Ergonomics Society*, *54*(5), 734–746.

Olson, W. A. and Sarter, N. B. (2000). Automation management strategies: Pilot preferences and operational experiences. *The International Journal of Aviation Psychology*, *10*(4), 327–341.

Onnasch, L., Wickens, C. D., Li, H., and Manzey, D. (2013). Human performance consequences of stages and levels of automation: An integrated meta-analysis. *Human Factors: The Journal of the Human Factors and Ergonomics Society*, *56*(3), 476–488.

Parasuraman, R. and Manzey, D. H. (2010). Complacency and bias in human use of automation: An attentional integration. *Human Factors: The Journal of the Human Factors and Ergonomics Society*, *52*(3), 381–410.

Parasuraman, R., Molloy, R., and Singh, I. L. (1993). Performance consequences of automation-induced "complacency". *The International Journal of Aviation Psychology*, *3*(1), 1–23.

Parasuraman, R. and Riley, V. (1997). Humans and automation: Use, disuse, and abuse. *Human Factors: The Journal of the Human Factors and Ergonomics Society*, *39*(2), 230–253.

Parasuraman, R., Sheridan, T. B., and Wickens, C. D. (2000). A model for types and levels of human interaction with automation. *IEEE Transactions on Systems, Man, and Cybernetics—Part A: Systems and Humans*, *30*(3), 286–297.

The President's Commission on the Accident at Three Mile Island. (1979). *The Need for Change: The Legacy of TMI*. Washington, DC: The White House.

Reichenbach, J., Onnasch, L., and Manzey, D. (2011). Human performance consequences of automated decision aids in states of sleep loss. *Human Factors: The Journal of the Human Factors and Ergonomics Society*, *53*(6), 717–728.

Sheridan, T. B. (2002). *Humans and Automation: System Design and Research Issues*. Santa Monica, CA: Wiley-Interscience.

Sheridan, T. B. and Parasuraman, R. (2006). Human-automation interaction. *Reviews of Human Factors and Ergonomics*, *1*, 89–129.

Sheridan, T. B. and Verplank, W. L. (1978). Human and computer control of undersea teleoperators (Report No. N00014-77-C-0256). Cambridge, MA: MIT Cambridge Man-Machine Systems Laboratory.

Slamecka, N. J. and Graf, P. (1978). The generation effect: Delineation of a phenomenon. *Journal of Experimental Psychology: Human Learning and Memory*, *4*(6), 592–604.

Smithsonian. (2009). Geese Involved in Hudson River Plane Crash Were Migratory. ScienceDaily. Retrieved March 21, 2014 from www.sciencedaily.com/releases/2009/06/090608125059.htm.

Sudeikat, J., Braubach, L., Pokahr, A., Lamersdorf, W., and Renz, W. (2007). On the validation of belief–desire–intention agents. In *Lecture Notes in Computer Science: Programming Multi-Agent Systems,* 4411, pp. 185-200, Heidelberg, Germany: Springer-Verlag.

Szalma, J. L. and Taylor, G. S. (2011). Individual differences in response to automation: The five factor model of personality. *Journal of Experimental Psychology: Applied,* *17*(2), 71–96.

Taylor, R. M. (2001a). *Cognitive Cockpit Systems Engineering: Pilot Authorisation and Control of Tasks.* In R. Onken (Ed), CSAPC'01. Proceedings of the 8th Conference on Cognitive Science Approaches to Process Control, Neubiberg, Germany, University of the German Armed Forces.

Taylor, R. M. (2001b). *Technologies for Supporting Human Cognitive Control,* RTO Meeting Proceedings 77, Human Factors in the 21st Century, pp. 18-1–18-15, Farnborough, UK: Defence Science and Technology Laboratory. In R. Onken (Ed), CSAPC'01. Proceedings of the 8th Conference on Cognitive Science Approaches to Process Control, Neubiberg, Germany, University of the German Armed Forces.

Taylor, R. M., Bonner, M.C., Dickson, B., Howells, H., Miller, C. A., Milton, N., Pleydell-Pearce, K., Shadbolt, N., Tennison, J., and Whitecross, S. E. (2001) Cognitive cockpit engineering: Coupling functional state assessment, task knowledge management and decision support for context sensitive aiding. *Human Systems IAC Gateway, 12(1), 20-21*

Vandoorne, S. (2011). Air France crash recovery ends with 74 bodies missing. Retrieved from http://www.cnn.com.

Vicente, K. J. (1990). Coherence- and correspondence-driven work domains: Implications for systems design. *Behaviour and Information Technology,* *9*(6), 493–502.

Vicente, K. J. (1999). *Cognitive Work Analysis.* Mahwah, NJ: Lawrence Erlbaum Associates.

Wang, H., Lewis, M., Velagapudi, P., Scerri, P., and Sycara, K. (2009). How search and its subtasks scale in n robots. In *Proceedings of the 2009 4th ACM/IEEE International Conference on Human-Robot Interaction (HRI),* pp. 141–147. March 11–13, San Diego, CA: ACM/IEEE.

Warm, J. S., Parasuraman, R., and Matthews, G. (2008). Vigilance requires hard mental work and is stressful. *Human Factors: The Journal of the Human Factors and Ergonomics Society,* *50*(3), 433–441.

Wickens, C. D. (1992). Introduction to engineering psychology and human performance. In J. G. Hollands and C. D. Wickens (eds), *Engineering Psychology and Human Performance,* 2nd edn., pp. 1–3. Upper Saddle River, NJ: Prentice Hall.

Wickens, C. D. and Dixon, S. R. (2007). The benefits of imperfect diagnostic automation: A synthesis of the literature. *Theoretical Issues in Ergonomics Science,* *8*(3), 201–212.

Wickens, C. D., Mavor, A. S., and McGee, J. P. (1997). *Flight to the Future: Human Factors in Air Traffic Control.* Washington, DC: National Academy Press.

Wickens, C. M., Mann, R. E., and Wiesenthal, D. L. (2013). Addressing driver aggression: Contributions from psychological science. *Current Directions in Psychological Science,* *22*(5), 386–391.

Woodson, W. E. and Conover, D. W. (1964). *Human Engineering Guide for Equipment Designers.* Berkeley, CA: University of California Press.

Section I

Theoretical Approaches

2 Overview of Intelligent Adaptive Systems*

2.1 OBJECTIVES

- Define interface and review the evolution of interface technologies
- Discuss automation types and review the evolution of automation technologies
- Discuss IASs as an evolution of intelligent adaptive interface (IAI) and intelligent adaptive automation (IAA) technologies
- Explain IAS design objectives and discuss the importance of coordination, cooperation, and collaboration in human–machine interaction

The Hudson River incident and the disaster in the Atlantic Ocean discussed in Chapter 1 serve as reminders of the difficulty of designing fully fail-proof systems. There are recurring themes within these stories that should be remembered when designing systems, such as keeping the operator in the loop, maintaining effective operator–machine interaction, and ensuring that the operator is in control of his or her machine partner. The Hudson River incident demonstrates humankind's brilliant ability to adapt to unknown situations. No matter how intelligent machines become, systems designers should not underestimate or forget human creativity and adaptability—the human cannot be replaced. Optimal human–machine interaction can only be achieved through the effective design of IASs.

Interface and automation are the two fundamental IAS technologies, and understanding how humans interact with machines and automation is crucial for IAS designers. A systems designer may start by sketching out how an interface for human interaction with a machine might look, or how the system architecture with different functions would be structured. They can then begin to determine which functions can be automated, or they might start by asking operators about their experience with a current system and what they expect from a new IAS. In either case, it is the relationship between human and machine that needs to be understood.

The maturation of computers, communication, artificial intelligence, and other technologies has led to increased automation in nearly all aspects of society. For example, through the use of artificial intelligence technology, software systems are gaining the ability to evaluate and make decisions on their own. This trend has led

---

to a shift in the role of the operator—from performing largely perceptual motor tasks, such as controlling a vehicle by directly manipulating a stick or wheel, to tasks that require greater amounts of cognitive processing, such as monitoring, reasoning, judging, decision making, and managing automated or semiautomated system operations. As both operator and machine adapt to their changing roles, effective interfaces become ever more critical to keep the operator not only on the loop (i.e., aware of what is happening), but also in the loop (i.e., able to take control) (Chen and Barnes, 2014).

Notably, many of today's conventional interfaces do not meet the interaction requirements imposed by complex IASs and often fail to properly reflect the tasks, goals, plans, and domain-specific problems of the operator (Sukaviriya et al., 1993; Bowman and Hodges, 1999; Sutcliffe et al., 2006). There is also very little guidance on how to evaluate the effectiveness of interface design and how to determine what an interface requires to aid decision making and facilitate user interaction. This is especially true for complex IASs, such as uninhabited aerial vehicle (UAV) control stations, where information flow is dynamic, operators have excessive workloads, and situation awareness is crucial for completing tasks.

This chapter begins by discussing the evolution of two basic components of an IAS—interface and automation. It addresses the concepts of IAI, IAA, and IASs in the context of human–machine interaction. It also notes the abilities of a well-designed system and discusses the range of human–machine interactions using coordination, cooperation, and collaboration.

2.2 THE EVOLUTION OF INTERFACE TECHNOLOGIES AND THE IAI CONCEPT

Figure 1.2 illustrates how the interface component of a human–machine system (HMS) acts as a bridge to facilitate interaction between human and machine subsystems. The necessity of the interface component is obvious; however, there is ambiguity about various interface terminologies. For example, human–computer interface (often abbreviated as HCI, but not to be confused with the similarly abbreviated human–computer interaction); human–machine interface (HMI), also known as man–machine interface (MMI) and user interface; and operator–machine interface (OMI) have different formal definitions, but all are loosely used to describe the same concepts.

A human–computer interface, or user interface, can be described as the point at which the operator, or user, interacts with the computer. The user interface design seeks to discover the most efficient way to design understandable electronic messages (Norman, 1988; Shneiderman et al., 2009). The user interface facilitates both input and output. Input allows the operator to enter information into the system, and output indicates to the operator any effects resulting from the input. For example, a keyboard is an input device, and a computer application that displays typed text is an output device.

HMIs and MMIs facilitate interaction between operator and machine. The terms *human–machine interface* and *man–machine interface* are typically used in industrial and process control applications, whereas user interface is common in the business information systems domain. An OMI is the aggregate of means by which an operator interacts with a machine. OMIs provide the means of input, allowing operators to manipulate the machine, and output, allowing the machine to produce the effects of the operators' manipulation.

2.2.1 DEFINING INTERFACE

Although there are many different ways to describe *interface*, the term ultimately has a singular definition: the means by which humans can interact with a machine to perform a task. Machine, as defined here, can be a tool, technology, a piece of hardware or software, or both. An interface is a subsystem of a HMS; it is the window (both metaphorically and literally) through which operators interact with the machine. For example, a computer mouse, keyboard, display, and operating system are individual components of a computer interface. Each stand-alone device is referred to as an interface device and as a whole they can be defined as "the interface." Because each definition of interface originates from a different domain, definitions are derived from different contexts. Within the HMS context, the OMI is a single component of an overall system.

2.2.2 CONVENTIONAL OPERATOR–MACHINE INTERFACES

Conventional OMIs support operator interaction with a particular machine, device, computer program, or other complex tool. In other words, the OMI facilitates interaction between the operator and the machine. For example, a computer operating system interface allows users to manipulate the system and perform tasks by producing outputs that correspond to operators' inputs. Using a computer without an operating system would be problematic.

The computer mouse is a good example of the iterative evolution of generic OMIs. In 1965, Douglas Engelbart developed the first mouse at the Stanford Research Laboratory as part of a research project working to augment human intellect and information processing using mechanical apparatuses. In the 1970s, Xerox introduced the mouse as a primary interface for operators interacting with their corporate computer systems. In the early 1980s, Apple Computer Inc. launched the first commercial computer system that included a graphical user interface (GUI) to be navigated using a mouse. The development of a pointer device, specifically the computer mouse, was crucial for complex GUIs, as it addressed a fundamental need to directly manipulate on-screen graphical objects (Myers, 1996).

To produce respondent outputs for operator inputs, operating systems need to have knowledge relating to the tasks performed by operators. Tasks are actions, and a series of multiple tasks make up an activity. Note that context is very important when using the word *task* as the definition varies. In some instances, task refers specifically to actions performed by the operator (as opposed to actions performed by the machine) and in other cases it may refer to actions performed by either the operator or the machine. It is also possible to find more precise definitions for the word task, such as its use in activity theory.

A task model is the system's knowledge of likely operations performed by an operator. It is technology-centered and focuses on the machine rather than the human. The overall purpose of technology-centered design is to reduce operator workload, increase task efficiency and accuracy, and create greater convenience. Figure 2.1 illustrates the evolution of interface design and shows the relationship of the technology-centered approach to conventional interfaces. The design of conventional interfaces requires only the use of task models.

Task models define how an operator uses a machine to achieve a goal, and how the machine supports the involved tasks and subtasks. They systematically capture the complete set of tasks that operators perform, as well as how the tasks relate to each other (Souchon et al., 2002).

The origin of most task modeling approaches can be traced back to activity theory (Kuutti, 1996), which maintains that operators carry out activities to change part of their environment in order to achieve a stated goal (Dittmar and Forbrig, 2003). Even simple activities, such as withdrawing funds from an automated teller machine (ATM), can be captured using a task model. Analytical techniques that are useful for building task models and understanding interaction needs are discussed in Chapter 4.

Figure 2.2 illustrates a task model that demonstrates the logic behind the design of an early version of an ATM, which is still available today in many places around the world. Newer models have been designed to eject the banking card before dispensing the bills. This advancement is to compensate for the potential human error of leaving the card in the machine after the withdrawal and was based on a better understanding of operator goals. People do not usually leave their money (i.e., the goal) behind, but it is common that they forget to retrieve their banking card after achieving their goal. As illustrated in Figure 2.1, the evolution of conventional interfaces to intelligent or adaptive interfaces shows the relationship of these interfaces to the user-centered approach that newer ATM designs are based on. The design of

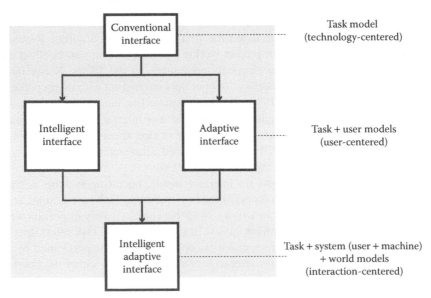

FIGURE 2.1 Evolution of interface technologies and their relationship to different design approaches.

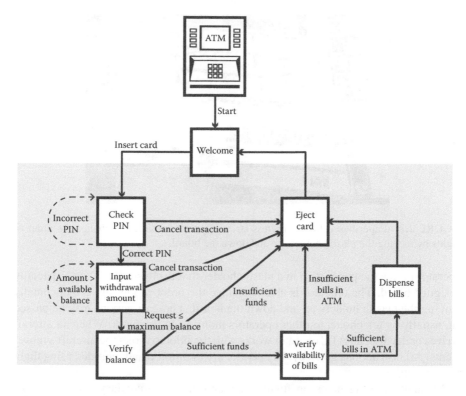

FIGURE 2.2 Task model for ATM withdrawal. (Redrawn from Wang, Y., Zhang, Y., Sheu, P., Li, X., and Guo, H., *International Journal of Software Science and Computational Intelligence*, 2, 102–131, 2010.)

these interface types requires the use of both task and user models. User models will be discussed in the next section.

Designers of effective HMSs need to understand not only what tasks an operator will typically perform using a particular machine, but also how the operator may interact with the machine in the context of human limitations, such as limited short-term memory capacity. HMSs need to be user-centered in order to facilitate optimal human–machine interaction. The need to assist operators in a flexible fashion has fostered the development of intelligent interface and adaptive interface technologies.

2.2.3 INTELLIGENT INTERFACES AND ADAPTIVE INTERFACES

Electronic flight strips (EFSs) are a good example of how the evolution of interface technology can improve the ratio of success to failure in safety-critical applications. The advances in flight strip technology have increased air traffic control (ATC) task efficiency and accuracy, improved situation awareness, and prevented human error.

Traditional flight progress strips are strips of paper that ATC operators use to coordinate flight progress and record the directions given to pilots before and after landing. Figure 2.3 illustrates how, after handwriting the flight details, an ATC

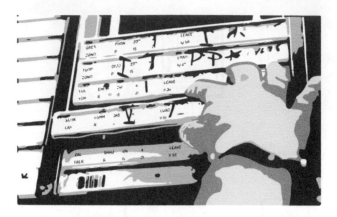

FIGURE 2.3   Paper-based flight progress strips. Operators manually change the order of flights by moving the plastic holders up and down the board.

operator places the paper strip in a plastic holder that is physically set on a specially designed board. The operator is able to change the order of the flights by manually moving the plastic holders up and down the board. These directions may be passed on, usually by telephone, to other operators monitoring the flight. When an aircraft arrives or departs, an ATC operator works with the pilot to direct the aircraft's movements. All instructions given by the operator are monitored and recorded using flight progress strips.

Manually writing notes on flight strips and calling secondary ATC operators cause a decrease in operator SA due to the volume of information that needs to be recorded and the limitations on the legibility of handwritten notes.

Software simulations of flight progress strips have been developed to improve on the speed and reliability of paper-based flight strips. Figure 2.4 illustrates EFSs, which offer many advantages to ATC operators, including improved situation awareness. Traditional flight progress strips required a new paper strip to be created for each parameter update, resulting in multiple flight progress strips to be sorted for each aircraft. EFS systems eliminate this problem (Eier and Fraenzl, 2006) by intelligently facilitating more efficient organization of strips. They automatically adjust to accommodate for varying traffic volumes, reduce waiting strip information, and limit the number of strips on the display to those deemed immediately important.

Another key advantage of EFS systems is their ability to digitize operators' notes. They mimic paper strips on a touch screen display, and provide intuitive and efficient strip markings that are clear and legible. Data can also be archived indefinitely. Well-designed EFS systems improve operator performance and facilitate cognitive processes by reducing the cognitive workload and maintaining accurate flight progress information (Manning, 1995).

EFS systems include a task model of ATC operations and all activities that are known to the system, such as recording, sorting, and coordinating. The systems design takes performance issues into consideration, such as limited human short-term memory capacity and the legibility of handwritten notes on strips of paper. It also takes a compensatory approach by allocating some of the system functions,

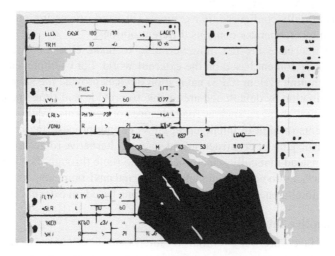

FIGURE 2.4 smartStrips® is an electronic flight strip software product that allows operators to digitally move, monitor, and record flight status on a touch screen. (Redrawn from Eier, D. and Fraenzl, T., *The Journal of Air Traffic Control*, 48, 41–43, 2006.)

such as recording, sorting, and coordinating, to automation technology instead of the operator. In short, good EFS systems are designed using both task and user models.

User models incorporate the system's knowledge of the capabilities, limitations, needs, and demands of the operator. To effectively assist the operator, a system must have the intelligence to trigger automated aids based on one or a combination of three conditions: (a) the evaluation of a task condition, including performance and error rate or duration of task accomplishment; (b) operator stress levels; and (c) operator workload level. Knowledge of these three conditions depends on the use of a user model.

Interface technology evolves into intelligent interface and adaptive interface technology by combining both task and user models. Systems designers must have a deep and detailed knowledge of potential operators and their capabilities, including their attention, short-term memory, and decision-making skills (Eggleston, 1992). As illustrated in Figure 2.1, using both task and user models, a user-centered design ensures that HMSs work so that people can use them, and can lead to the creation of safe HMSs that avoid failure.

2.2.3.1 Intelligent Interfaces

With embedded task and user models, interfaces can be personalized to enhance interaction and make systems more efficient, more effective, and easier to use (Hancock and Chignell, 1989; Maybury, 1993). Keeble and Macredie (2000) define an intelligent interface as "an interface where the appearance, function or content of the interface can be changed by the interface itself in response to the user's interaction." An intelligent interface can match its content to changing task-related circumstances. The machine controls whether an adaptation occurs, how it occurs, and the amount of adaptation that occurs, but the operator has control over how the machine adaptation is initially configured. For example, in word processing programs, each menu selected by an operator contains different options.

Intelligent interfaces are not, however, able to make changes to their appearance, function, or content based on the external world. For example, early versions of global positioning system (GPS) navigation devices provide the shortest possible route to a destination by default and are unable to provide real-time updates regarding environmental changes, such as road closures or accidents. Without knowledge of external events or built-in intelligence, GPSs cannot process environmental factors such as traffic jams or construction and suggest alternative routes to the operator. The operator instead manually selects an alternate route through the GPS interface. Intelligent interfaces do not have models for external environments, nor do they have the adaptation intelligence to deduct logical actions based on external context.

2.2.3.2 Adaptive Interfaces

Adaptive interfaces are very similar to intelligent interfaces. Originally, intelligent interfaces were known as adaptive user interfaces (Innocent, 1982; Browne et al., 1990; Benyon and Murray, 1993; Mitrovic et al., 2007; Abascal et al., 2008). Tomlinson et al. (2007) define adaptive interfaces as interfaces that seek to predict what features are desirable and customizable by operators. Adaptive interfaces are flexible tools intended to customize systems for individual operators (Höök, 2000). Customization enhances usability by matching content to changing circumstances, which results in increased efficiency and effectiveness.

Speech recognition tools that convert spoken words into computer-readable language are commonly used in telephony services as a way for customers to receive automated help from organizations. By speaking commands such as "account balance" or "billing inquiries" to a machine through a telephone, it is possible to receive the desired information in return. The system is able to recognize and adapt to the operator's spoken input- and output-relevant information. However, there are many instances when speech recognition fails because of factors such as noisy environments or the regional accent of the speaker. Speech recognition systems cannot recognize speech input in noisy environments without (a) knowledge of the noisy external environment and (b) the ability to discriminate between language and noise. System limitations mean that filtering out external noise is necessary for the interface to provide seamless instruction.

Although the definitions of intelligent interface and adaptive interface differ, both maintain a user-centered design approach. Typically, an intelligent interface decides when, what, how, and how much adaptation is needed, while an adaptive interface adjusts content output based on the changing situation of the task, operator, and environment.

It can be suggested that adaptive interface features are actually functions of intelligent interfaces, as both share the same goal of satisfying operators' needs by making the interface more effective, efficient, and easy to use. To facilitate adaptation, intelligent interfaces require an adaptive function that adjusts the way that the interface communicates results to the operator based on changes of the inputs taken from the operator, task information, machine status, and environmental information. An intelligent interface would not be intelligent without its adaptive properties, and an adaptive interface would have no purpose without its intelligent counterparts.

Additional models, such as machine and world models, are needed to provide a complete picture of the relationships between the task, user, machine, and environment. Note that user and machine models can also be referred to as a single unit, a system model. HMSs can adapt intelligently to changes in all four domains if they have knowledge of the task, operator, machine, and environment, as well as built-in adaptation mechanisms.

2.2.4 INTELLIGENT ADAPTIVE INTERFACES

The system's working environment is a critical element in designing effective IASs, and the dynamic nature of complex environments makes choosing the correct interface challenging. The right type of interface must adapt to rapidly changing tasks and operator cognitive states, as well as the working environment. For example, a jet-fighter radar system must communicate with satellites, ground control, the Internet, and all personnel monitoring the radar. Communication capabilities must act like a spider's web, interlinking the pilot, copilot, and airbase. When technology becomes intelligent, it is able to behave dynamically and make decisions based on environmental factors.

The progression from a conventional interface to an adaptive or intelligent interface to an IAI occurs through combining task, system (i.e., user and machine), and world models. An IAI requires more knowledge than either task or user model to appropriately allocate tasks between machine and human subsystems.

IAIs seek to maximize system performance by using technologies that intelligently adapt to changes in operator, machine, task, and environment. In other words, they support operators within both internal and external contexts. The machine's status and the operator's mental state are examples of internal context, while the system's working environment is an example of external context. A machine model consists of the machine's knowledge about itself and its ability to assist operators. For example, the machine model determines how the machine can help operators through offering advice, automation, or interface adaptation. IAIs can be defined as interfaces that react in real time to task and operator states and, as a result, dynamically change their physical presentation and control characteristics (Hou et al., 2007).

A world model allows the machine to monitor task status and environmental changes, calculate real-time decisions, and then relay the external status to the operator. It takes into account the machine's knowledge of its working environment, including the activities and objects that exist in the world, their properties, and the rules that govern them (Zhu and Hou, 2009). For instance, in the context of ATC, a world model would specify the principles of flight control and the principles of human behavior while under stress. Monitoring external factors also allows the machine to make suggestions to its operator based on those factors.

As computers become more powerful, more and more aiding tasks can be allocated to the computer, which makes the OMI more complicated. This makes it difficult for the systems designer to maintain operator situation awareness and may potentially place the human under the control of his or her machine partner. Although an IAI can be well-designed through the use of task, user, machine, and world models, an operator still needs to know what purpose an IAI is designed for and what activities the machine can perform. For safety-critical applications in the correspondence domain, it is especially important that systems designers have a deep knowledge of these activities and the way that operators conduct them. Note that the term *activity* is different from the term *task*.

These terms as used today are modified from the original ideals of activity theory that emerged from the Scandinavian and Russian cultural-historical research tradition (Kuutti, 1996; Norman, 2005). Figure 2.5 illustrates the hierarchy of activities according to activity theory. Activities are composed of actions (i.e., tasks), and actions are made up of operations (i.e., interactions) between the operator and the machine. According to Kuutti (1996), participating in an activity requires motive to perform conscious actions that have defined goals. Actions consist of chains of operations, which are executed in response to internal and external conditions faced when an action is performed. Thus, an activity can be defined as a coordinated, integrated set of tasks.

2.2.4.1 IAI Design Considerations

The 1979 meltdown of the Three Mile Island nuclear reactor discussed in Chapter 1 is an example of what can go wrong when operators are unaware of the full capabilities of an automated system. The discrepancies between the operators' mental model of how the plant worked and the actual activities being performed by the automated system meant that the operators took inappropriate actions when the alarms and indicators signaled abnormalities. Instead of asking an interface to adapt to the

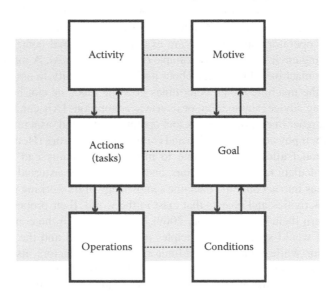

FIGURE 2.5 A hierarchical model of activity theory according to Kuutti (1996). Activity requires motive to perform conscious actions that have defined goals. Actions consist of chains of operations executed in response to internal and external conditions. (Redrawn from Kuutti, K., *Context and Consciousness: Activity Theory and Human-Computer Interaction*, MIT Press, Cambridge, MA, 1996.)

operator and the external world, the operators should have been intelligent enough to adapt to the changing machine activities.

Norman (2005) suggests that IAI design should use an activity-centered design approach, which is similar to user-centered design in that it requires a deep understanding of people. However, activity-centered design also requires a deep understanding of the technology that is used in the design, the tools that are used to facilitate interaction, and the reasons for the activities taking place (Norman, 2005). Successful IAIs fit gracefully into the requirements of the underlying activities, supporting them in a manner understood by the operator. Successful operators need to understand both the activity and the machine.

Substituting activity-centered design for user-centered design does not mean ignoring the human. Activities involve humans, and any system that supports activities must necessarily support the humans who perform them. However, human limitations and weaknesses may lead to negative consequences under abnormal circumstances. Neither user-centered design nor activity-centered design explicitly emphasizes the importance of understanding how humans act and react to events and how they communicate and interact with the machine to perform tasks. The oversight that the systems designers made regarding how operators might interact with a machine under abnormal circumstances contributed to both the catastrophic Three Mile Island event and the disastrous Flight 447 event that occurred 30 years later. HMS design, including interface design, considers not only human capabilities and limitations, the technologies, the working environment, and the activities themselves, but also, and perhaps more importantly, the interactions between operators and the machines through which operators conduct activities to achieve overall HMS goals (i.e., with the context). These factors are crucial to IAI design, and IAI design can best be described as interaction-centered design, which uses task, system, and world models, as illustrated in Figure 2.1.

Interaction-centered design is the design of HMSs for human communication and interaction, and is focused on finding ways to support human interaction within a HMS for applications primarily in the correspondence domain. The key concepts of interaction-centered design include both human–automation interaction and human–systems integration (i.e., the integration of multiple intelligent and adaptive components into an IAI). For example, UAVs have redefined the role of humans in warfare. The interaction between operators and UAVs has become increasingly important to accomplishing missions; operators generally remain in the loop and are paramount to an effective UAV operation (Pew and Mavor, 2007). To optimize operator interaction with various HMS components, systems designers must follow a coherent framework and coherent methodologies to understand not only operators' needs, but also how operators achieve system goals by interacting with the machine (both hardware and software) in its specific working environment under specific domain-context constraints. Interaction-centered design differs from the "interaction design" paradigm commonly found in the human–computer interaction (HCI) domain, which takes a user-centered design approach for applications primarily in the coherence domain. Interaction design focuses exclusively on satisfying the needs and desires of the user through effective interface design, and does not consider the needs and desires of other HMS components (Preece et al., 2002; Cooper, 2004; Cooper et al., 2012).

Although interaction design also uses analytical techniques, such as hierarchical task analysis, to identify user needs and interaction requirements, these techniques are not as comprehensive as other methods and processes that are used in interaction-centered design for correspondence domain applications. Furthermore, unlike most systems design approaches, which focus on system components as they are actually represented in the design, interaction design often focuses on imagining an ideal interface as it might be once the aforementioned techniques are applied (Cooper et al., 2012). *Interaction Design: Beyond Human–Computer Interaction* by Preece et al. (2002) and *About Face 3: The Essentials of Interaction Design* by Cooper et al. (2012) are recommended as introductory texts on interaction design.

IAI designers need to understand how operators conduct activities, how operators act, and how operators react to events. It is also important to understand operators' capabilities and limitations within assigned tasks and task contexts. Vital operator capabilities include attention, working memory, and decision-making skills (Eggleston, 1992). Typical IAIs are driven by software agents that intelligently aid decision making and operator actions under different workload levels and task complexity. IAIs should present the right information or present action sequence proposals or perform actions in the correct format, at the right time.

A successful IAI design should fit gracefully into the requirements of the underlying activity, facilitate interaction in a manner that is understandable to humans, and be able to optimize the interaction between the operator and machine partners to achieve overall system goals. The optimization of operator–machine interaction is further discussed in Chapter 5 in the context of operator–agent interaction.

2.2.4.2   IAI Design Example

The Microsoft Office Assistant, which was first included in Office 97 and removed after Office 2003, demonstrated both the positive and negative of commercial IAI applications. Clippy, an animated paperclip, presented various help search functions to users and offered advice based on mathematical algorithms or intelligence. Clippy automatically activated when it thought that the operator required assistance.

Clippy also used adaptation to modify the formatting of document and menu content. For example, typing an address followed by "Dear" would prompt Clippy to open and state, "It looks like you're writing a letter. Would you like help?" The embedded algorithms used a combination of task models (e.g., how a letter is formatted), user models (e.g., how many mistakes the operator made when writing a letter), machine models (e.g., how the Office software itself works), and world models (e.g., what the context of the task is and what the operator is trying to accomplish) to modify the interface to match the operator's needs and requirements. Clippy could be deactivated by the operator; otherwise, the system dynamically controlled the amount of adaptation that occurred.

The interaction-centered design logic of Clippy was an excellent example of IAI philosophy and many people welcomed it. However, Clippy was also an example of failure in IAS design because it misused automation. The inspiration for the design was a desire to assist people who become intimidated by complex software interfaces. The goal was for the software to observe and improve operator interaction with Office by providing valuable feedback and assistance using Bayesian methods

(Horvitz et al., 1998). The mathematics behind Bayesian methods are solid and have had many useful applications to date. However, a Bayesian inference engine never made it into Office 97. Instead, the inference engine was driven by a relatively simple rule-based method. Thus, the inference engine could only guess at the values of many variables using very little data, resulting in a small event queue with an emphasis on only the most recent operator interaction with the software interface. The inference engine did not have persistent operator profiles and could not reason about operator expertise. The separation between OMI events and word-based queries meant that the engine ignored any context and operator actions and focused exclusively on word-based queries. As a result, Clippy popped up all the time, causing Office users to dislike it and leading to its permanent removal from the program.

Clippy is a good example of an IAI feature that suffered because the designed automation did not reliably recognize operator goals and gradually caused the operator to lose trust (and patience) in its assistance capabilities. It is a reminder about the implications of poorly designed human–automation interaction. It also demonsrates the importance of the design of another essential IAS component, automation.

2.3 THE EVOLUTION OF AUTOMATION TECHNOLOGIES AND THE IAA CONCEPT

The realm of automation stems from human needs or, more specifically, the need to reduce labor and increase efficiency. Early forms of automation can be linked to water. In ancient China, horizontally oriented waterwheels powered trip hammers to crush ore and hull rice. Vitruvius, a Roman engineer from the Augustan Age, developed the first documented vertical waterwheel. In medieval Europe, social and economic catastrophe caused by events such as the Black Death pushed engineers to invent automated methods, including waterwheels, to do the work of the shrinking population (Hansen, 2009).

Waterwheels, and other early forms of automation that were capable of doing work without human intervention, are referred to as open-loop mechanical control systems (Wickens et al., 1997; Sheridan, 2002). The first completely automated industrial process was the centrifugal governor, which was originally designed to provide continuous self-corrective action to adjust the gap between millstones in flour mills, and was later used by James Watt to regulate the flow of steam into his steam engine (Bennett, 1979).

Automation is no longer dependent on water or steam for power. Since the Industrial Revolution, automation has been created using mechanics, hydraulics, pneumatics, electrics, electronics, and computers, usually in combination. Complex systems use a multitude of automation types. For example, an autopilot is a mechanism, or form of automation, that is used to guide an aircraft without pilot assistance. It is a collection of mechanical, hydraulic, pneumatic, electrical, electronic, and computing subsystems that help control various flight requirements, such as altitude, course, engine power, heading, and speed.

Automation, especially in the HF and HCI domains, can reduce human error, increase efficiency, and save lives. The extent to which any particular system should be automated depends on numerous factors. As a systems designer, it is important to

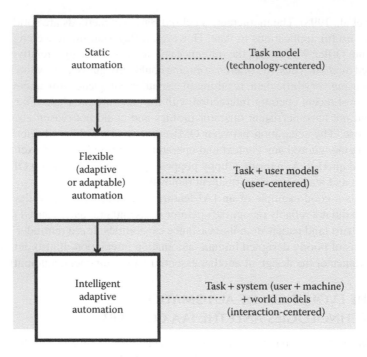

FIGURE 2.6   Evolution of automation technologies and their relationship to different design approaches.

understand the functions and tasks that automation can perform to assist operators as they conduct their activities. It is also important to understand the evolution of automation technologies and their impact on the allocation of functions and tasks between automation and operators.

Figure 2.6 illustrates the evolution of automation technologies, from static automation to flexible automation in the forms of adaptive automation, adaptable automation, and then IAA. Where static automation is often designed to improve efficiency by replacing human physical tasks, IAA seeks to augment and enhance human judgment, decision making, and responsibility.

2.3.1   STATIC AUTOMATION

Automation technologies are largely invisible and automated tasks are taken for granted. The general intention of static automation is to moderate the human workload to an appropriate level and increase efficiency and safety. ATMs are an example of static automation because they are unable to do anything beyond their designated activities: performing withdrawals, deposits, and transfers. Traffic lights are also an example of static automation as they repeat the same tasks in an infinite loop. Both are part of daily life in today's modern society and it would be difficult to imagine a busy intersection during morning rush hour without the presence of traffic lights.

Static automation design is technology-centered. Advances in automation technology have facilitated the moving of a wide variety of tasks to automated control, such as navigation, system health checks, and flight control.

The systems designer determines and assigns functions to both the operator and the automation and automation control is fixed for the duration of the task. Generally, if an automated machine can surpass human performance, the function is allocated to the machine. In other words, the function allocation takes a technology-centered approach using the identified task models illustrated in Figure 2.6. As technology advances, more functions can be considered for automation. It is important to note that because static automation is not context dependent, it is not sensitive to external factors. For example, even though an aircraft autopilot is reliable, human pilots will still need to monitor the system in case of sudden weather changes, wind speed changes, and so on and intervene in emergency situations.

Task allocation for static automation has to be predetermined at the HMS design stage. As discussed in Chapter 1, the allocation of tasks between the operator and the machine ranges from full operator control to full automation. Task allocation depends on system measurements and task efficiency rather than on satisfying operators' needs. As new tasks are assigned to static automation, issues emerge regarding how task allocation may affect operator and overall system performance. The effects on the cognitive abilities and deficiencies of the operator, such as motivation, tension, boredom, and fatigue, are extremely important considerations for the safe operation of human-in-the-loop systems. Studies show that the detection of automation failures substantially degrades in static automation systems where task allocation remains fixed over time (Parasuraman and Riley, 1997; Manzey et al., 2006; Flight Deck Automation Working Group, 2013). While automation can outperform humans for most monitoring tasks, humans excel in their ability to adapt when situations change.

There are concerns that the traditional static automation approach has not been entirely successful in the context of HMS design (Lee and Moray, 1992; Parasuraman et al., 2000; Wickens et al., 2012). Increased automation may make operators' tasks more difficult. Static automation can challenge operators by providing more information to interpret than is necessary. Operators may also experience difficulty maintaining alertness and vigilance over time without active involvement in the system's operation, which occurs as roles shift from operators being hands-on controllers to becoming system managers and system monitors.

A study conducted by Miller and Dorneich (2006) indicates that pilots must be the final authority in aircraft actions in order to keep operators in the loop and increase operators' acceptance of the automation system. This is especially vital for safety-critical correspondence domain applications. For example, when using cruise control, a vehicle maintains its predetermined speed without requiring the driver's foot on the gas pedal. After several hours, the driver may become detached from his or her role of driving the vehicle and lose situation awareness.

2.3.2 FLEXIBLE AUTOMATION

To minimize the negative effects of static automation and simultaneously achieve optimal HMS performance, automation needs to be adaptive to changes in both internal (i.e., human and machine) and external (i.e., working environment) states. The allocation of flexible automation tasks between human and automation cannot be predetermined at the HMS design stage as task allocation is not fixed for the duration of the task. An example of a simplistic, flexible automation system is a modern flight management system that automates the presentation and partial completion of operating checklists according to the phase of flight, or the detection of subsystem failures. More complex forms of flexible automation employ simplistic user models, such as simple behavioral indices of high workload, as mechanisms to control the onset and offset of automation.

Flexible automation dynamically, or adaptively, shares control of the onset and offset, and the degree of specific automated tasks distributed between human and machine (Barnes and Jentsch, 2010). In these systems, the human remains in the loop and the automation intervenes only when an increase or decrease in the human workload makes automation support necessary to meet operational requirements. In other words, flexible function allocation between automation and operators takes a user-centered approach with identified user and task models.

Keeping the operator in the loop maintains operator situation awareness. By providing adaptive support, flexible automation can reduce the perceived loss of control associated with static automation.

Flexible automation attempts to maximize operator interaction within a system in real time. The strategies for flexible task allocation may relate to operator workload levels. For example, ATC uses automated systems, such as an automated data synthesis module, to improve the quality of information received by ATC operators. This frees the operators from performing simple, routine activities (Wickens et al., 1997). As a result, ATC operators have less cognitive load, and can apply themselves to doing what humans do best—making decisions.

Flexible automation is a useful set of concepts for eliciting automation while minimizing workload, depending on the requirements of the specific tasks allocated to automation. There are various definitions for flexible automation (Kaber et al., 2006; Miller and Parasuraman, 2007; Parasuraman et al., 2007, 2009; Tecuci et al., 2007; Goodrich, 2010; Lewis et al., 2010; Wang et al., 2010; Chen et al., 2011), the most popular ones being adaptive automation and adaptable (or adjustable) automation.

The word *adaptive* describes flexible automation, but it also has multiple uses and understandings within systems design. Systems designers should be aware that many words that they encounter on a daily basis will carry different meanings depending on the context in which they are being used. For example, in the HF context, many people understand adaptive to mean the operator and automation sharing control when describing a HMS. In the HCI context, adaptive means that the machine has all the control.

In the HCI domain, the decision-making process stretches along a continuum between adaptive systems and adaptable systems. Adaptive systems rely on machine-initiated adaptivity and adaptable systems rely on human-initiated adaptivity. Figure 2.7 illustrates this continuum, developed by Oppermann and Simm (1994), and notes five categories for how human involvement and authority vary between the two systems. These categories are defined by (a) the amount of system status

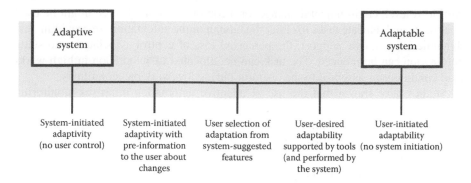

FIGURE 2.7 Adaptivity continuum illustrating systems with no user control to systems with full user control. (Drawn using data from Oppermann, R. and Simm, H., *Adaptive User Support: Ergonomic Design of Manually and Automatically Adaptable Software*, CRC Press, 1994.)

information given to the operator, and (b) the amount of control the machine and the operator has over the initiation of adaptation. The five categories are as follows:

1. *Adaptive*: The machine has total control over the adaptation.
2. *System-initiated adaptivity*: The machine notifies the operator of any changes made prior to execution. The operator has no control over the choice, timing, or implementation of the adaptation.
3. *Operator-selected adaptation*: The operator selects the adaptation based on the machine's suggestions. The machine performs the action.
4. *Operator-initiated adaptability*: The operator selects and initiates the adaptation without the machine's suggestions. The machine implements the actual change.
5. *Adaptable*: Complete operator control over the adaptation.

According to Oppermann and Simm's automation continuum, flexible systems cover the full range of adaptivity—flexible automation includes both adaptive automation and adaptable automation.

2.3.3 Adaptive Automation

The desire to improve the relationship between human and machine, and consequently overall system performance, has prompted the development of an alternative approach to automation. The goal of adaptive systems is to maintain a balance between reducing the workload for manual tasks and reducing the loss of situation awareness for automated tasks (Kaber et al., 2006).

Adaptive automation seeks to take advantage of the differences between human abilities and machine abilities by using strategies that allow for changes in task allocation. For example, one adaptive task allocation strategy might address how operator workload is handled. Emphasis is given to the prevention of task overload (or underload) on the operator. An adaptation trigger based on operator performance, workload level, environmental change, or a combination of all three might be used to allocate and relocate tasks to either the human or the automation. In providing this dynamic, or adaptive, support, the perceived loss of control associated with static automation can be reduced. The task can be allocated to automation in high-workload conditions or assigned to the operator during low-workload conditions.

Studies have shown that the use of adaptive automation improves monitoring and tracking performance, increases the detection rates of automation failures, and reduces task completion times (Chu and Rouse, 1979; Freedy et al., 1985; Rouse and Morris, 1986; Hilburn et al., 1993; Parasuraman et al., 1993). Parasuraman et al. (2009) investigated the effects of various types of automation for monitoring unmanned aerial and ground systems and concluded that superior situation awareness and change detection were demonstrated in adaptive automation rather than static automation conditions (Parasuraman et al., 2007). They also indicated that adaptive automation is most effective for multitasking situations where fully automated tasks may be neglected if an operator becomes complacent or allocates his or her cognitive resources to other tasks.

Adaptation can be triggered by objective measurements of operator workload. Both human behavioral measurements, such as operator error rate, and task difficulty have been used as triggering conditions to improve situation awareness and secondary task performance for the supervisory control of heterogeneous robotic systems (Parasuraman et al., 2007). Psychophysiological monitoring technologies such as electroencephalography (EEG), functional magnetic resonance imaging, and heart rate variability (HRV) have also been used successfully in a number of laboratory settings as triggering conditions related to different workload levels (Prinzel et al., 2000; Wilson and Russell, 2003). Psychophysiological monitoring technologies are further discussed in Chapter 6.

Adaptive automation improves the operator's monitoring performance by reducing task management workload so that the operator does not need to decide when to initiate automation, or what should trigger automation. This is more important in situations where safety and time constraints can become issues. For example, when monitoring an incoming missile in flight, an operator may not have time to give permission before the automated machine is activated to respond and avoid the missile attack (Barnes and Grossman, 1985). However, adaptive automation introduces complexity during task allocation that can result in new problems of awareness regarding system functional status and automation failure detection. A key design issue is optimizing triggering conditions for task reallocation, which usually involves monitoring operators' behavior, operators' mental state, or situation and task events. If triggering conditions are not optimized, sudden changes in the task or the operator's state may be annoying to the operator or dangerous to the system as a whole, assuming authorization of task allocation has been delegated to the machine.

Parasuraman et al. (2007) concluded that adaptive automation should be set to moderate levels in high-risk environments because humans may make fatal errors if they are not included in the decision-making process. Brief periods of manual task performance help operators maintain their situation awareness and, as a result, increase their detection rate of automation failures.

2.3.4 ADAPTABLE AUTOMATION

According to Oppermann and Simm's automation continuum, automation is adaptable if the human initiates the adaptivity of joint human–machine tasks (Miller and Parasuraman, 2007; Langan-Fox et al., 2009; Goodrich, 2010). Adaptable automation ensures that operators play an active role in the control loop, which is important as reductions in operator workload occur when the system improves, rather than replaces, the operator's decision-making ability.

A good example of an adaptable automated system is Playbook™ developed by Miller et al. (2005) to facilitate the effective and flexible tasking of multiple automation agents while keeping the operator in the decision-making loop. Plays are called by an operator to execute specific automated tasks, much like a football quarterback would call specific plays to his team. The operator has control over agents and Playbook facilitates active human–machine interaction.

Playbook is designed to give the operator the flexibility to modify the constraints placed on automation agents based on the required solution. This flexibility is beneficial as constraints ensure a plan that is more precisely in keeping with the operator's intent, even though it takes more time. Having fewer constraints on agents can be quicker, but it gives more freedom to the automation and less predictability in the results. Playbook achieves the goal of keeping the operator in the decision-making loop without increasing the operator's mental workload, and allows the operator and automation agents to successfully adapt to unpredictable changes in the environment (Miller et al., 2005). The Playbook approach has been successfully applied to complex operations, such as the control of a group of simulated robots (Squire et al., 2006) and the simulation and field testing of multiple UAVs (Miller et al., 2005, 2011; Parasuraman and Miller, 2006; Fern and Shively, 2009; Miller and Parasuraman, 2007). Playbook and other agent design approaches are further discussed in Chapter 5.

A number of researchers have examined the effectiveness of adaptable automation versus adaptive automation. Valcro-Gomez et al. (2011) compared automation types when controlling multiple robots in a simulated search-and-rescue tasking environment. Their results indicated that when the operator could authorize automation execution in the adaptable automation condition, human–robot team performance was better than in the adaptive automation condition, which had fixed levels of automation. In a similar study for a multirobot control task, Kidwell et al. (2012) found that when a higher workload was associated with adaptable automation, the change detection performance of participants was better than in the adaptive automation condition.

Adaptable automation ensures that the human maintains control of his or her machine partner by providing the operator with the ultimate authority to execute automation. However, the dynamic nature of task allocation may not be practical if automated tasks become more complex and more time is required to make decisions. This is critical in time-constrained and high-workload environments under emergency conditions. Additionally, the operator may have to be well trained to understand the behavior of all automated tasks and their influences on human–machine interaction and the overall HMS performance.

Systems designers should not build too much complexity or too much automation into the machine component, as too much automation may be met with resistance by the operator. For example, operators who depend on their "manual control expertise" may prefer to have little or no automation (Parasuraman et al., 2007). Systems designers must address issues such as (a) the ideal balance of manual control and automation, and (b) at what point should the operator take control over automation.

Both adaptive automation and adaptable automation are considered dynamic as the loci of control, or the extent that individuals believe that they can control the events that surround them, constantly change (Rotter, 1954, 1966). During task allocation and reallocation, this dynamic complexity can result in new problems that may affect awareness of system functional status and automation failure detection. A key design issue is to optimize automation triggering conditions for task reallocation, such as monitoring operators' behavior or monitoring situations or task events. Ideally, systems designers will incorporate a task model, user model, machine model, and world model into any HMS design.

2.3.5 INTELLIGENT ADAPTIVE AUTOMATION

Traditionally, automation design took a technology-centered approach and focused on maximizing technological performance. Flexible automation, which includes both adaptive and adaptable automation, aims to reduce the negative effects of static automation by dynamically shifting tasks between human and machine. Its goal is to keep the operator in the loop and maintain situation awareness. However, the nature in which flexible automation handles changes in task allocation results in the separation of HMS functions. It invites issues regarding the operator's loss of authority and situation awareness in adaptive automation, and high workload and potential system failure in response to abnormal events in adaptable automation. In addition, the validity of applying the concept of functional separation of tasks between human and automation is questionable when tasks become more mental than physical. This is especially true when a HMS design is based strictly on the considerations of human capabilities and limitations rather than on the overall system performance and efficiency.

The US Defense Science Board suggested that the design and operation of autonomous systems need to be considered in terms of human-machine collaboration (Defense Science Board, 2012). Automation within HMSs needs to be designed in such a way that it not only keeps humans on the loop but also in the loop when necessary. Automation should not only provide feedback to the operator but also intelligently assist in maintaining situation awareness and making timely decisions by knowing when and how to assist the operator.

Intelligence is necessary as functional integration, rather than function allocation, is a key characteristic of IAA. With functional integration, the behaviors required by the domain are shared across multiple functional components, including the operator and automation. Thus, the same behavior can be performed by one of several components. Functional integration creates robust systems that are better able to handle unexpected events. For example, safety redundancies are built into aircraft control systems to allow an alternate course of action if a key component fails.

The aim of IAA is to augment and enhance the operator's judgment, decision making, and responsibility while mitigating the operator's limitations. IAA systems are intelligent insofar as they exhibit behavior that is consistent with the humanlike characteristics defined by Taylor and Reising (1998): (a) actively collecting information, (b) being goal driven, (c) being capable of reasoning at multiple levels, and (d) learning from experience. IAA systems seek to restore the operator to the role of the decision maker. IAA systems also seek to provide safeguards for situations where time constraints or problem complexity restrict the operator's problem-solving ability.

2.3.6 Design Basics for IAA

IAA is considered a higher level of automation than either adaptive or adaptable automation because it is context dependent and is therefore capable of implementing changes based on the environment (Wickens et al., 1997). For example, the autopilot system on a Boeing 777 monitors airspeed, temperature, and weather, and warns pilots when a potential problem occurs. Vehicle cruise control functions are also context dependent, to a lesser degree. Cruise control will maintain a set speed when it encounters environmental changes, such as steep hills; it will also maintain speed if it encounters unexpected events, such as an animal darting in front of the vehicle. This partial adaptation can lead to negative results (e.g., the animal may be hit by the vehicle) and demonstrates why it is vital for operators to be aware of automation, and have authority over automation at all times.

IAA systems can be further distinguished in terms of the tasks and roles that they perform and the knowledge that they manipulate (Geddes and Shalin, 1997). Geddes and Shalin (1997) divided IAA systems into three categories:

- *Assistant*: An assistant performs specific tasks when instructed by the operator. For example, an assistant system can provide a fighter pilot with an assessment of a threatening aircraft when asked.
- *Associate*: An associate automatically recognizes that the operator requires assistance and provides some level of support. For instance, an associate system can recognize a threatening situation and automatically provide the pilot with all threat information.
- *Coach*: A coach uses task, situation, and operator knowledge to determine the need for automation in order to achieve a task objective, and provides instructions to the operator on how to achieve the objective. In the case of a coach system, a pilot would be presented with the most threatening aircraft first, in accordance with the higher-level goal of maximizing own-ship survivability.

The technology-centered and user-centered design approaches that are used in static and flexible automation are not robust enough to accommodate IAA design requirements. IAA design requires an interaction-centered approach that relies on detailed and comprehensive task, system, and world models, as illustrated in Figure 2.6. By focusing on the interaction between humans and machine, IAA design seeks to enhance the operator's judgment, decision making, and responsibility, and maximize overall HMS performance.

For example, the rain-sensing wiper control systems seen in many General Motors cars demonstrate interaction-centered design. Using an advanced sensing system, analog signal processing, and a control algorithm, this IAA technology provides drivers with added safety, convenience, and comfort in various weather conditions.

Traditional windshield wipers are difficult to adjust properly and the operator (i.e., driver) is continually switching between fast and slow speeds to accommodate for external conditions. Rain-sensing wiper control systems work using multiple built-in models. A sensor detects the amount of moisture on the windshield. The electronics and software in the sensor activate the wipers when the amount of moisture reaches a specific level devised using a world model. Using a task model, the software sets the speed of the wipers based on how quickly the moisture builds up on the window. The system adjusts the wiper speed as often as necessary to match the rate of moisture accumulation. The change in the amount of moisture is a result of the driver's change in car speed (which reflects a user model) or of the weather conditions (as seen in a world model). Whenever necessary, the system can be overridden or turned off. In other words, the wiper system intelligently determines when to activate the wipers and how fast to run the wiper motor based on inputs from both the driver and the weather conditions. Without diverting the driver's attention, it keeps the windshield clear, allowing the driver to focus on the road and maintain the authority of overall interaction.

2.4   UNDERSTANDING INTELLIGENT ADAPTIVE SYSTEMS

Systems designers must be aware of operator requirements and operator preferences when designing conventional automation and interfaces. Operator requirements are the functionality and capabilities that are required to perform a task. Operator preferences are the functionality and capabilities that improve the perceived quality of an operator's interaction with the machine. For example, customizing operator preferences, such as changing the brightness level of a car dashboard or the display properties of aircraft radar, can improve operator–machine interaction.

Conventional automation and conventional OMIs are both designed using a technology-centered approach based on task models. They incorporate a combination of operator requirements and operator preferences and place an emphasis on how technology advances can help allocate more tasks to automation. Their purpose is to reduce workload, increase task efficiency and accuracy, and provide greater convenience. The task model preplans and predesigns all automation functions.

Automation capability is derived from the leftover principle and the compensatory principle. The leftover principle states that operators are left with functions that have not been automated, or could not be automated. The compensatory principle states that function allocation is based on the strengths and weaknesses of human and machine (Fitts, 1951). With the advance of computerized aiding technologies, more and more tasks are being allocated to automation. However, operators may have difficulty maintaining situation awareness when limited to the static nature of conventional automation and OMIs (Billings, 1991; Cook et al., 1991; Endsley and Kiris, 1995). To mitigate human information processing limitations and maintain situation awareness, HMSs need to know how operators are doing. Furthermore, interface and automation technologies need to be flexible in order to adapt to changes in operators' state.

2.4.1 THE EVOLUTIONARY NATURE OF INTELLIGENT ADAPTIVE SYSTEMS

The need to assist the operator in a flexible fashion has fostered the development of adaptive and adaptable automation, as well as adaptive and intelligent interface technologies. To assist the operator, the machine must trigger automated aids based on one or more of the following: (a) evaluation of work conditions, (b) operator arousal and stress levels, and (c) operator workload levels. Knowledge of these triggers requires the use of a user model that encompasses the machine's understanding of the capabilities, limitations, needs, and demands of the operator. A good user-centered design approach ensures that HMSs, including IASs, work effectively and that people can use them safely. To facilitate this, it is imperative that systems designers have a deep and detailed knowledge of potential operators.

It is also critical that systems designers take into account the working environment of the HMS. Complex working environments make it challenging to choose the correct type of automation or interface. Generally speaking, the right type of automation for IAS design must adjust its machine processes and adapt to rapidly changing tasks, operator cognitive states, and working environments.

IASs require knowledge beyond their built-in user and task models to intelligently allocate tasks between operators and automation. When a technology becomes intelligent, it is able to behave dynamically and make decisions based on its environment as well as self-knowledge. The progression from adaptive or intelligent interfaces and adaptive or adaptable automation to IAIs and IAAs is possible through combining task, user, machine, and world models.

An IAS is essentially the unified evolution of an IAI and IAA into a hybrid system that features state-of-the-art automation and interface technologies. Figure 2.8 illustrates the existence of conceptual parallels in the evolution processes of both interface and automation. It shows the consistencies in their evolution and their eventual amalgamation into an IAS.

2.4.2 IAS DESIGN OBJECTIVES

IASs are a synergy of IAI and IAA technologies that are capable of context-sensitive communication with an operator. These technologies are possible through technological advances in artificial intelligence and the physiological monitoring of operators (Schmorrow, 2005). One of the goals of IAS systems is augmented cognition, which involves using technology to monitor operator cognitive state and trigger automation aids to reduce and augment cognitive loads and increase overall operator situation awareness. Operator state monitoring concepts and approaches are further discussed in Chapter 6.

Unlike conventional automation systems that rely on discrete control centers, IASs are fully integrated, intelligent systems that can adopt agent-like properties. In this

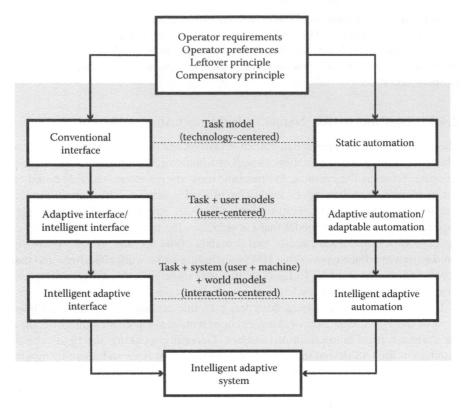

FIGURE 2.8 Evolution of intelligent adaptive systems (IASs) from intelligent adaptive interface (IAI) and intelligent adaptive automation (IAA) technologies.

context, an agent is a subsystem within a HMS that is authorized to act on behalf of the operator. Agent-based design methods are further discussed in Chapter 5. According to Eggleston (1997), a well-designed IAS should enable the system to

- Respond intelligently to operator commands and provide pertinent information to operator requests
- Provide knowledge-based state assessments
- Provide execution assistance when authorized
- Engage in dialogue with the operator either explicitly or implicitly, at a conceptual level of communication and understanding
- Provide the operator with a more usable and nonintrusive interface by managing the presentation of information in a manner that is appropriate to mission content

In addition, IASs should be able to provide support for the basic functions of assessment: planning, coordinating, and acting. In most cases, IASs will propose a solution for the operator. In extreme cases, IASs will propose, select, and execute the solution for the operator.

IASs should provide several of the following functional capabilities:

- *Situation assessment*: This function supports the organization of large amounts of dynamic data into concepts at varying levels of aggregation and abstraction, provides the context in which the system operates, and relays information to other machine processes and the operator, using task, system (user and machine), and world models.
- *Planning*: Plans, which may cover different periods of time at different abstraction levels, are formulated based on the current situation as determined by the IAS's situation assessment capabilities. With information from the operator and the environment, an IAS can independently formulate and propose plans to the operator, and can also complete the details of partial plans provided by the operator.
- *Acting*: IASs are active HMSs that are capable of acting on behalf of their operators. Given a set of plans and an evolving situation, the system may issue commands directly to active system elements, such as sensors, flight controls, and secondary support systems.
- *Coordination of behaviors*: An IAS must be able to coordinate its behavior at four distinct levels. At the lowest level, an IAS must coordinate its own assessment, planning, and executing processes to produce coherent behaviors. At the planning level, it must coordinate with operators to receive direction and provide useful recommendations. At the action level, it must coordinate with operators so that the operator and automation act in concert. Finally, it must coordinate with other independent participants to avoid undesirable conflicts and satisfy the requirements of higher-level system goals.

The IAS functional capabilities necessitate an interaction-centered design approach with an emphasis on augmenting human judgment, decision making, and responsibilities to maximize overall system performance. Details regarding IAS conceptual architecture and design guidance are provided in Chapters 3 and 7.

2.4.3 Design Basics for IAS

IASs seek to enhance HMS performance by leveraging technologies that intelligently adapt the OMI, task automation, and support provided to the operator in accordance with internal and external contexts. Given the closed-loop nature of IASs, the operator is an intrinsic part of the system. As a system monitors and adapts to operator state, the nature of how tasks are performed changes, which requires the system to again monitor and adapt to operator state.

For example, a pilot flying a top-secret aircraft over enemy territory scans the controls at cruising altitude. The aircraft's safety warning engages and an electronic voice says, "Engine bay two and three have malfunctioned. Perform emergency landing maneuvers?" The pilot freezes, the aircraft's engines stall, and the plane quickly descends. "Performing emergency landing procedures." The aircraft banks hard to one side, and the pilot can see a landing strip in the distance. The pilot attempts to take control but the safety system detects that his or her psychological state is not calm enough to provide sound judgment. "Override automatic landing procedures?" the voice asks. "Override," confirms the pilot. As he or she attempts to set the aircraft on the proper trajectory, the safety system calculates that there is not enough speed to make the landing strip. The electronic voice states, "Commencing evacuation procedures. Eject aircraft in thirty seconds."

In decomposing this scenario, a number of factors come to light. The IAS recognizes mission parameters—because the aircraft is top secret, the system instructs the aircraft to self-destruct if it is about to crash in enemy territory. The IAS monitors the pilot's physiological and psychological states (i.e., internal operator context), as well as the speed required to make it to the landing strip (i.e., internal machine context). Note that the system recognizes the pilot's stress and the machine status, and engages in emergency landing maneuvers. IASs recognize internal and external contexts, giving them the ability to calculate the best course of action for the operator. Despite automatic landing procedures, the pilot maintains full control and is able to perform landing procedures manually as needed. However, the pilot must understand all high-level automation activities and must be confident that the system will make the right decision, particularly in safety-critical situations, such as the aircraft's decision to eject the pilot rather than allow an attempted landing.

Operators are an intrinsic part of closed-loop IASs and monitoring and adapting to operator state changes are two functions that IASs must perform. For example, if an aircraft operator's state shifts from relaxed to tense, the system will recognize the change in pilot physiology. The nature of how a pilot performs tasks may also change. If pilots are tense, their decision-making ability can be affected. Therefore, the system must understand that a tense pilot will perform tasks differently than a relaxed pilot. The system must also continuously sample the operator's state to ensure that the operator is kept in the loop and is able to maintain situation awareness.

A real-world example of an IAS that acts as an associate to the operator and a coach to subordinate robots is the RoboLeader system for multirobot control in multitasking environments (Chen and Barnes, 2012a,b). RoboLeader was developed to investigate human performance issues associated with the control of multiple intelligent systems in both civilian environments (e.g., the operation of autonomous farm equipment) and military environments (e.g., explosive device detection and disposal). The systems design was based on (a) mixed-initiative human–machine interaction, where operators have ultimate decision authority to ensure system safety and flexibility while incorporating implicit high-level system goals that automation may not be aware of (Guerlain, 1995; Woods et al., 2004; Linegang et al., 2006); (b) supervisory control to ensure that the operator maintains a manageable workload and has adequate situation awareness for controlling multiple robotic systems (Chen et al., 2011); and (c) automation transparency, where both the operator and the autonomous subsystems are aware of each other's role within the system (Lee and See, 2004).

Figure 2.9 illustrates the strategy of operator interaction with RoboLeader-based systems. The operator manages one entity, instead of interacting with multiple robots. This allows operators to concentrate on other tasks requiring their attention. RoboLeader can collect information from subordinate robots with limited autonomy, such as collision avoidance and self-guidance to reach target locations; make

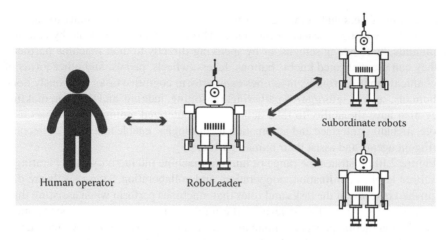

FIGURE 2.9 Operator interaction with a RoboLeader-based system. RoboLeader acts as an interface between the operator and subordinate robots and can make some decisions autonomously.

tactical decisions; and coordinate subordinate robots through means such as issuing commands, setting waypoints, and plotting motion trajectories (Snyder et al., 2010). Typically, RoboLeader recommends route revisions when environmental circumstances require robot redirection. In response, the operator either accepts the recommended revisions or modifies them as appropriate.

The hierarchical characteristics of this system enable RoboLeader to serve as an interface between its human supervisor and subordinate robots. RoboLeader is a mixed-initiative system that consists of both human and machine decision-making components. It supports collaborative decision making between human and IAS, and resembles a human collaborating with a subordinate but autonomous associate (Tecuci et al., 2007; Hardin and Goodrich, 2009; Cummings et al., 2012). RoboLeader automation has proved ideal for circumscribed solutions and as a means to reduce the operator's multitasking burden. It is, however, essential to maintain human decision-making authority without affecting automation awareness of the unfolding task situation (Chen and Barnes, 2014).

2.4.4 Automation and Collaboration

It is vital to understand the connections in every human–machine interaction at a detailed level. If too much automation occurs, the human is left out of the loop. If humans are out of the loop, they are less aware of what automation does and what changes are made in their working environment as a result of automation action. Thus, to optimize human–machine and human–automation interaction, there must be a balance between human and automation.

As discussed previously, the introduction of computers has changed the way that humans interact with machines, as well as the tasks that machines perform to assist their human partners. Exponential increases in computing power since the introduction of personal computers in the 1970s have enabled humans to interact with machines using a variety of interfaces. Humans can communicate by entering commands, by touching screens, or by speaking directly to their machine partners, or they can use dedicated knobs, buttons, levers, wheels, pedals, and other forms of mechanical input. Currently, machines can perform cognitive tasks previously done by humans, such as information gathering, reasoning, judging, and decision making. They also have the ability to take actions. Computerized aiding technologies as a whole, including interface and automation technologies, enable machines to become intelligent agents and assist their human partners.

Figure 2.10 illustrates the range of human–machine interactivity by separating it into three levels: coordination, cooperation, and collaboration. These levels are distinguished in terms of the tasks and roles that machines perform when assisting their human partners. At the lowest level of the interactivity continuum is coordination, where the interaction between human and technology is passive. Passive interaction occurs when operators and technologies work independently to achieve the same

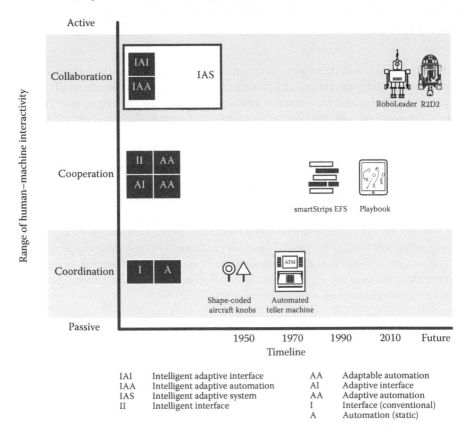

IAI	Intelligent adaptive interface	AA	Adaptable automation
IAA	Intelligent adaptive automation	AI	Adaptive interface
IAS	Intelligent adaptive system	AA	Adaptive automation
II	Intelligent interface	I	Interface (conventional)
		A	Automation (static)

FIGURE 2.10 A historical view of interface and automation technologies and the range of human–machine interactivity. The *x*-axis displays a timeline that begins with the conventional interfaces and automation explored in the 1950s, and ends with current and future IASs, such as RoboLeader and R2D2.

goal. To produce coherent behaviors, technology must coordinate between its own assessment, planning, acting, and coordinating processes. Through coordination, interfaces and automation perform their specific tasks according to preplanned programs and predefined protocols embedded in the system and do not take over operator tasks. Likewise, operators do not take over the automation's tasks. Coordination is particular to conventional interface and static automation systems that provide specific support using basic task and situational knowledge, which is a technology-centered design approach. Technologies at the coordination level act as an assistant (rather than an associate or coach) to the operator. For example, an ATM does its own job to support a person's financial transactions when asked.

The next level on the interactivity continuum is cooperation, where interaction can be active as operator and technology work together to achieve system goals. Cooperation is achieved through user-centered design and the technology must coordinate with the operator at the planning level to receive direction and provide useful recommendations. The technology must also cooperate with the operator at the action level to work in unison, and each should offer help to the other if necessary. Cooperation also allows the technology to take over operator tasks if given permission. For example, vehicle cruise control systems exhibit cooperation because they can take over the task of maintaining vehicle speed from the operator, and the operator can end the automation assistance and accelerate the vehicle at any time.

Cooperation also applies to intelligent or adaptive interfaces and adaptive or adaptable automation systems that provide different solutions and act as an associate to the operator (rather than as an assistant or coach). A computerized aiding system that acts as an associate can automatically recognize the needs of the operator based on complex task and situational knowledge, as well as basic user model and operator coordination knowledge.

Collaboration facilitates active human–machine interactivity at the highest level. Technology must collaborate with other independent participants in order to avoid undesirable conflicts and satisfy higher authority requirements within the domain. In collaborative system environments, the technology automatically augments operator tasks. Collaborative systems support the operator proactively, acting as a coach. They explain what is happening, offer available options, and then suggest an appropriate course of action. By using built-in protocols and relying on mutual understanding with operators, a collaborative system may take over operator tasks without seeking permission. This is extremely important in emergency situations. For example, the autopilot that presented available options, but ejected the pilot without authorization after it realized that the aircraft would not reach the landing strip.

Proactive support is inherent to IASs. These machines can aid and instruct to better assist operators by using collaboration knowledge (for decision making, dynamic function allocation, etc.), coupled with complex task, system, and world models. The design of collaborative aiding systems takes an interaction-centered approach. Note that collaboration and cooperation continuums can be applied to any system, from high-end mission-critical systems to low-end everyday applications.

Computerized aiding technologies at different stages of their evolution provide different levels of interaction and assistance to operators. For conventional interfaces and static automation, automated assistance is only applied to preplanned system functions. These automated functions are either defined before actions begin or are designed to describe current actions to the operator. Based on these predefined functions, conventional automation and interfaces take on a coordination role as assistants and respond to operators' needs, as do the shape-coded aircraft knobs and the ATM in Figure 2.10.

Intelligent interfaces and adaptive interfaces, as well as adaptable automation and adaptive automation, have automated functions that not only inform the operator about current actions, but also provide solutions in response to changes in system, tasks, and working environment. These types of interfaces and automation can dynamically allocate functions and act as an associate in a cooperation role to support operator tasks, as do the smartStrips and Playbook in Figure 2.10.

For an IAS combining both IAI and IAA, automated functions are proactive and focus on the operator. Interfaces and automation take on a collaborative role as coach, associate, or assistant to enhance operator capabilities and maximize overall system performance, as do RoboLeader and R2D2 in Figure 2.10. Without asking for human permission, proactive functions can make their own plans and decisions using their own assessment of the task, operator state, machine status, and working environment. By clearly understanding the definitions of interface and automation in the context of IAS design, as well as the nature of operator, interface, and automation interaction, systems designers have the potential to realize safe and effective IAS design.

2.5 SUMMARY

This chapter introduced the two most basic components of IASs: interface and automation. It discussed both concepts from their conventional beginnings to their current intelligent adaptive states. Specific interface terminology—notably, HMI, MMI, and OMI—was addressed. EFSs were used to show the evolution of interface technology and the benefits of using it, and Microsoft Office's Clippy was highlighted as a software agent that was both an excellent IAI concept and, ultimately, a failed IAI implementation.

The ideas of static and flexible automation were examined, and the differences between adaptive, adaptable, and IAA were explained. The evolution of automation design from technology-centered to user-centered to interaction-centered was explored. IAA technologies were identified as assistants, associates, or coaches depending on their tasks, roles, and knowledge; and a rain-sensing wiper control system was explored as an example of current IAA technology.

The leftover principle and compensatory principle were addressed as starting points for automation design; and the importance of operator situation awareness in the context of IASs was reiterated. The evolutionary nature of IASs as a hybrid of IAI and IAA technologies was explored, including their ability to perform situation assessment, plan, and act. RoboLeader was provided as an example of an IAS that acts as both an associate (to humans) and a coach (to subordinate automation). Human–machine interaction was illustrated as an interactivity continuum focusing on the concepts of coordination, cooperation, and collaboration, and the importance of operators staying in-the-loop and on-the-loop was explained.

Further background on early IASs is explored in Chapter 3, which also outlines conceptual architectures for IAS design that can be used and adapted by systems designers.

REFERENCES

Abascal, J., De Castro, I. F., Lafuente, A., and Cia, J. M. (2008). Adaptive interfaces for supportive ambient intelligence environments. In K. Miesenberger, J. Klaus, W. Zagler, and A. Karshmer (eds), *Computers Helping People with Special Needs*, vol. 5105, pp. 30–37. Berlin: Springer.

Barnes, M. J. and Grossman, J. (1985). The intelligent assistant concept for electronic warfare systems (Report No. NWCTP5585). China Lake, CA: Naval Weapons Center.

Barnes, M. J. and Jentsch, F. (eds). (2010). *Human-Robot Interactions in Future Military Operations*. Farnham Surrey, UK: Ashgate Publishing.

Bennett, S. (1979). *A History of Control Engineering 1800–1930*. London, England: Peter Peregrinus.

Benyon, D. and Murray, D. (1993). Adaptive systems: From intelligent tutoring to autonomous agents. *Knowledge-Based Systems*, 6(4), 197–219.

Billings, C. E. (1991). Human-centered aircraft automation: A concept and guidelines. (NASA Technical Memorandum 103885.) Moffett Field, CA: NASA-Ames Research Center.

Bowman, D. A. and Hodges, L. F. (1999). Formalizing the design, evaluation, and application of interaction techniques for immersive virtual environments. *Journal of Visual Languages and Computing*, 10(1), 37–53.

Browne, D., Totterdell, P., and Norman, M. (1990). *Adaptive User Interfaces*. London, England: Academic Press.

Chen, J. Y. C. and Barnes, M. J. (2012a). Supervisory control of multiple robots effects of imperfect automation and individual differences. *Human Factors: The Journal of the Human Factors and Ergonomics Society*, 54(2), 157–174.

Chen, J. Y. C. and Barnes, M. J. (2012b). Supervisory control of multiple robots in dynamic tasking environments. *Ergonomics*, 55(9), 1043–1058.

Chen, J. Y. C. and Barnes, M. J. (2014). Human-agent teaming for multi-robot control: A review of human factors issues. *IEEE Transactions on Human-Machine Systems*, 44(1), 13–29.

Chen, J. Y. C., Barnes, M. J., and Harper-Sciarini, M. (2011). Supervisory control of multiple robots: Human-performance issues and user-interface design. *IEEE Transactions on Systems, Man, and Cybernetics, Part C: Applications and Reviews*, 41(4), 435–454.

Chu, Y. Y. and Rouse, W. B. (1979). Adaptive allocation of decision making responsibility between human and computer in multitask situations. *IEEE Transactions on Systems, Man and Cybernetics*, 9(12), 769–778.

Cook, R. I., Woods, D. D., McColligan, E., and Howie, M. B. (1991). Cognitive consequences of "clumsy" automation on high workload, high consequence human performance. In R. T. Savely (ed.), *Fourth Annual Workshop on Space Operations, Applications and Research, SOAR '90* (Report No. CP–3103). Washington, DC: NASA.

Cooper, A. (2004). *The Inmates are Running the Asylum: Why High Tech Products Drive Us Crazy and How to Restore the Sanity*, 2nd edn. Indianapolis, IN: SAMS Publishing.

Cooper, A., Reimann, R., and Cronin, D. (2012). *About Face 3: The Essentials of Interaction Design*. Hoboken, NJ: Wiley.

Cummings, M. L., Marquez, J. J., and Roy, N. (2012). Human-automated path planning optimization and decision support. *International Journal of Human-Computer Studies*, 70(2), 116–128.

Defense Science Board. (2012). Task Force Report: The role of autonomy in DoD systems [online]. Washington, DC: Office of the Secretary of Defense. Retrieved July 20, 2014 from http://www.fas.org/irp/agency/dod/dsb/autonomy.pdf.

Dittmar, A. and Forbrig, P. (2003). Higher-order task models. In *Proceedings of the Thirteenth International Workshop on Design, Specification, and Verification of Interactive Systems*, pp. 187–202. June 11–13, Berlin, Germany: Springer.

Eggleston, R. G. (1992). Cognitive interface considerations for intelligent cockpits. In *Proceedings of the AGARD Conference on Combat Automation for Airborne Weapon Systems: Man/Machine Interface Trends and Technologies* (Report No. AGARD-CP-520). October 19–22, Edinburgh: NATO.

Eggleston, R. G. (1997). Adaptive interfaces as an approach to human-machine co-operation. In M. J. Smith, G. Salvendy, and R. J. Koubek (eds), *Design of Human-Computer Systems: Social and Ergonomic Considerations*, pp. 495–500. Amsterdam: Elsevier Science.

Eier, D. and Fraenzl, T. (2006). Human centered automation in NGATS terminal environment. *The Journal of Air Traffic Control*, 48(4), 41–43.

Endsley, M. R. and Kiris, E. O. (1995). The out-of-the-loop performance problem and level of control in automation. *Human Factors: The Journal of the Human Factors and Ergonomics Society*, 37(2), 381–394.

Fern, L. and Shively, R. J. (2009). A comparison of varying levels of automation on the supervisory control of multiple UASs. In *Proceedings of AUVSI's Unmanned Systems North America*, pp. 1–11. Washington, DC: AUVSI.

Fitts, P. M. (1951). *Human Engineering for an Effective Air-Navigation and Traffic-Control System*. Oxford, England: National Research Council.

Flight Deck Automation Working Group. (2013). *Operational Use of Flight Path Management Systems: Final Report of the Performance-Based Operations Aviation Rulemaking Committee/Commercial Aviation Safety Team Flight Deck Automation Working Group*. Washington, DC: Federal Aviation Administration.

Freedy, A., Madni, A., and Samet, M. (1985). Adaptive user models: Methodology and applications in man-computer systems. In W. B. Rouse (ed.), *Advances in Man-Machine Systems Research: A Research Annual*, Vol. 2, pp. 249–293. Greenwich, CT: JAI Press.

Geddes, N. D. and Shalin, V. L. (1997). Intelligent decision aiding for aviation (Report No. 91-7219-125-2). Linkoping, Sweden: Centre for Human Factors in Aviation.

Goodrich, M. A. (2010). On maximizing fan-out: Towards controlling multiple unmanned vehicles. In M. J. Barnes and F. Jentch (eds), *Human-Robot Interactions in Future Military Operations*, pp. 375–395. Farnham Surrey, UK: Ashgate Publishing.

Guerlain, S. (1995). Using the critiquing approach to cope with brittle expert systems. In *Proceedings of the Human Factors and Ergonomics Society Annual Meeting*, vol. 39, no. 4, pp. 233–237. New York, NY: SAGE Publications.

Hancock, P. A. and Chignell, M. H. (eds). (1989). *Intelligent Interfaces: Theory, Research, and Design*. Amsterdamn, NL: Elsevier Science.

Hansen, R. D. (2009). Water-related infrastructure in medieval London [online]. Retrieved July 20, 2014 from http://waterhistory.org/histories/london/london.pdf.

Hardin, B. and Goodrich, M. A. (2009). On using mixed-initiative control: A perspective for managing large-scale robotic teams. In *Proceedings of the Fourth ACM/IEEE International Conference on Human Robot Interaction*, pp. 165–172. March 11–13, New York, NY: ACM.

Hilburn, B., Molloy, R., Wong, D., and Parasuraman, R. (1993). Operator versus computer control of adaptive automation. In *The Adaptive Function Allocation for Intelligent Cockpits (AFAIC) Program: Interim Research and Guidelines for the Application of Adaptive Automation*, pp. 31–36. Warminster, PA: Naval Air Warfare Center—Aircraft Division.

Höök, K. (2000). Steps to take before intelligent user interfaces become real. *Interacting with Computers*, 12(4), 409–426.

Horvitz, E., Breese, J., Heckerman, D., Hovel, D., and Rommelse, K. (1998). The Lumiere project: Bayesian user modeling for inferring the goals and needs of software users. In *Proceedings of the Fourteenth Conference on Uncertainty in Artificial Intelligence*, pp. 256–265. Burlington, MA: Morgan Kaufmann Publishers.

Hou, M., Gauthier, M. S., and Banbury, S. (2007). Development of a generic design framework for intelligent adaptive systems. In J. Jacko (ed.), *Human-Computer Interaction. HCI Intelligent Multimodal Interaction Environments*, pp. 313–320. Berlin: Springer.

Innocent, P. R. (1982). Towards self-adaptive interface systems. *International Journal of Man-Machine Studies*, *16*(3), 287–299.

Kaber, D. B., Perry, C. M., Segall, N., McClernon, C. K., and Prinzel, L. J. (2006). Situation awareness implications of adaptive automation for information processing in an air traffic control-related task. *International Journal of Industrial Ergonomics*, *36*(5), 447–462.

Keeble, R. J. and Macredie, R. D. (2000). Assistant agents for the world wide web intelligent interface design challenges. *Interacting with Computers*, *12*(4), 357–381.

Kidwell, B., Calhoun, G. L., Ruff, H. A., and Parasuraman, R. (2012). Adaptable and adaptive automation for supervisory control of multiple autonomous vehicles. In *Proceedings of the Human Factors and Ergonomics Society Annual Meeting*, vol. 56, no. 1, pp. 428–432. New York, NY: SAGE Publications.

Kuutti, K. (1996). Activity theory as a potential framework for human-computer interaction research. In B. A. Nardi (ed.), *Context and Consciousness: Activity Theory and Human-Computer Interaction*, pp. 17–44. Cambridge, MA: MIT Press.

Langan-Fox, J., Canty, J. M., and Sankey, M. J. (2009). Human–automation teams and adaptable control for future air traffic management. *International Journal of Industrial Ergonomics*, *39*(5), 894–903.

Lee, J. D. and Moray, N. (1992). Trust, control strategies and allocation of function in human-machine systems. *Ergonomics*, *35*(10), 1243–1270.

Lee, J. D. and See, K. A. (2004). Trust in automation: Designing for appropriate reliance. *Human Factors: The Journal of the Human Factors and Ergonomics Society*, *46*(1), 50–80.

Lewis, M., Wang, H., Chien, S. Y., Velagapudi, P., Scerri, P., and Sycara, K. (2010). Choosing autonomy modes for multirobot search. *Human Factors: The Journal of the Human Factors and Ergonomics Society*, *52*(2), 225–233.

Linegang, M. P., Stoner, H. A., Patterson, M. J., Seppelt, B. D., Hoffman, J. D., Crittendon, Z. B., and Lee, J. D. (2006). Human-automation collaboration in dynamic mission planning: A challenge requiring an ecological approach. In *Proceedings of the Human Factors and Ergonomics Society Annual Meeting*, vol. 50, no. 23, pp. 2482–2486. New York, NY: SAGE Publications.

Manning, C. A. (1995). Empirical investigations of the utility of flight progress strips: A review of the VORTAC studies. In *Proceedings of the Eighth International Symposium on Aviation Psychology*, pp. 404–409. April 24–27, Columbus, OH: Ohio State University.

Manzey, D., Bahner, J. E., and Hueper, A. D. (2006). Misuse of automated aids in process control: Complacency, automation bias and possible training interventions. In *Proceedings of the Human Factors and Ergonomics Society Annual Meeting*, vol. 50, no. 3, pp. 220–224. New York, NY: SAGE Publications.

Maybury, M. T. (1993). Communicative acts for generating natural language arguments. In *Proceedings of the Eleventh National Conference on Artificial Intelligence*, pp. 357–364. July 11–15, Palo Alto, CA: AAAI Press.

Miller, C., Funk, H., Wu, P., Goldman, R., Meisner, J., and Chapman, M. (2005). The Playbook™ approach to adaptive automation. In *Proceedings of the Human Factors and Ergonomics Society Annual Meeting*, vol. 49, no. 1, pp. 15–19. New York, NY: SAGE Publications.

Miller, C. A. and Dorneich, M. C. (2006). From associate systems to augmented cognition: 25 years of user adaptation in high criticality systems. In D. D. Schmorrow, K. M. Stanney, and L. M. Reeves (eds), Foundations of Augmented Cognition, 2nd Edition, pp. 344–353. Arlington, VA: Strategic Analysis.

Miller, C. A. and Parasuraman, R. (2007). Designing for flexible interaction between humans and automation: Delegation interfaces for supervisory control. *Human Factors: The Journal of the Human Factors and Ergonomics Society*, *49*(1), 57–75.

Miller, C. A., Shaw, T., Emfield, A., Hamell, J., Parasuraman, R., and Musliner, D. (2011). Delegating to automation performance, complacency and bias effects under non-optimal conditions. *In Proceedings of the Human Factors and Ergonomics Society Annual Meeting*, vol. 55, no. 1, pp. 95–99. New York, NY: SAGE Publications.

Mitrovic, A., Martin, B., and Suraweera, P. (2007). Intelligent tutors for all: The constraint-based approach. *IEEE Intelligent Systems*, *22*(4), 38–45.

Myers, B. A. (1996). User interface software technology. *ACM Computing Surveys (CSUR)*, *28*(1), 189–191.

Norman, D. A. (1988). *The Design of Everyday Things*. New York, NY: Basic Books.

Norman, D. A. (2005). Human-centered design considered harmful. *Interactions*, *12*(4), 14–19.

Oppermann, R. and Simm, H. (1994). Adaptability: User-initiated individualization. In R. Oppermann (ed.), *Adaptive User Support: Ergonomic Design of Manually and Automatically Adaptable Software*, pp. 14–66. Hilldale, NJ: Lawrence Erlbaum Associates.

Parasuraman, R., Barnes, M., and Cosenzo, K. (2007). Adaptive automation for human-robot teaming in future command and control systems. *The International C2 Journal*, *1(2)*, 43–68.

Parasuraman, R., Cosenzo, K. A., and De Visser, E. (2009). Adaptive automation for human supervision of multiple uninhabited vehicles: Effects on change detection, situation awareness, and mental workload. *Military Psychology*, *21*(2), 270–297.

Parasuraman, R. and Miller, C. (2006). Delegation interfaces for human supervision of multiple unmanned vehicles: Theory, experiments, and practical applications. *Advances in Human Performance and Cognitive Engineering Research*, *7*, 251–266.

Parasuraman, R., Molloy, R., and Singh, I. L. (1993). Performance consequences of automation-induced "complacency". *The International Journal of Aviation Psychology*, *3*(1), 1–23.

Parasuraman, R. and Riley, V. (1997). Humans and automation: Use, misuse, disuse and abuse. *Human Factors: The Journal of the Human Factors and Ergonomics Society*, *39*(2), 230–253.

Parasuraman, R., Sheridan, T. B., and Wickens, C. D. (2000). A model for types and levels of human interaction with automation. *IEEE Transactions on Systems, Man, and Cybernetics—Part A: Systems and Humans*, *30*(3), 286–297.

Pew, R. W. and Mavor, A. S. (2007). *Human-System Integration in the System Development Process: A New Look*. Washington, DC: National Academies Press.

Preece, J., Rogers, Y., and Sharp, H. (2002). *Interaction Design: Beyond Human-Computer Interaction*. New York, NY: Wiley.

Prinzel, L. J., Freeman, F. G., Scerbo, M. W., Mikulka, P. J., and Pope, A. T. (2000). A closed-loop system for examining psychophysiological measures for adaptive task allocation. *The International Journal of Aviation Psychology*, *10*(4), 393–410.

Rotter, J. B. (1954). *Social Learning and Clinical Psychology*. Englewood Cliffs, NJ: Prentice Hall.

Rotter, J. B. (1966). Generalized expectancies for internal versus external control of reinforcement. *Psychological Monographs: General and Applied*, *80*(1), 1–28.

Rouse, W. B. and Morris, N. M. (1986). On looking into the black box: Prospects and limits in the search for mental models. *Psychological Bulletin*, *100*(3), 349–363.

Schmorrow, D. D. (2005). *Foundations of Augmented Cognition*. New York: Lawrence Erlbaum.

Sheridan, T. B. (2002). *Humans and Automation: System Design and Research Issues.* Hoboken, NJ: Wiley.

Shneiderman, B., Plaisant, C., Cohen, M., and Jacobs, S. (2009). *Designing the User Interface: Strategies for Effective Human-Computer Interaction,* 5th edn. Englewood Cliffs, NJ: Prentice Hall.

Snyder, M. G., Qu, Z., Chen, J. Y. C., and Barnes, M. J. (2010). RoboLeader for reconnaissance by a team of robotic vehicles. In *Proceedings of the 2010 International Symposium on Collaborative Technologies and Systems (CTS),* pp. 522–530. May 17–21, Washington, DC: IEEE.

Souchon, N., Limbourg, Q., and Vanderdonckt, J. (2002). Task modelling in multiple contexts of use. In P. Forbrig, Q. Limbourg, J. Vanderdonckt, and B. Urban (eds), *Interactive Systems: Design, Specification, and Verification,* pp. 59–73. Berlin: Springer Verlag.

Squire, P., Trafton, G., and Parasuraman, R. (2006). Human control of multiple unmanned vehicles: Effects of interface type on execution and task switching times. In *Proceedings of the First ACM SIGCHI/SIGART Conference on Human-Robot Interaction,* pp. 26–32. March 2–3, New York, NY: ACM.

Sukaviriya, P., Foley, J. D., and Griffith, T. (1993). A second generation user interface design environment: The model and the runtime architecture. In *Proceedings of the INTERACT'93 and CHI'93 Conference on Human Factors in Computing Systems,* pp. 375–382. April 24–29, In P. Forbrig, Q. Limbourg, J. Vanderdonckt, and B. Urban (eds) ACM.

Sutcliffe, A. G., Kurniawan, S., and Shin, J. E. (2006). A method and advisor tool for multimedia user interface design. *International Journal of Human-Computer Studies, 64*(4), 375–392.

Taylor, R. M. and Reising, J. R. (1998). The human-electronic crew: Human-computer collaborative team working. In *Proceedings of the First NATO RTO Human Factors and Medical Panel Symposium on Collaborative Crew Performance in Complex Operational Systems,* pp. 1–17. April 20–22, Brussels: North Atlantic Treaty Organization (NATO).

Tecuci, G., Boicu, M., and Cox, M. T. (2007). Seven aspects of mixed-initiative reasoning: An introduction to this special issue on mixed-initiative assistants. *AI Magazine, 28*(2), 11–18.

Tomlinson, G. B., Baumer, E., Yau, M. L., and Renz, W. (2007). Dreaming of adaptive interface agents. In *Proceedings of the ACM Conference on Human Factors in Computing Systems,* pp. 2007–2012. April 30–May 3, New York, NY: ACM.

Valero-Gomez, A., De La Puente, P., and Hernando, M. (2011). Impact of two adjustable-autonomy models on the scalability of single-human/multiple-robot teams for exploration missions. *Human Factors: The Journal of the Human Factors and Ergonomics Society, 53*(6), 703–716.

Wang, Y., Zhang, Y., Sheu, P., Li, X., and Guo, H. (2010). The formal design model of an automatic teller machine (ATM). *International Journal of Software Science and Computational Intelligence, 2*(1), 102–131.

Wickens, C. D., Hollands, J. G., Banbury, S., and Parasuraman, R. (2012). *Engineering Psychology and Human Performance,* 4th edn. Upper Saddle River, NJ: Pearson.

Wickens, C. D., Mavor, A. S., and McGee, J. P. (1997). *Flight to the Future: Human Factors in Air Traffic Control.* Washington, DC: National Academies Press.

Wilson, G. F. and Russell, C. A. (2003). Operator functional state classification using multiple psychophysiological features in an air traffic control task. *Human Factors: The Journal of the Human Factors and Ergonomics Society, 45*(3), 381–389.

Woods, D. D., Tittle, J., Feil, M., and Roesler, A. (2004). Envisioning human-robot coordination in future operations. *IEEE Transactions on Systems, Man, and Cybernetics, Part C: Applications and Reviews, 34*(2), 210–218.

Zhu, H. and Hou, M. (2009). Restrain mental workload with roles in HCI. In *Proceedings of the 2009 IEEE Toronto International Conference Science and Technology for Humanity (TIC-STH),* pp. 387–392. September 26–27, Toronto, ON: IEEE.

3 Conceptual Architecture for Intelligent Adaptive Systems*

3.1 OBJECTIVES

- Discuss the importance of conceptual architecture in IAS design
- Review previous IAS projects and discuss their key system components
- Describe a generic conceptual architecture for IAS design, including critical system components and the interrelationships among them
- Describe how an IAS adapts to, or is triggered by, explicit commands from the operator or implicit inferences about operator state

The history and definitions of human–machine systems (HMSs), intelligent adaptive interfaces (IAIs), intelligent adaptive automation (IAA), and the IAS discussed in Chapters 1 and 2 serve as the background for the conceptual IAS architecture explored in this chapter, which embodies many of the notions, assumptions, and ideas that researchers from a broad cross section of domains and disciplines have developed over the past 20 years or so.

Conceptual architectures are important as they allow systems designers and developers to understand what the critical components of an IAS are at the very start of a project or work activities. Figure 3.1 illustrates the four-step process involved in a typical IAS development project:

1. *Conceive*: Conceive and articulate the conceptual architecture that will guide IAS development by identifying the operational priorities and the critical IAS components needed to support them.
2. *Analyze*: Select appropriate analytical techniques to capture the requirements for the display, communication, and control of the operator–machine interface (OMI), and the task and functional decomposition of the domain conceived for the IAS. Analyzing also involves the identification of the means to capture more detailed knowledge from subject matter experts (SMEs) for function allocation between the operator and machine. Function allocation assigns the performance of each function to the element (e.g., human, hardware, or software) best suited to perform the function (Nemeth, 2004). This provides information

---

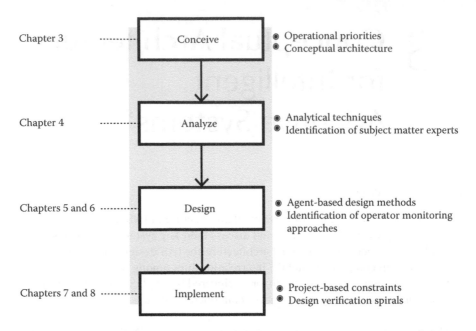

FIGURE 3.1   Overview of the IAS design process. Each chapter of this book is correlated to a specific step.

about the objective states of the IAS within a specific application or context, and provides a basis for the interface to intelligently support the operator through adaptation. Analytical techniques are discussed in Chapter 4 and additional issues that should be considered are discussed in Chapter 7.

3. *Design*: Optimize human–machine interaction for the IAS in terms of OMI design and automation design. Optimizing necessitates following agent-based design methods, which are discussed in Chapter 5. Additionally, IAS design requires information about the objective and subjective states of the operator within a mission or task context. Design also involves the identification of behavioral-based and psychophysiological-based monitoring approaches for the evaluation of operator behavioral, psychological, and physiological states. Knowledge about the internal context (i.e., operator state) can provide the basis for an intelligent adaptation of the system to support the operator and achieve system goals. Operator state monitoring approaches are further discussed in Chapter 6.

4. *Implement*: The final step is to implement the fully planned IAS design.

Fortunately for systems designers today, there are a number of conceptual architectures that can be reviewed and modified to meet the needs and constraints of a project design plan. Specific design concepts can then be developed and evaluated (i.e., design verification spirals) to capture the scope of the entire project.

This chapter reviews several conceptual architectures that have been used in a number of different applications and shows commonalities among them. The aim is to describe the basic anatomy (i.e., conceptual architecture) of an IAS so that it may be adapted and extended to meet particular research or development needs.

Systems designers can use this conceptual architecture to assist in the implementation of a wide range of adaptation technologies to design solutions for specific applications. The selection of an appropriate development framework affords the development team a number of advantages: (a) reducing development time and costs by leveraging previous research; (b) benefitting from lessons learned during past projects; and (c) providing insight into the potential operational impact of the developed system.

3.2 EXISTING CONCEPTUAL ARCHITECTURES

This review of conceptual architectures for IASs starts with two of the first adaptive systems that developed from large, well-funded defense projects, which have inspired the development of IASs ever since. Several more recent IASs that have been developed for a broad range of applications are also reviewed, including intelligent tutors, intelligent interfaces, and intelligent decision support tools. This is not a comprehensive summary of IAS development that has been conducted to date, but it captures a general sense of the scope of work that has been undertaken. From this review, some common system components can be extracted to provide the basic anatomy of IASs.

3.2.1 EARLY INTELLIGENT ADAPTIVE SYSTEMS

In 1988, the first workshop on the "human-electronic crew" was held in Ingolstadt, Germany. The objective of the workshop was to bring together internationally renowned artificial intelligence and cockpit design experts to exchange ideas, concepts, and data on the potential development and impact of an aircraft crew composed of human and electronic team members, and to answer the question "can they work together?" (Emerson et al., 1989). The second workshop, held 2 years later, continued the theme by asking "is the team maturing?" (Emerson et al., 1992). The organizers reported that significant progress had been made between the two workshops in terms of examining what it actually takes to build an electronic crew member. Additionally, more emphasis was placed on understanding how human and electronic teams could work together, particularly in the context of building pilot trust in the electronic team member. The final workshop was held in 1994 and again focused on issues of pilot trust in automated systems by asking "can we trust the team?" (Taylor and Reising, 1995). The main outcome of the final workshop was to reiterate the importance of assisting, rather than replacing, the pilot. However, it was also acknowledged that the complexity of military missions was increasing to the point where automated decision assistance would likely be needed.

The papers presented at these workshops make it apparent that there were two different strategies for developing electronic crew members. The first strategy was more short term and sought the development of stand-alone pilot support tools such as the fusion and management of sensor information, mission planning, and tactical advisors. The second strategy was more long term and sought the development of integrated systems that use artificial intelligence technologies to adapt the support provided to the pilot in a manner that is sensitive to both the external world and the internal (i.e., cognitive) state of the pilot. Of the second strategy, the main themes common across many of the research papers presented at the workshop were:

- *OMI*: An interface that could adapt content to support pilot needs
- *Situation assessment*: The assessment of threat from entities in the environment such as other aircraft
- *Operator state estimation*: The assessment of the workload level of pilots, their goals, and the tasks they attempt to undertake

The Pilot's Associate (PA; United States) and the Cognitive Cockpit (United Kingdom) research programs were two of the projects undertaken during this time that emphasized promoting a more integrated human–electronic crew member team.

3.2.1.1 Pilot's Associate Research Program

The US Air Force PA program (1986–1992) sought to exploit artificial intelligence technologies to develop integrated, real-time intelligent systems that supported the pilot in monitoring the external situation, assessing threats, and planning reactions to events. The technology was dubbed "associate" because the intent was to have the system give advice to the pilot. The intent of the PA program was to let the human pilot remain in charge of the aircraft.

The PA was an IAS in that it provided recommendations to the pilot that were responsive to changes in the current offensive or defensive situation, changes to pilot objectives, and changes in the ability of the aircraft to complete the current plan due to failures, battle damage, or the status of consumable resources such as fuel, weapons, or ammunition. This level of responsiveness was achieved using the following functional modules:

- *Systems status*: This module is responsible for assessing the ability of the aircraft to perform its mission.
- *Situation assessment*: This module is responsible for determining the status of the external environment (e.g., other entities and threats).
- *Pilot–vehicle interface (PVI)*: This module is responsible for managing the presentation of information to the pilot by monitoring and interpreting pilot actions.

Other modules included mission and tactics planning, and mission management.

Although the PA program ended in 1992, it was succeeded by the Rotorcraft PA (RPA) program (1993–1998), whose objective was to continue the development of "associate technologies" (or IASs) to the point of implementation and flight demonstration. Of particular interest is the RPA program's emphasis on intelligent information management of cockpit behaviors to support task-sensitive, dynamically generated cockpit configurations (Miller et al., 1999). In other words, the cockpit displays of the RPA were configured and adapted intelligently by the system to support the pilot in a manner that was sensitive to the pilot's immediate task.

From an IAS perspective, the most important component of the RPA was the cognitive decision aiding system. The cognitive decision aiding system comprised a number of modules, including (a) an external situation assessment module that considered the significance of entities in the external world to known mission goals; (b) an internal situation assessment module that performed a similar function, but was concerned with the internal health of the aircraft; and (c) the cockpit information manager, which comprised the intelligent adaptive OMI (i.e., an IAI) component of the system. To provide the cockpit information manager with the intelligence to adapt to the needs of the pilot, it was supported by a crew intent estimator that was responsible for interpreting pilot actions and world events against preestablished mission plans. The crew intent estimator did this by monitoring the external environment and the crew actions and matching them against a hierarchical knowledge base of pilot tasks. On the basis of this comparison, the RPA could offer support to the pilot through varying the amount (i.e., level) of automation depending on the severity of the situation.

To summarize the conceptual architecture of both the PA and the RPA, three main functional components were required: (a) an IAI, (b) a module that assesses external (i.e., the outside world) status, and (c) a module that assesses internal (i.e., the mental state of the crew) state.

3.2.1.2 Cognitive Cockpit Research Program

The Cognitive Cockpit research program began in the early 1990s and was originally a large multidisciplinary project funded by the UK Ministry of Defence conducted at the Defence Evaluation and Research Agency (DERA). Its objective was to specify the cognitive requirements for building the next generation of cockpit intelligent aiding systems for use in the 2010–2015 time scale (Taylor et al., 2001). This was to be achieved by developing an integrated system that focused on the pilot's requirement of being in control of the system and not being overwhelmed with system control information. After the year 2000, the program continued under funding from the US Defense Advanced Research Projects Agency (Kollmorgen et al., 2005).

Not only did the Cognitive Cockpit program develop several functional modules similar to the PA and RPA, such as the situation assessment module and the pilot intent estimation module, but it also integrated expertise in the psychophysiological monitoring of the pilot's cognitive state, which is further discussed in Chapter 6. The Cognitive Cockpit had three main functional modules:

- *Cognition monitor (COGMON)*: This module was concerned with real-time analysis of the psychophysiological and behavioral states of the pilot. The primary functions of this system included continuous monitoring of pilot workload, and making inferences about the current focus of pilot attention and intentions. The COGMON was also responsible for detecting excessively high and low levels of pilot arousal. The information that the

COGMON provided about the cognitive state of the pilot within a mission context, taken as a whole, was used as a basis for the implementation of aiding and support tools.

- *Situation assessor (SASS)*: This module was concerned with real-time analysis of the mission and the provision of pilot aiding and support provided by decision support systems. It collected information about the status of the current mission, other aircrafts in the vicinity (e.g., heading, altitude, and threat), and environmental status (e.g., weather conditions). The module also comprised extensive tactical, operational, and situational knowledge. Overall, the SASS provided information about the objective status of the aircraft within a mission context and used extensive knowledge-based tools to aid and support the pilot.

- *Tasking interface manager (TIM)*: This module was concerned with real-time analysis of high-level outputs from the COGMON, SASS, and other aircraft systems. A central function for this module was the maximization of the goodness of fit of the aircraft status, "pilot-state," and tactical assessments provided by the SASS. These integrative functions meant that the system was able to influence the prioritization of pilot tasks and determine the means by which information was communicated to and from the pilot. In other words, the cockpit displays were intelligently configured and adapted to the current needs of the pilot. The TIM also allowed pilots to manage their interaction with the cockpit automation by controlling the allocation of tasks to the automated systems.

The same three main functional components seen in the PA and RPA programs were present in the Cognitive Cockpit: (a) an OMI through which intelligent adaptation takes place, (b) a module that assesses external status, and (c) a module that assesses internal state.

3.2.2 OTHER INTELLIGENT ADAPTIVE SYSTEMS

Four of the more recent IASs, which support the operator over a range of functions, including document formatting, data retrieval and analysis, stock market trading, and tutoring, will now be addressed in order to identify the main functional components of these IASs and determine the commonalities among them.

3.2.2.1 Work-Centered Support System

The Work-Centered Support System (WCSS) is a job performance aid developed by Young and Eggleston (2002) that provides support for operators so that they can concentrate on the task at hand. Unlike traditional stand-alone software applications, the WCSS supports operators by conducting a connected series of work activities that are directed toward accomplishing a goal. Each activity, or work-thread, usually consists of multiple steps, which often require operators to quickly shift their cognitive frame of reference, resulting in increased mental workload and increased probability of error. Operators are supported by the WCSS, which allows them to focus on a primary task, such as mission planning, by reducing the time needed to locate and integrate necessary information.

WCSS comprises a number of intelligent agents that work as team players side by side with the operator (agent concept will be discussed in Chapter 5). These agents are used in three layers:

- *Acquisition agents*: These agents contain knowledge on how to find and retrieve data. They automatically monitor and access data sources for the operator and notify other agents when new data have been retrieved or received.
- *Analysis agents*: These agents contain the knowledge required to transform data into information that supports operator decision making. They use the data collected by the acquisition agents to continually appraise the situation, identify possible problems, and generate a prioritized list of potential operator actions.
- *Presentation agent*: This agent is part of a communications and dialogue module that controls the information presented to the operator, the responses to operator requests, and the provision of alerts to the operator to identify potential problems and opportunities. This is achieved through knowledge of who the operator is and what his or her current requirements are.

In other words, the main functional components of the WCSS are modules pertaining to knowledge about the external world (e.g., where to find and retrieve data and how to transform these data into information that can support the operator's decision making), knowledge about the operator (e.g., preferences for information presentation and dialogue and his or her profile, requests, and requirements), and an OMI that presents the information to the operator.

3.2.2.2 Stock Trader System

The Stock Trader system is an adaptive recommendation system for stock market trading that uses models of an operator's investment style and preferences to tailor its advice on whether to buy or sell particular stocks (Yoo et al., 2003). The system architecture of Stock Trader comprises the following elements:

- *Data processing*: This module converts raw data, such as current stock readings and historical trading information, into reports that contain buy and sell recommendations for the operator.
- *Recommendation*: Using the reports generated by the data processing module, this module creates tailored recommendations for each stock based on individual operator trading profiles.
- *User modeling*: This module constructs a trader profile based on operator responses to previous system recommendations.
- *Information manager*: This module records all operator interactions with the system, including the buying and selling of stocks, and maintains an awareness of their investment portfolios.
- *Graphical user interface (GUI)*: This module presents all reports to the operator, and accepts commands such as the buying and selling of stocks and viewing portfolios.

Similar to the WCSS, the main functional components of Stock Trader are (a) an OMI, (b) a module pertaining to knowledge about the external world (e.g., retrieving stock market data), and (c) a module containing knowledge about the operator (e.g., preferences accepting recommendations, and information and requirements). These functional components allow Stock Trader to transform stock market data into recommendations to support decision making based on the trader's profile.

3.2.2.3   Generalized Intelligent Framework for Tutoring

IAS technologies have also been implemented within computer-based training applications to improve their effectiveness. Such intelligent tutoring systems (ITSs) capture feedback from the learner's performance, prior exposure to knowledge, and learning rate to deliver, evaluate, and react according to pedagogical principles and educational goals. In doing so, an ITS is able to provide the learner with the same benefits as one-on-one instruction from a human instructor. One such system is the Generalized Intelligent Framework for Tutoring (GIFT) developed by Sottilare et al. (2012). Their goal was to enhance the learning experience by better aligning ITS instructional strategy with learner needs. This alignment was facilitated by evaluating the cognitive, emotional, and physical states of the learner, together with his or her learning performance, so that the most appropriate instructional strategy could be deployed. GIFT comprises several functional modules that support the delivery of training to the student:

- *Sensor module*: This module receives data from several psychophysiological sensors attached to the student (e.g., heart rate and activity) and filters, segments, or extracts features in these data to determine the cognitive state of the student.
- *Learner (user) module*: This module supports the assessment of the learner's cognitive, affective (i.e., emotional), and physical state based on a combination of historical data (e.g., performance, traits, preferences) and data from the sensor module. A secondary function of this module is to track relevant learner trait data (e.g., personality preferences and learning style) and lesson data (e.g., lesson start/completion, lesson scores).
- *Domain module*: This module defines and structures domain knowledge, such as instructional content, domain-relevant tasks, conditions, standards, problem sets, and common questions.
- *Pedagogical module*: This module queries learner state data from the learner module and the assessment of the learner's performance (and associated feedback) from the domain module and uses this information to determine the content, order, and flow of the instruction. Furthermore, the pedagogical module adapts the instructional material presented to learners through the OMI. For example, the elements within a training scenario can be manipulated, hints or feedback can be provided, or the pace and difficulty of the interaction can be changed.

There are many similarities among the GIFT conceptual architecture and the previously described systems. The main functional components comprise (a) an

OMI that presents information in a manner that is best suited to the learning context, (b) a module pertaining to knowledge about the world (in this case, the subject matter knowledge and instructional content), and (c) modules pertaining to knowledge about the learner (e.g., cognitive, physical, and emotional state, and learning performance).

3.2.2.4 Edwards' Generic Framework

One of the most comprehensive attempts to generate a design and development framework for IASs is Edwards' generic framework (2004). To investigate the efficacy of an IAS for the control of multiple uninhabited aerial vehicles (UAVs), Edwards examined a variety of theoretical approaches to produce a generic, integrated, and comprehensive framework to provide guidance for the development of UAV ground control stations. Accordingly, an intelligent, adaptive, agent-based IAS was designed, implemented, and evaluated (Hou et al., 2007a,b). Edwards' generic framework comprised a number of design approaches that can be used to specify several models that are used to construct IASs:

- *Task model*: This model contains knowledge pertaining to the tasks that an operator is expected to perform and is represented as a hierarchy of actions, goals, and plans. Satisfying low-level goals allows for the attainment and achievement of high-level goals. The pathway from a low-level goal to a high-level goal identifies and defines a plan for attaining that goal. The techniques used to hierarchically decompose operator tasks are discussed in Chapter 4.
- *System (i.e., machine) model*: This model is also characterized by a goal hierarchy, and comprises a description of the tasks, goals, and plans that a system completes in order to support an operator. When multiple system agents are involved, the system model can also comprise a distinct goal hierarchy for each agent. The system model also defines the type or level of assistance provided by the system to aid an operator, including advice, automation, interface adaptation, and so on. Note that Edwards' description of a system model is different from the version used throughout this book; system model in this book does not refer exclusively to a machine model, but is a combination of both machine and user models.
- *User (i.e., operator) model*: This model is generated from information obtained from the monitoring of operator activities or system requests. The system model can be built to facilitate the system's recognition of each operator's unique profile.
- *Dialogue (i.e., interaction) model*: This model identifies the communication and interaction that take place among the operator, machine, and other system agents.
- *World (i.e., environment) model*: This model defines the external world according to the objects that exist in the environment, their properties, and the rules that govern them. These rules can be varied, such as physical rules, psychological rules, and cultural rules. For example, the model incorporates physical (e.g., principles of flight controls), psychological (e.g.,

principles of human behavior under stress), and cultural (e.g., rules associated with tactics adopted by hostile forces) knowledge of the external world
relevant to UAV control. Edwards' framework is discussed in the context of
explicit models design in Chapter 5.

3.3 THE BASIC ANATOMY OF INTELLIGENT ADAPTIVE SYSTEMS

Based on this review of existing design frameworks for IASs, the following section
describes the critical components that are common to all IASs. Each component is
described in detail to enable systems designers to gain a high-level technical understanding of the scope and capabilities of each of the components and how they interact with one another.

3.3.1 CRITICAL COMPONENTS OF INTELLIGENT ADAPTIVE SYSTEMS

From this review of conceptual architectures that have been developed for a broad
range of applications, a number of fundamental IAS characteristics have been
extracted:

- Tracking operator goals, plans, or intents, and the progress made toward
  them
- Monitoring and inferring the internal state of the operator (e.g., cognitive,
  emotional, physical, and behavioral)
- Monitoring and inferring the external state of the world (e.g., other entities,
  threats, environmental conditions, and domain knowledge)
- Monitoring the effects of system status, advice, automation, and adaptation
  on operator and world state (i.e., closed-loop feedback)
- A customized OMI to handle the interaction/dialogue between the operator
  and the system agents

Figure 3.2 illustrates the critical components of a generic IAS and the interrelationships among them. It includes modules pertaining to situation assessment and
operator state assessment, an adaptation engine, and an OMI. The models that support each of these modules as described by Edwards (2004) and Hou et al. (2007a,
2011) are also included in this figure.

3.3.1.1 Situation Assessment Module

The situation assessment module is concerned with the assessment of the external
world, or the "situation," and comprises functionality relating to the real-time analysis of the mission (i.e., the sequence of activities needed to achieve a specific goal or
learning objective), automation, and decision support. A situation assessment module monitors and tracks the current progress toward a specific mission, goal, or status through the sensing and fusion of internal (i.e., system status) and external data
sources using the knowledge bases of task, goal, tactical, operational, and situational
knowledge. Using this knowledge, the module is able to assist the operator through

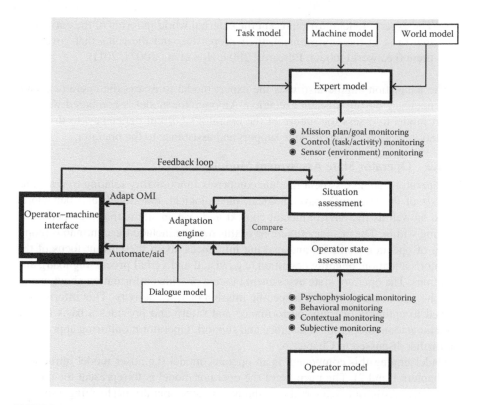

FIGURE 3.2 Critical components of an IAS. (Redrawn from Hou, M., Gauthier, M.S., and Banbury, S., *Proceedings of Human-Computer Interaction International Conference, Part III*, pp. 313–320, Springer, Berlin, 2007.)

decision support and automation, or by adapting the information that is presented to the operator.

Underpinning this module is an expert model. The purpose of the expert model is to represent the knowledge, skills, or behaviors that embody the desired state of the operator within the context of a specific mission, work activity, or learning experience. The contents of the expert model are mission-specific or work activity–specific and remain static during mission execution—the operator's interaction with the system does not alter its representation of knowledge, skills, or behaviors. The expert model encapsulates three aspects of representation:

- The expert model must represent the required operator proficiency within the specifics of the mission tasks. In other words, the "what" that the operator must do to successfully complete the mission or work activity (i.e., task model: Edwards, 2004; Hou et al., 2007a, 2011).
- The expert model must represent expert behaviors and skills—the "how" that is used by the system to provide feedback and assistance to the operator in the form of advice or support (i.e., system model: Edwards, 2004; Hou et al., 2007a, 2011).

- The expert model must represent the external world in terms of the entities that exist in the environment, their properties, and the rules that govern them (i.e., world model: Edwards, 2004; Hou et al., 2007a, 2011).

The adaptation engine also uses the expert model to assess the operator's state of competency during mission execution. An operator model is compared with the expert model to assess the nature of the operator's current deficiency, to drive the adaptation process, and to provide support and assistance to the operator.

3.3.1.2   Operator State Assessment Module

The operator state assessment module comprises functionality relating to real-time analysis of the operator's psychological, physiological, emotional, and behavioral states, as well as the environmental context and system status that the operator is working within. The primary functions of this module include the continuous monitoring of operator workload and making inferences about the current focus of the operator's attention, ongoing cognition (e.g., visual and verbal processing load), and intentions. The operator state assessment module provides information about operator state within the context of a specific mission or work activity. This information is used to optimize operator performance and safety and provides a basis for the implementation of operator assistance and support. Operator monitoring approaches are further discussed in Chapter 6.

Underpinning this component is an operator model (i.e., user model introduced in Chapters 1 and 2). The purpose of the operator model is to represent the current knowledge, skills, or behaviors that embody a specific operator performing a specific task based on models of human cognition, control abilities, and communication. The contents of the operator model are both task and operator specific, and they evolve dynamically during the execution of the task. In other words, the operator's interactions with the system update the modeled state of the operator's performance and competence. The adaptation engine uses the operator model to compare the current operator model with the expert model to assess the operator's deficiencies. This information is used to drive the adaptation of operator assistance and support to enhance and mitigate human processing capabilities and limitations.

The operator model can be segregated into information directly related to the mission or work activity and information that is more general and unrelated to the mission. The mission- or work-related information would have the same composition as the expert model, but it would contain an assessment of current operator state rather than desired operator state. The information considered more general and unrelated to the mission would include knowledge about cognitive abilities, operator preferences, and training background. This aspect of the operator model can also be populated, where possible, with historical information about the operator such as training received, previous mission performance assessments, and so on.

The operator model can also be updated dynamically during the performance of the mission or work activity. In this case, the operator's interaction with the system and the operator's performance can be used to update the mission-related aspects of the operator model. The operator state assessment module must therefore be able to use operator state measurements (e.g., eye tracking, psychophysiological data, and operator behavior and self-assessments) and contextual measurements (e.g., ambient conditions and system status) to make assessments of high-level operator state (e.g., workload, stress, performance, proficiency) and update the operator model accordingly.

3.3.1.3 Adaptation Engine Module

The adaptation engine module uses the high-level outputs from the operator state assessment and situation assessment modules. This module seeks to maximize the fit of the operator state provided by the operator state assessment module to the machine status and situational assessments provided by the situation assessment module.

These integrative functions require that the adaptation engine module be able to influence the prioritization and allocation of tasks (i.e., IAA) and determine the means by which information is presented to the operator (i.e., IAI). For example, the adaptation engine module can offer a range of authority between adaptive systems (i.e., system-initiated adaptation) and adaptable systems (i.e., operator-initiated adaptation). In other cases, the adaptation engine module can also provide advice to the operator (e.g., suggesting how to accomplish a specific task that the operator is having difficulty with) or adapt the information available for presentation (e.g., reducing the amount of information presented to the operator by presenting only critical information during periods of excessively high-workload levels).

Figure 3.3 illustrates the three functional components that comprise the adaptation engine module:

- *Adaptation assessment component*: This component is responsible for producing system-initiated adaptation demands by comparing the operator model with the expert model. Some adaptation aspects might depend only on the operator model. For example, workload is one factor that influences

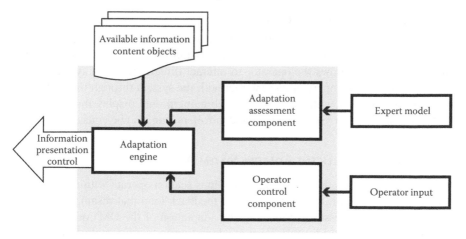

FIGURE 3.3 Components of the adaptation engine module. The adaptation assessment component produces system-initiated demands, while the operator control component produces operator-initiated demands. The adaptation engine considers the demands and controls the module output.

adaptation that depends only on the operator model. Other adaptation demands are based on the differences between the operator model and the expert model. For example, task accomplishment or mission performance is assessed by a comparison of the operator's actual performance with the hypothetical expert performance and performance standards captured in the expert model.

- *Operator control component*: This component is responsible for handling operator-initiated adaptation demands (i.e., operator input). For example, during a mission the operator might produce adaptation demands through the OMI, similar to the Cognitive Cockpit's TIM. The operator control component coordinates operator demands that are received through the OMI to the adaptation engine.
- *Adaptation engine*: This component is responsible for both the translation of system- and operator-initiated adaptation demands into changes in the information presented to the operator through the OMI and the level of machine support that can be given to the operator. The adaptation engine considers the adaptation demands, the available information content that can be presented to the operator (e.g., information parameters, automation status, and system-generated advice), and the range of adaptations possible (e.g., information presentation options, levels of machine automation, or the availability of decision support). From this, the adaptation engine module controls the OMI in its delivery of the adapted content or support and presents relevant information to the operator accordingly (i.e., dialogue model: Edwards, 2004 or interaction model: Hou et al., 2011). Based on the characteristics of the information content, this adaptation could either be continuous during the execution of a task or there could be a dynamic alteration of the sequence of information objects that are presented to the operator.

3.3.1.4   Operator–Machine Interface Module

The final component of an IAS is the OMI. It is the means by which the operator interacts with the system in order to satisfy mission tasks and goals. The OMI is also the means that allows the operator to interact directly with the system. There are many ways that an operator can interact with the system through the OMI: keyboard, mouse, headset with microphone, web camera, and display monitor. Other input devices, such as joysticks and game controllers, can also be considered.

3.3.2   How an Intelligent Adaptive System Adapts

All four of the modules described in the previous section operate within the context of a closed-loop system insofar as there is a feedback loop that resamples operator state and situation assessment following the adaptation of the OMI or automation. The goal of an IAS is to adjust the level of adaptation so that optimal operator states (e.g., performance and workload) are attained and maintained. As was previously discussed, adaptation can be initiated (or triggered) directly by the operator through explicit commands, or indirectly by the operator through implicit inferences made about the operator's performance against task or mission objectives by the IAS.

As such, there are two main modes of control over function allocation (Rieger and Greenstein, 1982). Explicit allocation refers to situations where the operator has allocation control over whether tasks are to be performed automatically by the system or manually by the operator (in other words, an adaptable rather than an adaptive system). Implicit allocation refers to system-initiated allocation of tasks (Tattersall and Morgan, 1996). Research comparing explicit and implicit modes of adaptive automation indicates that although most operators prefer explicit control, implicit adaptive automation is superior in terms of overall system performance (Greenstein et al., 1986; Lemoine et al., 1995). Although implicit adaptive automation affords lower levels of operator workload, a trade-off has to be made against an increased risk of problems associated with increased monitoring load, out-of-the-loop performance problems, loss of skills, overtrust and undertrust, and increased system complexity when using this type of adaptive system. For a recent review, see Wickens et al.' s (2012) *Engineering Psychology and Human Performance.*

Many approaches to adaptive automation assume that operator control over function allocation (i.e., adaptable systems) is preferable to the system having control (i.e., adaptive systems). At the least, consent should be mandatory. This assumption may reflect the safety-critical nature of the correspondence domains that are researched (e.g., aviation, warfare, and process control). Indeed, returning to the early conferences on human-electronic crews discussed toward the beginning of the chapter, there are many references to the pilot wishing to remain in control of the electronic crew member. Harris et al. (1994) looked at the comparative effectiveness of operator- and system-initiated automation during anticipated and unanticipated increases in task load. When participants received written warnings that workload increases were likely to occur, performance during the operator- and system-initiated automation did not differ. However, when there was no warning before workload increase, resource management error was greater during periods of operator-initiated automation. The results suggest that system-initiated automation is most beneficial when rapid workload increase occurs without warning, and that when operator initiation is necessary, responses to rapid task load increases improve when warnings are provided.

Two main modes of control over function allocation have been described: (a) explicit allocation, which refers to situations where the operator has control over the allocation of tasks; and (b) implicit allocation, which refers to the system allocation of tasks automatically. All implicit adaptive systems require a mechanism to initiate, or trigger, changes in the OMI and levels of automation. A number of criteria can be used as triggers for adaptation.

3.3.2.1   Critical Events

Critical events are an important trigger for adaptation, and can be defined as situations during a mission that have a direct and significant impact on the successful accomplishment of mission or work-related goals. In terms of adaptation triggers, if a critical event occurs, the system adaptations are invoked. For example, if a specific predefined critical event occurs, such as the sudden appearance of a hostile aircraft, the appropriate defensive measures are performed by an automated system.

One problem is that it is often difficult, particularly in a complex multitask environment, to adequately specify *a priori* all eventualities that may occur in real settings. For example, the series of eventualities that led to the Air France incident discussed in Chapter 1 could not have been anticipated. Another related problem arises from potential insensitivity to the current needs of the operator, as there is an assumption *a priori* that the occurrence of a critical event necessitates automation of some functions because the operator cannot efficiently carry out these functions and deal effectively with the critical event. For example, it might not always be the case that a pilot will want the automation of particular functions during an emergency situation, and would instead want to retain manual control in order to troubleshoot the problem.

3.3.2.2   Operator State and Behavior

The dynamic assessment of the cognitive state (e.g., workload) and behavior (e.g., performance) of the operator is achieved using a combination of psychophysiological and behavioral measurements. As discussed earlier, adaptation to the OMI or automation is based on the real-time assessment of operator state or behavior; the violation of one or more predetermined criteria triggers the adaptation. The main advantage of this approach is that the measurement is taken in real time and is reasonably sensitive to unpredictable changes in operator cognitive states. However, this approach is only as accurate as the sensitivity and diagnostic capabilities of the measurement technology. Furthermore, due to the time lag inherent in these kinds of closed-loop systems (i.e., feedback), state and behavioral measurement usually occur after the fact. As a result, the actual adaptation might occur after the point in time when it may have been required due to changes in task demands or operator behavior. This is known as closed-loop instability (Wickens et al., 2012), which might lead to situations where an inference of a sudden onset of high workload is made after a significant delay, leading to the adaptation being implemented at the very moment that the operator's workload is back to a normal level. Finally, in highly automated systems that do not require the operator to make many observable responses during task accomplishment, the measurement of behavior might not, in fact, be practical. In this case or similar cases where not all behavior is directly observable, assessing the cognitive or affective state of the operator is the only alternative.

3.3.2.3   Operator Modeling

To mitigate the disadvantages of using solely operator state and behavioral assessment, this approach can be used in conjunction with operator modeling, whereby the adaptation strategy is enhanced by knowledge of human cognition and action. The system uses models of human cognition to predict (i.e., feed-forward) the operator's performance on a task, and adapt the information, support, or advice presented to operators when they are not able to cope with task demands. Modeling techniques are advantageous in that they can be incorporated into real-time operator state and behavior adaptation systems. However, this approach is only as good as the theory behind the model, and many models may be required to deal with all aspects of operator performance in complex task environments. Many approaches to adaptation invoke automation or OMI adaptation on the basis of impending performance degradation, as predicted by a human performance model. Models can be classi-fied broadly as either optimal performance models, such as signal detection, infor-mation, and control theories, or information processing models, such as multiple resource theory. For example, an adaptation strategy based on multiple resource the-ory (Wickens, 1984) would predict performance degradation whenever concurrent tasks place excessive demands on finite common cognitive resources, such as visual or auditory modality. When the combination of information from these different sources exceeds a threshold in the algorithm, tasks are allocated to the system or the OMI is adapted to reduce task demands.

Finally, adaptation based on cognitive models is dependent on their predictive validity. In other words, adaptation is dependent on whether the model predicts the actual cognitive state of the operator across a wide range of situations. Unfortunately, these models are descriptive. For example, they do not quantify the degree of cog-nitive degradation as a result of task factors. Rather, they indicate that degradation might be present. Another difficulty associated with operator modeling is that it requires real-time assessment of what the operator intends to do (i.e., goal tracking). This assessment is especially problematic in multitask situations where the activi-ties and intentions of operators at any moment in time might be difficult to infer accurately. Nevertheless, research will continue to improve both the fidelity and predictive validity of human cognitive models, and the accuracy of measurement techniques.

3.3.2.4 Combination-Based Approach

Of course, it is possible for IASs to combine performance and modeling, critical-event and performance, or other combinations of these methods, in order to opti-mize their relative benefits (Hilburn et al., 1993). Hilburn et al. (1993) propose that a hybrid system incorporating more than one of these methods might optimize ben-efits and minimize drawbacks. For example, the Cognitive Cockpit described in Section 3.2.1.2 compares pilot behavior with inferences about their goals and intent, which then drives decision aiding and OMI adaptation, a concept similar to the belief–desire–intention design method discussed in Chapter 5.

A hybrid approach to triggering adaptation within an IAS has enormous appeal for systems designers as it helps mitigate the significant challenges that are faced when implementing these kinds of systems. These challenges include obtaining artifact-free measurements of operator state and behavior in complex and sometimes hostile work settings (e.g., a military aircraft cockpit) and user acceptance and trust (Cummings, 2010). A very real challenge to systems designers is the time required by the system to capture and integrate a sufficient amount of data to allow accurate inferences about operator state and behavior to be made. If adequate time and data are not allowed, inferences about operator state and behavior are likely to be inaccu-rate, leading to inappropriate adaptations to the automation, OMI, or system advice, which will have a negative impact on operator trust in the system and acceptance of it (Wickens et al., 2012). Clearly, assessments of the time available for closed-loop systems to function should be made early on in the design process to establish what types of data sources can be used by the operator state assessment module. These issues are further discussed in Chapters 6 and 7.

3.4   SUMMARY

This chapter has sought to provide guidance to systems designers by describing IAS architectures at the conceptual level (not the systems design level). This facilitates an understanding of the functional components that an IAS needs to achieve high-level system goals. The four-step process involved in a typical IAS development project was outlined, and the importance of considering operator state during the design phase was noted.

A number of conceptual architectures were reviewed, and architectures from the PA, RPA, and Cognitive Cockpit research programs were examined in detail. Other IASs explored include the WCSS, Stock Trader system, and GIFT.

Edwards' generic framework was described, and his interpretations of task model, system model, user model, dialogue model, and world model were explained. The critical components of IASs were then discussed: situation assessment module, operator state assessment module, adaptation engine, and OMI. IAS adaptation and triggering mechanisms were examined, and the difference between explicit and implicit function allocation was clarified.

A hybrid approach to triggering adaptation was then suggested. Analytical techniques that can be used to populate the individual modules and models of the IAS conceptual architecture are explored in Chapter 4.

REFERENCES

Cummings, M. L. (2010). Technology impedances to augmented cognition. *Ergonomics in Design, 18*(2), 25–27.

Edwards, J. L. (2004). A generic agent-based framework for the design and development of UAV/UCAV control systems (Report No. CR 2004-062). Toronto, Canada: Defence Research and Development.

Emerson, J., Reinecke, M., Reising, J., and Taylor, R. M. (eds). (1992). The human-electronic crew: Is the team maturing? (Report No. WL-TR-92-3078). In *Paper Presented at Second International Workshop on Human-Electronic Crew Teamwork, Ingolstadt, Germany*. Dayton, OH: Wright Air Force Research Laboratory.

Emerson, J., Reising, J., Taylor, R. M., and Reinecke, M. (eds). (1989). The human-electronic crew: Can they work together? (Report No. WRDC-TR-89-7009). In *Paper Presented at the First International Workshop on Human-Electronic Crew Teamwork, Ingolstadt, Germany*. Dayton, OH: Wright Air Force Research Laboratory.

Greenstein, J. S., Arnaut, L. Y., and Revesman, M. E. (1986). An empirical comparison of model-based and explicit communication for dynamic human-computer task allocation. *International Journal of Man-Machine Studies, 24*(4), 355–363.

Harris, W. C., Goernert, P. N., Hancock, P. A., and Arthur, E. J. (1994). The comparative effectiveness of adaptive automation and operator initiated automation during anticipated and unanticipated taskload increases. In M. Mouloua and R. Parasuraman (eds), *Human Performance in Automated Systems*, pp. 40–44. Hillsdale, NJ: Lawrence Erlbaum Associates.

Hilburn, B., Molloy, R., Wong, D., and Parasuraman, R. (1993). Operator versus computer control of adaptive automation. In J. G. Morrison (ed.), *The Adaptive Function Allocation for Intelligent Cockpits (AFAIC) Program: Interim Research and Guidelines for the Application of Adaptive Automation*, pp. 31–36. Warminster, PA: Naval Air Warfare Center, Aircraft Division.

Hou, M., Gauthier, M. S., and Banbury, S. (2007a). Development of a generic design frame-work for intelligent adaptive systems. In J. Jacko (ed.), *Proceedings of Human-Computer Interaction International Conference, Part III*, pp. 313–320. Berlin: Springer.

Hou, M., Kobierski, R., and Brown, M. (2007b). Intelligent adaptive interfaces for the control of multiple UAVs. *Journal of Cognitive Engineering and Decision Making*, *1*(3), 327–362.

Hou, M., Zhu, H., Zhou, M., and Arrabito, G. R. (2011). Optimizing operator-agent interaction in intelligent adaptive interface design: A conceptual framework. *IEEE Transactions on Systems, Man, and Cybernetics, Part C: Applications and Reviews*, *41*(2), 161–178.

Kollmorgen, G. S., Schmorrow, D., Kruse, A., and Patrey, J. (2005). The cognitive cockpit—State of the art human-system integration (Paper No. 2258). In *Proceedings of the Interservice/Industry Training, Simulation and Education Conference (I/ITSEC)*. Arlington, VA: National Training Systems Association.

Lemoine, M. P., Crevits, I., Debernard, S., and Millot, P. (1995). Men-machines cooperation: Toward an experimentation of a multi-level co-operative organisation in air traffic control. In *Proceedings of the Human Factors and Ergonomics Society 41st Annual Meeting*, pp. 1047–1051. Santa Monica, CA: HFES.

Miller, C. A., Guerlain, S., and Hannen, M. D. (1999). The Rotorcraft Pilot's Associate cockpit information manager: Acceptable behavior from a new crew member. In *Proceedings of the American Helicopter Society 55th Annual Forum*, pp. 1321–1332. Alexandria, VA: American Helicopter Society.

Nemeth, C. P. (2004). *Human Factors Methods for Design: Making Systems Human-Centered*. Oxford, England: Taylor & Francis.

Rieger, C. A. and Greenstein, J. S. (1982). The allocation of tasks between the human and computer in automated systems. In *Proceedings of the IEEE 1982 International Conference on Cybernetics and Society*, pp. 204–208. New York, NY: IEEE.

Sottilare, R. A., Brawner, K. W., Goldberg, B. S., and Holden, H. K. (2012). *The Generalized Intelligent Framework for Tutoring (GIFT)*. Orlando, FL: US Army Research Laboratory-Human Research and Engineering Directorate (ARL-HRED).

Tattersall, A. J. and Morgan, C. A. (1996). Dynamic Task Allocation. Literature Review and Plan of Studies (Milestone Deliverable 1 Ref No. PLSD/CHS/(HS3)/6/5/RO3/7/17/5/01). Farnborough, England: Defence Evaluation and Research Agency.

Taylor, R. M., Bonner, M. C., Dickson, B., Howells, H., Miller, C. A., Milton, N., Pleydell-Pearce, C. W., Shadbolt, N., Tennison, J., and Whitecross, S. E. (2001). Cognitive cockpit engineering: Coupling functional state assessment, task knowledge management and decision support for context sensitive aiding. *Human Systems IAC Gateway*, *13*(1), 20–21.

Taylor, R. M. and Reising, J. (eds). (1995). The human-electronic crew: Can we trust the team? (Report No. WL-TR-96-3039). In *Paper Presented at Third International Workshop on Human-Electronic Crew Teamwork, Cambridge, United Kingdom*. Dayton, OH: Wright Air Force Research Laboratory.

Wickens, C. D. (1984). Processing resources in attention. In R. Parasuraman and D. R. Davies (eds), *Varieties of Attention*, pp. 63–102. San Diego, CA: Academic Press.

Wickens, C. D., Hollands, J. G., Banbury, S., and Parasuraman, R. (2012). Automation and human performance. In C. D. Wickens, J. G. Hollands, S. Banbury, and R. Parasuraman (eds), *Engineering Psychology and Human Performance*, 4th edn., pp. 377–404. Boston, MA: Pearson.

Yoo, J., Gervasio, M., and Langley, P. (2003). An adaptive stock tracker for personalized trading advice. In *Proceedings of the 8th International Conference on Intelligent User Interfaces*, pp. 197–203. New York, NY: ACM.

Young, M. J. and Eggleston, R. G. (2002). Work-centered decision support. In *Proceedings of RTO Human Factors and Medicine Panel (HFM) Symposium*, pp. 12-1–12-12. Neuilly sur Seine, France: NATO Research and Technology Organisation.

Section II

---

*Analysis and Design of
Intelligent Adaptive Systems*

4 Analytical Techniques for IAS Design*

4.1 OBJECTIVES

- Discuss analytical techniques for understanding operator tasks or goals
- Compare analytical techniques and how they generate different requirement types
- Discuss analytical techniques for OMI and IAS design
- Describe the creation of a hybrid analytical technique for a specific IAS project

Systems designers must understand certain key factors when designing products to be used by operators, including the work that the operator will do, the complexities and challenges that the operator will face, and the context in which the operator will work. The deeper the understanding of the roles, jobs, and environmental complexities of a particular operator and a potential system, the more likely a designer will be able to develop an innovative product that increases both the quality and the effectiveness of the operator's work.

In large-scale systems design or redesign efforts, it can be daunting to generate appropriate requirements; thus, a systematic approach that explores the problem space with a plan is critical. The analytical techniques that are discussed in this chapter have all been developed for large-scale industrial systems. First, the general concept of each technique is explained, including the background and the context in which the technique was developed. Next, the primary steps are outlined. Typical artifacts or models produced and the resources required for a particular technique are then discussed. In many cases, there are similarities in the resources used in and the artifacts produced by different techniques, which is not unusual as they have the common goal of facilitating an understanding of complex human work requirements. The strengths and limitations of each technique are then outlined, and an example is provided whenever possible.

This chapter is an introduction to systems design analysis and it targets key industry techniques; systems designers should perform deeper, more detailed study of the considered techniques, and references are provided to help facilitate this.

4.2   REVIEW OF COMMON ANALYTICAL TECHNIQUES

All of the analytical techniques described in this chapter provide potential useful input for IAS design. They act as a basis for the information that IASs must handle through the interface presentation to the operator, or through automation. Techniques such as operational sequence diagrams (OSDs) can be useful to predict what tasks and information will be presented to operators under different states of adaptation. By identifying periods of high workload or high operator stress, triggers for initiating different adaptive states can also be identified. Performance simulations provide an opportunity to test and evaluate an IAS structure before its implementation, resulting in cost savings, time savings, and refined user requirements. IAS components can be developed using a number of analytical techniques:

- *Situation assessment*: Situation assessment refers to an assessment of the external world and the functionality related to the analysis, automation requirements, and decision support of a given situation. This component requires both an external model and an expert model, which were discussed in Chapter 3. Certain analytical techniques can contribute to developing these models. In particular, cognitive work analysis (CWA) can inform the external model by identifying environmental constraints and complexities. Mission, function, and task analysis (MFTA); hierarchical task analysis (HTA); cognitive task analysis (CTA); and several other techniques can be used to build an expert model, as they identify mission goals, information requirements, and expected courses of action. Hierarchical goal analysis (HGA) can be used to develop discrete event simulations of expected operator behavior.
- *Operator state assessment*: In operator state assessment, current operator behavior is monitored for workload and inferred focus of attention. Analytical techniques can define the information and actions that may help systems designers make good choices in monitoring operator motor, perceptual, and cognitive states. Operator state monitoring approaches are further discussed in Chapter 6.
- *Adaptation engine*: The adaptation engine compares the operator model with the expert model and decides if intervention is required. Understanding operator work analytically can help identify likely situations where adaptation may be required. Furthermore, work analysis can suggest possible functionality or information changes that may suit the operator.
- *Operator–machine interface*: Joint application design (JAD) and ecological interface design (EID) are potential design frameworks for OMI design in IASs. These techniques can help define OMI characteristics that aid the operator, such as information content and form. In particular, previous studies have shown EID to be effective in improving operator understanding of automation behavior, limitations, and future actions (Furukawa and Parasuraman, 2003; Furukawa et al., 2004; Seppelt and Lee, 2007; Cummings and Bruni, 2010).

The working environment analytical techniques discussed in this chapter reflect techniques that have been previously applied in complex decision-making environments where IASs might typically be used. In each case, the general concept behind the technique is discussed, the major components of the technique are summarized, and, where available, examples are provided. This list is not exhaustive and other techniques may also be useful. Lee and Kirlik's (2013) *The Oxford Handbook of Cognitive Engineering*, and other referenced texts are a good starting point for further study on the techniques noted here.

4.2.1 MISSION, FUNCTION, AND TASK ANALYSIS

MFTA is a top-down analytical framework that has been developed for military environments, but the technique and its concepts can be applied elsewhere. As its name suggests, MFTA begins with an analysis of high-level mission goals and user-developed scenarios. Top-level functions are determined, and are then successively decomposed to operator task levels. Low-level functions are then allocated to the operator, hardware, software, or automation. In MFTA, functions that are allocated specifically to the operator are referred to as tasks, and an analysis of these tasks completes the MFTA (US Department of Defense, 2011). At each stage, requirements can be extracted to inform the design of a decision support system.

Ideally, MFTA should be applied early in the design process, as it tends to answer questions about scope and capability. In order to obtain detailed design information, MFTA should be augmented using other analytical techniques (Darvill et al., 2006). Figure 4.1 illustrates the primary components of MFTA:

- *Mission and scenario analysis*: Based on the concept of operations, this phase provides the scope requirements and defines the major objectives of the system design. The types of missions or scenarios to be considered are outlined; major roles and operators are described; and other stakeholders, conditions of use, and design constraints are noted.
- *Function analysis*: The functional capabilities of the system are outlined without consideration of the specific design. Functions are connected to mission phases, and are then decomposed to subfunctions as needed.
- *Function allocation*: Function allocation is often described according to the mission phases and can play a critical role in determining various elements of the operator interface, the number and capabilities of the crew or operators, and the optimal workload levels of those operators. Functions described at the lowest level of the function analysis are allocated to either operator or machine (hardware, software, or automation). Functions allocated to technology and automation are considered system functions, and functions allocated to operators are tasks.
- *Task analysis*: This phase of MFTA can be completed using any common form of task analysis, such as HTA. The key aspect of a task analysis is the decomposition of necessary tasks into low-level subtasks. The overall execution of tasks should be analyzed while focusing on efficiency to ensure that the steps are clear and effective. At each task step, the requirements

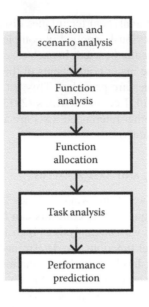

FIGURE 4.1 The primary components of mission, function, task analysis (MFTA). MFTA should be applied early in the design process to help answer questions about scope and capability.

for operator information, critical decisions, and expected interaction are outlined.

- *Performance prediction*: Using the gathered information, analysts predict system performance under various conditions. Analysts may predict task completion times, performance accuracy, workload levels, crew utilization, and task completion success rate. This phase can be used to validate the conceptual design to ensure that it will likely be successful if implemented.

4.2.1.1   Results and Artifacts of MFTA

- *Mission analysis report*: This document should list all the expected missions, key baseline scenarios (usually produced by the customer), environments, related systems, operation constraints, operator roles and responsibilities, and the expected scope of personnel and equipment in the system being analyzed.
- *Composite scenario*: A composite scenario is used to show the key mission segments for the system in terms of mission phases, sequencing, events, constraints, and environmental conditions. The segments included in the composite scenario are those that are determined to be design drivers in the mission analysis report.
- *Function flow diagrams*: These diagrams present a hierarchy of system functions as they are employed over time in various mission phases.
- *Task data*: These data include a decomposition of tasks and their sequencing, and the cognitive and physical requirements for execution. Important data items to document include (a) information requirements, to support the

design of display formats; (b) action requirements, to support the design of controls; (c) initiating conditions; and (d) appropriate feedback, to confirm that the task has been completed successfully.

- *Operational sequence diagrams*: These are often used as part of a task analysis to show how tasks change at various decision points or due to communication between operators if more than one operator is involved in the mission. OSDs represent the logical sequence of tasks along a mission time line. They usually represent the composite scenario and they normally incorporate multiple ongoing tasks. OSDs are used to conduct workload analyses, as well as information flow and cognitive processing analyses. In order to produce an OSD, the task functions must be allocated to a specific operator.
- *Performance predictions*: These predictions illustrate or document the results of performance simulations.

4.2.1.2 Tools and Resources Required

MFTA uses typical human factors engineering techniques, such as conducting documentation reviews, operator and subject matter expert (SME) interviews, and observations. Other MFTA-specific tools have also been developed to help keep track of task analysis information, such as the integrated performance modeling environment (IPME), which has been developed to provide the simulation that is required for performance prediction (Alion Micro Analysis and Design, 2009). The IPME is further discussed in Chapter 8.

The following is a list of the resources that are commonly used for MFTA:

- Documents
- Customer-produced baseline scenarios
- Procedures
- Checklists
- History with current systems
- Training material used for current systems (if this is for a systems redesign)
- Interviews
- Observations
- SME sessions
- Discrete event simulation

4.2.1.3 Strengths and Limitations

MFTA was designed explicitly to support a military human–systems integration process and it is therefore well suited to IAS design. Since the analysis covers both functions and tasks, it provides a rich view of the work that is required in the environment being analyzed. This is similar to CWA, which models both functions and tasks, and CTA, which breaks down tasks and can use OSDs. However, the baseline scenarios and composite scenarios that are developed are unique to MFTA and can be very useful when evaluating a new system.

The MFTA process is rigorous and it should be conducted in a consistent manner. To ensure that the task system is modeled correctly, MFTA requires access to numerous SMEs. In large systems, an MFTA can scale up very quickly by generating many tasks to analyze. Past operator performance data or access to current operators for subtask data collection are required to complete simulation models and create accurate performance predictions. Subtask data include the visual, auditory, cognitive, and psychomotor ratings for each task, task completion time, and task conflict prediction ratings. The tools that support MFTA are complex and have a significant learning curve.

4.2.1.4   MFTA Applied to a Halifax Class Frigate

Figure 4.2 illustrates part of an MFTA that was conducted for a Canadian Halifax class frigate. The frigate is a multifunction ship deployed as part of a larger naval task force. Many of its functions are related to surveilling contacts within a specified sector. In this case, the composite scenario combined the frigate's peacetime and wartime missions. In Figure 4.2, the MFTA is shown starting from the top level, a mission to conduct transit. Various functions are identified, such as conducting surveillance and detecting contact, and a task-level decomposition, which includes determining distance and observing contact, is provided. To understand the scale of this analysis, note that three missions were analyzed, and these were subsequently decomposed into approximately 2600 tasks (Chow et al., 2007). OSDs were created for eight key task sequences. The analysis was used to identify the information and action requirements for operators and to look at task allocation across multiple operators. The analysis could then be used to inform interface design and crew selection decisions.

1. Conduct transit (mission, top-level function)
   1.1. Proceed on watch
   1.2. Conduct continuous surveillance (mission segment, first-level function)
      1.2.1. Maintain picture
      1.2.2. Detect contact (second-level function)
         1.2.2.1. SWC task (third-level function, by operator)
            1.2.2.1.1. Determine distance (task, fourth-level function)
            1.2.2.1.2. Observe contact, information displayed
         1.2.2.2. ASWC task
            1.2.2.2.1. Optimize radar for detection
      1.2.3. Classify contact
      1.2.4. Localize contact
      1.2.5. Manage friendly track
      1.2.6. Manage hostile track
      1.2.7. Manage unknown track
   1.3. Monitor, manage systems
   1.4. Maintain communications
2. Conduct peacetime operations (mission, top-level function)
3. Conduct warfare operations (mission, top-level function)

FIGURE 4.2   Mission, function, task analysis (MFTA) for a Halifax class frigate. The mission, or top-level function, is decomposed into individual tasks. (Redrawn from Chow, R., Kobierski, R. D., Coates, C. E., and Crebolder, J., *Proceedings of the 50th Annual Meeting of the Human Factors and Ergonomics Society*, pp. 520–524, SAGE Publications, New York, 2006.)

4.2.2 Hierarchical Task Analysis

HTA is arguably the oldest human factors approach, dating back to the earliest studies on improving the way humans work (Annett and Duncan, 1967; Annett and Stanton, 1998). Almost all other forms of task or work analysis build from HTA, using the basic concept of decomposing work to some degree, while modifying the approach to explicitly address other needs.

The fundamental concept behind HTA is the decomposition of complex tasks to successively smaller and smaller subtasks. Task breakdowns are most often expressed in a hierarchical structure diagram, but may also appear in table format in cases where richer descriptions are required. OSDs are also used in HTA to help show decision points and loops.

4.2.2.1 Results and Artifacts of HTA

*   *Structure model*: The structure model begins at the top level with an overall task. Each level decomposes the task further, from subtasks at Level 2 to sub-subtasks at Level 3. Task sequences are maintained and a hierarchical numbering system is often used to help keep the model readable and traceable. The decomposition continues until the analyst reaches a level where the tasks have become individual actions, or where variance in the actions is not consequential or meaningful to the overall task. The structure model can be used comparatively to look at different task designs for efficiency, or informatively to examine operator needs at each task step. Figure 4.3 illustrates what a task structure model might look like. The arrows indicate that the structure represents a sequence of subtasks that constitute a mission segment, as opposed to the decomposition of an overall task type into increasingly specific examples of the task type. For example, "land aircraft" decomposes into "conduct normal landing; conduct short

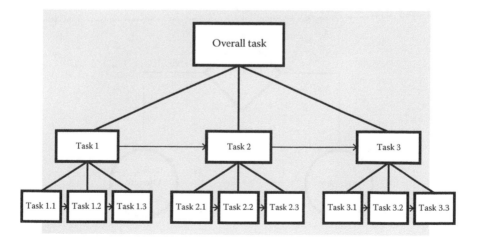

FIGURE 4.3 A task structure model for hierarchical task analysis (HTA). Sequences of subtasks constitute a mission segment. The arrows indicate variations of a specific task type.

field landing; conduct soft field landing," and the individual subtasks may decompose into "conduct landing with no crosswind; conduct landing with crosswind; conduct landing with a barrier," and so on.

- *Task analysis table*: A task analysis can also be expressed in table format. In tables, numbering becomes the main method for retaining task hierarchy. The table format is useful for providing a richer task description, as well as organizing information that is relevant to the task, such as operator needs or task complexity.
- *Operational sequence diagram*: The OSD was developed for task analysis as a response to the rise in control-theoretic perspectives. The OSD shows decision points and loops and can be particularly useful when modeling tasks that do not follow a linear path. The symbols used in an OSD are consistent with similar approaches found in engineering and design. Figure 4.4 illustrates a simple OSD.

4.2.2.2   Tools and Resources Required

The best way to gather information for a HTA is by using any approach that clearly shows the decomposition of tasks. Observations and previously documented procedures are good sources for confirming task decomposition. However, interviews and

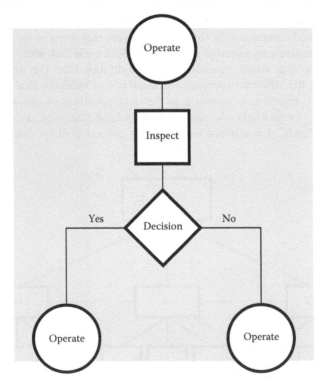

FIGURE 4.4   A simple operational sequence diagram (OSD). The symbols used in OSDs, such as a diamond shape for a decision, are consistent with symbols used in engineering and design.

SMEs are useful when preferred sources are unavailable. The following is a list of the resources that are commonly used for HTA:

- Documents
- Procedures
- Checklists
- Training material used for current systems (if this is for a systems redesign)
- Interviews
- Observations
- SME sessions

4.2.2.3 Strengths and Limitations

The main strength of HTA is that it is well known and easy to learn. In the hands of a skilled analyst, HTA can be flexible enough to capture much of what HTA-derivative approaches seek to capture. One of the limitations of HTA is that, if performed in a rote or unskilled manner, it can generate near meaningless sequences of operational steps. A HTA's strength lies in analyzing the operational steps to extract the information and cognitive demands required. A skilled analyst begins a project by analyzing a very short period of the scenario, and completes all steps of the anticipated analysis just to determine the fidelity required of each step through to the drawing of the project's conclusions.

4.2.2.4 HTA Applied to Medication Management

Drug administration can be a lengthy and complex process where human error may be introduced with ill effects. For these reasons, it is a good candidate for HTA and, where possible, simplification and error-reducing steps (Lane et al., 2006). Figure 4.5 illustrates a HTA performed by a hospital pharmacy technician skilled in HTA that was reviewed by other hospital personnel. The top-level goal (or task) is "to administer drugs to the patient." At the next level, four tasks are described and these are

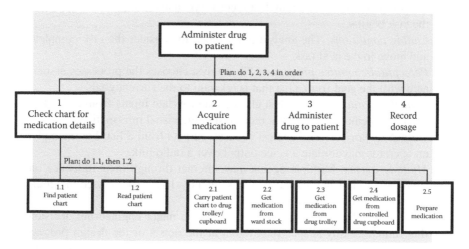

FIGURE 4.5 Hierarchical task analysis (HTA) for drug administration. (Redrawn from Lane, R., Stanton, N., and Harrison, D., *Applied Ergonomics,* 37, 669–679, 2006.)

then further decomposed. Following the task analysis, the systematic human error reduction and prediction approach (SHERPA; Embrey, 1986; as cited in Harris et al., 2005) was applied to identify points of risk for human error.

4.2.3 HIERARCHICAL GOAL ANALYSIS BASED ON PERCEPTUAL CONTROL THEORY

Perceptual control theory (PCT; Powers, 1973) takes a control-theoretic view of human behavior. The essential proposition behind PCT is that people do not respond to stimuli or compute an output, but rather they observe their own behavior and input variables and compare these inputs with their desired inputs, thus controlling their output behavior until they have achieved the desired observed inputs. This theory takes advantage of control theory but stands in stark contrast to behavioral psychology, where organisms respond to stimuli, and information processing theory, where organisms compute outputs. The advantage of PCT is that the focus is on goal states, input variables, and actions, which are all very relevant to design. HGA (Hendy et al., 2001) is an analytical technique built on PCT. HGA identifies goals and builds control loops that identify output interfaces, input interfaces, and influencing variables at each goal level. It should be noted that there is a special case of PCT called layered protocol theory (LPT; Farrell et al., 1999; Taylor et al., 1999) that relates specifically to issues of communication. PCT and LPT are further discussed in the context of interaction-centered design in Chapter 5.

4.2.3.1 Results and Artifacts of HGA

- *Goal structure model*: The goal structure model requires specific cognitive and perceptual information related to each goal loop at every hierarchical level.
- *Required knowledge states*: The analyst identifies both declarative knowledge (e.g., procedures) and situational knowledge.
- *Initiating conditions*: The analyst records when and under what conditions the task begins.
- *Ending conditions*: The analyst notes when to consider the task complete and move to the next task.
- *Perceptual/cognitive processes*: The analyst chooses the processes associated with the goal from a list that is relevant to the current goal.
- *Inputs/sensations*: The analyst chooses appropriate inputs from a list. For example, memory tasks might require specific visual information.
- *Outputs/behaviors*: The analyst chooses outputs from a list. For example, an operator may initiate a voice output over a radio link.
- *Multiple agents*: Multiple agents may interact through their influence on shared environmental variables. Consequently, HGA can be used in a multiperson or team environment.
- *Objectives*: The assignment of objectives is a major engineering decision that fundamentally shapes the system at the outset of the design process (Hendy et al., 2001). Therefore, no assignments to human or machine are made using HGA until the goal-related information has been collected for all system goals within the hierarchy.

- *Performance predictions*: Using the information from the goal structure model, HGA creates a structure of control loops. These can be simulated using discrete event simulation, such as the IPME, in order to examine the sequencing and timing of activities.

4.2.3.2 Tools and Resources Required

HGA uses typical human factors engineering techniques, such as observation and interviews of operators and SMEs, but also includes the use of focus groups and introspection. Focus groups and introspection can help transform typical operator observations into a goal structure model. Additionally, the IPME includes features to assist with HGA and the execution of goal structure model performance predictions. The following is a list of the resources that are commonly used for HGA:

- Interviews
- Observations
- SME sessions
- Focus groups
- Introspection
- Discrete event simulation

4.2.3.3 Strengths and Limitations

Although HGA requires similar information and a similar level of effort as other analytical techniques discussed in this chapter, it is more specialized and relies on analysts who are knowledgeable about both the analytical technique and the system domain. HGA requires sufficient dedicated resources to collect all information for hierarchical decomposition and establish performance prediction exercise parameters, generally using PCT constructs. HGA also requires that feedback associated with low-level goal achievement be transmitted upward to high-level goals so that these goals may be perceived as being complete. This unique bottom-up feature complements the appropriate feedback data element of a task analysis. HGA may provide insight into tasks that are complex and control based, or tasks where human error has previously been a problem. HGA can be started at any point of human interaction with the system, and can provide useful information for the smaller parts of a complex system when it is used in expanding iterations. After executing the models in the IPME, analysts must interpret the results for use in the design and development phases.

4.2.3.4 HGA Applied to a Halifax Class Frigate

Figure 4.6 illustrates part of an HGA for a Halifax class frigate. This example was chosen for comparison with the MFTA example. The difference between the techniques is apparent from the first step. In MFTA, the first step is a mission-level "conduct transit." In HGA, the first step is a goal: "I want to perceive the conduct of sea denial operations." The MFTA is action oriented, where the HGA is goal oriented and begins at a higher conceptual level. The HGA proceeds to decompose the goals, going into finer and finer detail until it reaches the lowest-level control loops,

Goal: I want to perceive	Operator
The conduct of sea denial operations (top-level goal)	CO
1. Current mission is received and acknowledged	CO
2. Predeployment preparations are complete	CO
3. Ship is ready to undertake critical operational taskings	CO
4. Combat organization and resources are managed effectively	CO
5. Optimal level of SA is being maintained (first-level goal)	CO
5.1. Accurate recognized maritime picture is created and maintained (second-level goal)	ORO
5.1.1. Accurate tactical air picture is compiled (third-level goal)	SWC
5.1.1.1. Coordination with other units is effective	SWC
5.1.1.2. Effective display configuration	SWC
5.1.1.2.1. Display settings are appropriate	SWC
5.1.1.2.2. Alarms are accurate	SWC
5.1.1.2.3. Display overlays are implemented	SWC
5.1.1.2.4. Effective radar manipulation occurs	SWC
5.1.1.3. Effective air track management	ARRO
5.1.1.4. Effective visual watch is maintained	ORO
5.1.2. Accurate tactical surface picture is compiled	SWC
5.1.3. Accurate tactical subsurface picture is compiled	ASWC
6. Ongoing operational tasks are being actioned effectively	CO
7. Mission follow-up action is completed	CO
The conduct of peace operations (top-level goal)	CO
The provision of assistance to other government departments (top-level goal)	CO

FIGURE 4.6 Hierarchical goal analysis (HGA) for a Halifax class frigate. Top-level goals are decomposed into more manageable subgoals. (Redrawn from Chow, R., Kobierski, R. D., Coates, C. E., and Crebolder, J., *Proceedings of the 50th Annual Meeting of the Human Factors and Ergonomics Society*, pp. 520–524, SAGE Publications, New York, 2006.)

ensuring that display settings are appropriate. The corresponding MFTA would be used to detect a contact, and then optimize radar for detection. Both techniques should extract the same information, even though the process and perspective are different.

4.2.4 Goal-Directed Task Analysis

Endsley and Jones (2011) developed goal-directed task analysis (GDTA) to reveal situation awareness (SA) requirements. GDTA begins with a definition of the goals and progresses through a decomposition to identify the perceptual variables that are needed to support SA. The SA requirements that have been identified by GDTA have been shown to improve operator performance in complex environments (Bolstad et al., 2002; Endsley et al., 2003).

4.2.4.1 Results and Artifacts of GDTA

- *Goal structure model*: In contrast to the previous techniques, GDTA focuses on decomposing goals into a goal hierarchy. It does not break down to specific tasks. The information variables required to support the SA of each goal are outlined in detail. GDTA is similar to HGA in that both techniques use goal decomposition, but they differ in perspective. For example, GDTA follows a treelike breakdown similar to HTA and does not particularly look for control loops.

- *SA requirements*: Each low-level goal should be analyzed to identify SA requirements. The requirements should be classified by the level of SA as supporting perception, supporting understanding, or predicting future states or actions. A sample breakdown of SA requirements could include (a) goal, (b) subgoal, and (c) SA variables.

4.2.4.2 Tools and Resources Required

GDTA can be conducted from interviews and observations, and does not require any special tools. Sessions with SMEs are useful to generate the goal decomposition and variable lists. The following is a list of the resources that are commonly used for GDTA:

- Interviews
- Observations
- SME sessions

4.2.4.3 Strengths and Limitations

GDTA is a good technique for obtaining information requirements that support SA in complex environments. GDTA is domain focused and is normally able to generate requirements independent of the existing interface or information system. This allows GDTA to support the development of new technologies. GDTA does not provide particularly useful information on task sequencing and does not support performance prediction.

4.2.4.4 GDTA Applied to Army Brigade Officer Teams

Bolstad et al. (2002) applied GDTA to determine the SA requirements for army brigade officers. Interviews were held with three brigade officers per position. After the interviews, goal hierarchies and SA requirements were developed, and then reviewed by the brigade officers for accuracy. The analysis revealed shared SA requirements and SA requirement differences among the positions. For example, some positions required detailed information about the effect of terrain on enemy assets, while other positions required greater information about friendly assets. General information about troops and courses of action needed to be shared among positions. The GDTA was a useful exercise to identify these requirements. Table 4.1 shows a subset of the SA requirements for officers.

4.2.5 Cognitive Task Analysis

CTA is the decomposition of cognitive work into low-level subtasks to determine work requirements. There are various approaches to CTA. In general, however, all CTA techniques have the same two steps: (a) decomposition of the work into smaller task units, and (b) analysis of the cognitive requirements or demands of each task unit. This section focuses on applied CTA (ACTA; Militello and Hutton, 1998), as it is representative of most CTA techniques. ACTA offers a set of streamlined techniques for performing efficient CTA. The critical decision method and concept mapping, two other CTA techniques, have distinct differences from both hierarchical CTA and ACTA and will be discussed elsewhere.

TABLE 4.1

Situation Awareness Requirements Generated from a Goal-Directed Task Analysis (GDTA) of Army Brigade Officers

SA Level	Intelligence	Operations	Logistics	Engineer
1	• Areas of cover/concealment	• Areas of cover/concealment	• Areas of cover/concealment	• Type
	• Enemy boundaries	• Key terrain	• Potential choke points due to terrain	• Conditions
	• Engagement areas	• Type	• Type	• City plan
	• Location of restrictive terrain	• Conditions	• Conditions	• Map of area
	• Map of the area	• City plan	• City plan	• Subsurface
	• Restrictive points	• Map of area	• Map of area	• Features
	• Significant terrain characteristics	• Subsurface	• Subsurface	• Vegetation
	• Type	• Features	• Features	• Hydrology
	• Conditions	• Vegetation	• Vegetation	• Location
	• City plan	• Hydrology	• Hydrology	• Swamps
	• Map of area	• Location	• Location	• Lakes
	• Subsurface	• Swamps	• Swamps	• Wet lands
	• Features	• Lakes	• Lakes	• Rivers
	• Vegetation	• Wet lands	• Wet lands	• Locations
	• Hydrology	• Rivers	• Rivers	• Conditions
	• Location	• Bank slopes	• Bank slopes	• Bank
	• Swamps	• Water tables	• Stream beds/drainage	• Slopes
	• Lakes	• Obstacles	• Water tables	• Condition
	• Wet lands		• Obstacles	• Water tables
	• Rivers		• Contour/elevation	• Obstacles
	• Bank slopes		• Firmness of ground	• Type
	• Water tables		• Grade	• Location
	• Obstacles			• Quantity
				• Rocks

- Enemy limitations/advantages due to terrain
- Friendly limitations/advantages due to terrain
- Effect of terrain on enemy and friendly assets
- Effect of terrain on anticipated troop movement time
- Effect of terrain on system detection capability

- Accessibility of routes
- Effect of terrain on movement times/time to position troops
- Effect of terrain on rate of enemy closure
- Effect of terrain on visual capabilities
- Effect of terrain on communication capabilities
- Effect of terrain on route difficulty

- Suitability of land for unit
- Effect of terrain on ability to access location with each vehicle type
- Effect of terrain on type of vehicles to be supported

- Houses
- Terrain
- Roads
- Vehicles
- Villages
- Buildings
- Trees
- People
- Mines
  - Location enemy
  - Location friendly
- Potential approaches and exiting areas
- Potential staging areas
- Potential terrain suppression areas
- Traffic ability
- Visibility of the locations
- Critical obstacle information
- Past enemy usage of obstacles
- Effect of terrain on location of enemy counter attacks

(continued)

2

TABLE 4.1 (Continued)
Situation Awareness Requirements Generated from a Goal-directed Task Analysis (GDTA) of Army Brigade Officers

SA Level	Intelligence	Operations	Logistics	Engineer
3	• Predicted effects of terrain on enemy courses of action (COAs) • Projected effects of terrain on friendly COAs • Projected terrain • Projected effect of terrain on troop movements	• Predicted effects of terrain on enemy COAs	• Projected effect of terrain on usage rates per item per unit • Projected effect of terrain on security of resources	• Estimated obstacle effectiveness • Predicted most secure location for assets, soldiers, vehicles • Predicted most survivable routes

Note: Data from Bolstad, C. A., Riley, J. M., Jones, D. G., and Endsley, M. R., *Proceedings of the 46th Annual Meeting of the Human Factors and Ergonomics Society*, pp. 472–476, 2002. New York, NY: SAGE Publications.

4.2.5.1   Results and Artifacts of CTA

ACTA suggests using the following CTA components:

*   *Task diagram*: Similar to HTA, SMEs decompose tasks into subtasks. While decomposing the task, the analyst asks the SME to discuss the breakdown, identifying cognitive skills and areas of difficulty.
*   *Knowledge audit*: A knowledge audit probes for domain knowledge and skills that are relevant to diagnosis, prediction, SA, perception, and solution development. In a knowledge audit, experts may be asked to identify areas where novices have difficulty.
*   *Simulation interview*: In a simulation interview, a scenario is presented and the SME is walked through the scenario. Key probes look at major events, decision-making points, critical cues, and potential errors. The intention of the simulation interview is to capture requirements in a situation that is more closely representative of actual operations.
*   *Cognitive demands table*: A cognitive demands table organizes and collates CTA results. Potential columns are (a) cognitive element, (b) reasons for difficulty, (c) common errors, (d) cues needed, and (e) strategies used.

CTA may result in a task breakdown and tables describing the cognitive challenges and information requirements to complete the associated tasks. These tasks are matched to the motor, perceptual, cognitive, and multimodal aspects of the operations. Knowledge audits and simulation interviews result in tables of cognitive demands, difficulties, skills, and strategies. Thus, a cognitive demands table may be used to summarize demands, difficulties, cues, strategies, and errors.

4.2.5.2   Tools and Resources Required

CTA requires access to SMEs for interviews and scenario walkthroughs. Observations are also helpful. The following resources are commonly used for CTA:

*   Interviews
*   Observations
*   SME sessions

4.2.5.3   Strengths and Limitations

CTA can lead to the development of a rich list of cognitive requirements to help define a new support system. By identifying both difficult areas and expert strategies, CTA helps ensure that the new system offers improvements over the previous system. However, CTA is limited to understanding current tasks and may not be very good at revealing the requirements for unanticipated tasks. Functional techniques, such as MFTA or CWA, may better suit unanticipated situations.

4.2.5.4   CTA Applied to Fireground Commanders

Militello and Hutton (1998) propose ACTA as an efficient CTA technique. They use the example of fireground commanders, previously used by Klein et al. (1989) to

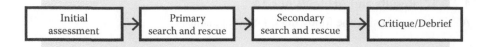

FIGURE 4.7 An example task diagram for fireground commanders. (Redrawn from Militello, L. and Hutton, R., *Ergonomics*, 41, 1618–1641, 1998.)

show the usefulness of ACTA in studying a complex work situation where experts have deep knowledge that novices lack. Militello and Hutton demonstrate that ACTA has similar requirements to CTA, but it is able to provide more structure to aid the process. The results of their fireground commander analysis are provided as an example of the outputs of ACTA: Figure 4.7 illustrates a task diagram; Table 4.2 shows a knowledge table; Table 4.3 shows a simulation interview table; and Table 4.4 shows the resulting cognitive demands and provides an overall summary of the ACTA, noting priority areas for system support.

4.2.6 Concept Mapping

Concept mapping decomposes cognitive work using ideas rather than tasks (Novak and Canas, 2008). Concept mapping is useful for learning about complex tasks and

---

TABLE 4.2
Knowledge Audit Table for an Applied Cognitive Task Analysis (ACTA) of Fireground Commanders

Aspects of Expertise	Cues and Strategies	Why Difficult?
• Past and future (e.g., explosion in office strip; search the office areas rather than source of explosion)	• Material safety data sheets (MSDS) tell you that explosion is in the area of dangerous chemicals and information about chemicals • Start where most likely to find victims and own safety considerations	• Novice would be trained to start at source and work out • May not look at MSDS to find potential source of explosion and account for where people are most likely to be found
• Big picture includes source of hazard, potential location of victims, ingress/egress routes, other hazards • Noticing breathing sounds of victims	• Senses, communication with others, building owners, MSDS, building preplans • Both you and your partner stop, hold your breath, and listen • Listen for crying, victims talking to themselves, victims knocking things over	• Novice gets tunnel vision, focuses on one thing (e.g., victims) • Noise from own breathing in apparatus, fire noises • Do not know what kinds of sounds to listen for

Source: From Militello, L. and Hutton, R., *Ergonomics*, 41, 1618–1641, 1998. Reprinted by permission of the publisher (Taylor and Francis Ltd, http://www.tandf.co.uk/journals).

TABLE 4.3

Simulation Interview Table for an Applied Cognitive Task Analysis (ACTA) of Fireground Commanders

Events	Actions	Assessment	Critical Cues	Possible Errors
On scene arrival	• Account for people (names) • Ask neighbors (but do not take their word for it, check it out yourself) • Must knock on or knock down doors to make sure people are not there	• It is a cold night, need to find place for people who have been evacuated	• Night time • Cold $\pm > 15°$ • Dead space • Add-on floor • Poor materials: wood (punk board) and metal girders (buckle and break under pressure) • Common attic in whole building	• Not keeping track of people (could be looking for people who are not there)
Initial attack	• Watch for signs of building collapse • If signs of building collapse, evacuate and throw water on it from outside	• Faulty construction, building may collapse	• Signs of building collapse include: what walls are doing (cracking); what floors are doing (groaning); what metal girders are doing (clicking, popping) • Cable in old buildings holds walls together	• Ventilating the attic (this draws the fire up and spreads it through the pipes and electrical system)

Source: From Militello, L. and Hutton, R., *Ergonomics*, 41, 1618–1641, 1998. Reprinted by permission of the publisher (Taylor and Francis Ltd, http://www.tandf.co.uk/journals).

the relationships that are important to understanding them. Concept maps have been used in education—seeing the relationships between ideas can help students learn the more complex overall system. They are also useful in interface design to show the relationships between different ideas.

A concept map can be developed from interviews with experts or from a group brainstorming session. Concept maps are generally free form and display unstructured associations of ideas. In the map, a concept is a stabilized idea connected to other concepts by propositions that describe the relationship between the concepts. For example, a possible concept unit would be "a dog (concept) is an example of (proposition) an animal (concept)."

4.2.6.1   Results and Artifacts of Concept Mapping

A concept map is the artifact from a concept mapping exercise.

TABLE 4.4

Cognitive Demands Table for an Applied Cognitive Task Analysis (ACTA) of Fireground Commanders

Difficult Cognitive Element	Why Difficult?	Common Errors	Cues and Strategies Used
Knowing where to search after an explosion	• Novices may not be trained in dealing with explosions • Other training suggests you should start at the source and work outward • Not everyone knows about the material safety data sheets (MSDSs), which contain critical information	• Novices would be likely to start at the source of the explosion • Starting at the source is a rule of thumb for most other kinds of incidents	• Consider the type of structure and where victims are likely to be • Start where you are most likely to find victims, keeping in mind safety considerations • Refer to material safety data sheets (MSDSs) to determine where dangerous chemicals are likely to be • Consider the likelihood of further explosions • Keep in mind the safety of your crew
Finding victims in a burning building	• There are lots of distracting noises • If you are nervous or tired, your own breathing makes it hard to hear anything else	• Novices sometimes do not recognize their own breathing sounds; they mistakenly think they hear a victim breathing	• Both you and your partner stop, hold your breath, and listen • Listen for crying, victims talking to themselves, victims knocking things over, and so on

Source: From Militello, L. and Hutton, R., Ergonomics, 41, 1618–1641, 1998. Reprinted by permission of the publisher (Taylor and Francis Ltd, http://www.tandf.co.uk/journals).

4.2.6.2   Tools and Resources Required

The following is a list of the resources that are commonly used for concept mapping:

- Experienced operators
- Concept mapping interviews or focus group exercises

4.2.6.3   Strengths and Limitations

Concept mapping is a useful technique when trying to uncover operator associations. Concept maps can help reveal complicated relationships or unexpected associations that are understood by operators in the work domain.

4.2.6.4   Concept Map of Seasons

Figure 4.8 illustrates a concept map of the seasons, originally provided by Novak and Canas (2008). Note the rich connectivity between concepts and the natural flow of association that results in a concept map.

4.2.7   THE CRITICAL DECISION METHOD

Originated by Klein et al. (1989), the critical decision method is a technique for eliciting cognitive task requirements. It builds on Flanagan's (1954) critical incident technique, which focuses on probing operators along particular influential event time lines. The technique has been very successful in probing highly experienced personnel working in dynamic, fast-paced environments where observation or real-time verbal protocol techniques may be inappropriate for the following reasons:

- Events happen rarely and observations cannot be organized to capture a meaningful sample of events. The observation of accidents is an example of this situation.

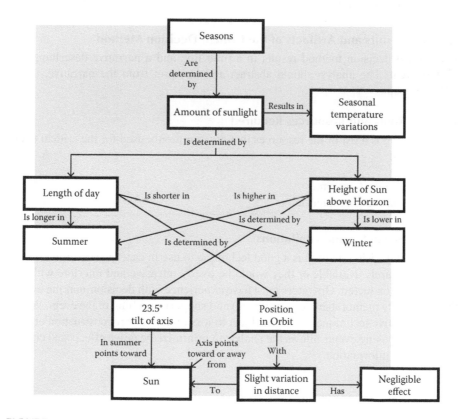

FIGURE 4.8   Concept map illustrating the concept of seasons. (Redrawn from Novak, J.D. and Canas, A.J., The theory underlying concept maps and how to construct and use them (Report No. IHMC Technical Report Cmap Tools 2006-01 Rev 01-2008), Florida Institute for Human and Machine Cognition, Pensacola, FL, p. 10, 2008.)

- Observing operators, asking them questions during the task, or asking them to generate a verbal protocol would interfere with their ability to perform the task. Users in safety-critical and high-pressure tasks, such as emergency workers, are an example of operators in this situation.
- The high expertise level of operators tends to make it difficult for them to articulate their reasoning in detail. As operators gain expertise, in many cases their knowledge becomes tacit, rather than explicit. This makes it difficult for them to articulate the details of their knowledge, prompting answers like "I just know." By reconstructing a past event, the analyst may be able to obtain more explicit knowledge.

The critical decision method relies almost exclusively on probed interviews surrounding an event that was memorable for its intensity, criticality, or lessons learned. The interviewer must determine a suitable event, and then begin probing before or after the event. The critical decision method is a core technique for understanding naturalistic decision making (NDM), or how people work in real situations under intense time constraints.

4.2.7.1   Results and Artifacts of the Critical Decision Method

The critical decision method results in a time line and a narrative describing the related event. The analyst should abstract and interpret from the narrative, using NDM concepts.

4.2.7.2   Tools and Resources Required

The following is a list of the resources that are commonly used for the critical decision method:

- Experienced operators
- Critical incident interviews

4.2.7.3   Strengths and Limitations

The critical decision method is a good technique to use in cases where direct observations are rarely available or they would be overly intrusive and interfere with the work being conducted. Operators must have experience with decision-making events that are highly memorable (i.e., significant) and support a probing of the event. While all retrospective techniques may be subject to loss of detail or reconstruction errors, carefully choosing events allows the analyst to minimize negative effects and collect the required information.

4.2.8   Team Cognitive Task Analysis

Team CTA is an extension of CTA that considers the team as a single cognitive entity (i.e., more than a collection of individuals; Klein, 2000). By using a technique that is designed explicitly for team use, specific team processes, such as collaboration and communication, can be studied. Techniques that are designed for individual task analysis may not reveal these requirements. Team CTA uses the concept of "team mind" (Klein, 2000), which views the team as a single whole with specific cognitive activities. This technique is closely tied to understanding "macrocognition," which originated at the same time in Klein et al. (2000). There are a number of key team processes related to macrocognition that should be considered in a Team CTA:

- Sensemaking
- Planning
- Learning
- Coordinating
- Managing uncertainty
- Managing risk
- Detecting problems
- NDM

Understanding these processes and viewing the team as a whole, a Team CTA should reveal the cognitive skills that are required by the team, the types of expertise held by the team, and the decision-making requirements of the team. There are many techniques for obtaining this information, including observations, interviews, examination of critical incidents experienced by the team, and team concept mapping. While the knowledge elicitation methods are similar to those found in CTA, analysts will employ slightly different strategies, such as interviewing team members individually or together, or introducing scenarios and asking one team member to comment on another from his or her perspective. The analyst may focus on communication, how information is shared, how the team makes sense of complicated situations, how roles are coordinated, or how the team learns or plans its activities.

4.2.8.1 Results and Artifacts of Team CTA

Team CTA produces similar artifacts to CTA, but focuses on team decision-making and macrocognitive processes.

4.2.8.2 Tools and Resources Required

The following is a list of the resources that are commonly used to perform Team CTA:

- Interviews
- Observations
- Scenario walkthroughs
- Focus groups with individuals, teams, or pairs of team members

4.2.8.3 Strengths and Limitations

Team CTA is an important technique to consider where team performance and coordination improvements are the objectives of the systems redesign. By looking at team decision-making requirements, this technique can help analysts understand how information should be distributed across various team members. Team CTA is limited in that individual requirements may not be fully considered.

4.2.9   Cognitive Work Analysis

CWA evolved in the early 1990s in response to large-scale accidents that resulted from systems so complex that operator tasks and accident-causing events could not be anticipated. The technique was designed to elicit the knowledge required for problem solving and the construction of effective solutions. CWA works best in environments so complex that operators often cannot fully explain them; conducting a task analysis would not give an appropriate view of the constraints needed for effective problem solving in these environments. The CWA perspective is best understood through Vicente's (1999) text, which outlines the formative nature of the analytical technique and the five primary analytic phases. While a full CWA includes all five phases, in many cases analysts have been successful using only one or two phases of the technique. A successful variant of CWA is applied CWA, which results in explicit decision and cognitive work requirements that can be helpful in bridging the gap between analysis and design (Elm et al., 2004).

4.2.9.1   Results and Artifacts of CWA

- *Work domain analysis*: Easily the best-known phase of CWA, a work domain analysis creates a functional decomposition of the work environment. This is distinctly different from an operator-centered technique, such as CTA, as the operator is not represented in the model. CWA focuses on developing a functional understanding of the world that the operator must interact with, monitor, or control. This functional understanding begins with exploring the designed-for purposes of the world; the priorities, values, and principles of the world; and, at the lowest levels, the components of the world and how those components contribute to the world. The analysis is constraint-based and identifies capability limitations and fundamental laws. A work domain analysis most often results in a work domain model, a link explanation table, and a narrative describing the rationale for constructing the model.
- *Control task analysis*: A control task analysis examines the information processing requirements to complete fundamental tasks within the system. Designed for generic classes of tasks that involve distinctly different control patterns, control task analysis has also been used in a format similar to CTA to provide descriptions of how operators perform tasks. A control task analysis is typically performed using a decision ladder (i.e., a folder template of information processing steps on which operator actions can be mapped). A unique feature of the decision ladder is its ability to show leaps between stages, which illustrate the shortcuts that expert operators may take.
- *Strategies analysis*: A strategies analysis examines the different ways that operators can perform tasks. Typically, the strategies are triggered by initiating conditions, such as (a) low- versus high-workload, (b) novice versus expert operator, or (c) level of available work domain information. A strategies analysis typically results in a narrative description of the various possible strategies, though in some cases information flow maps (Vicente, 1999) have been used

to explicitly show the connection between strategies and domain information. Time lines (Burns et al., 2008a) have also been used to show how strategies may be adopted or abandoned at different stages of problem resolution.

- *Social organizational analysis*: A social organizational analysis seeks to identify organizational constraints on decision making. Several different techniques have been used to describe this phase, notably work domain models focused on organizational purposes, contextual activity templates, and, more recently, Team CWA.
- *Worker competency analysis*: A worker competency analysis examines the competency requirements for successful workers in the work domain. Most often, the skills–rules–knowledge framework that was developed by Rasmussen (1983) is used for this analysis. The most common result is a table outlining the expected skills, rules, and knowledge requirements.

4.2.9.2   Tools and Resources Required

Performing CWA requires access to SMEs, interviews, and observations. Typically, performing a work domain analysis requires access to manuals or the ability to talk with trainers to develop an understanding of complex relationships that system operators may not fully understand. The following is a list of the resources that are commonly used for CWA:

- Documents
- Customer-produced baseline scenarios
- Procedures
- Checklists
- History with current systems
- Training material used for current systems (if this is for a systems redesign)
- Interviews
- Observations
- SME sessions
- Discrete event simulation

4.2.9.3   Strengths and Limitations

Developing a work domain analysis requires experience in functional decomposition. As an artifact, a work domain model is not easy to confirm with operators and is best evaluated by returning to the original knowledge elicitation methods, as opposed to presenting operators with the model itself and asking for verification. However, CWA provides a rich and multifaceted view of operator requirements and is well suited for use in complex environments where operators must detect unexpected events and develop novel solutions.

4.2.9.4   CWA Applied to Cardiac Triage

A CWA was conducted in the cardiac patient health work domain to aid in the design of a triage system for a hospital. The CWA phases used in this project were (a) work domain analysis, (b) control task analysis, and (c) strategies analysis. Shortened examples of the primary artifacts of these analyses are provided: Table 4.5 shows

TABLE 4.5

Work Domain Analysis Illustrating the Relationships Important for Patient Healing

	Body	Systems	Organs
Purposes	• Maintain and improve patient health	• Regulate circulatory and respiratory systems • Improve circulatory and respiratory function	• Regulate and improve organ function
Principles, priorities, and balances	• Balance recovery and restore cardiac workload to normal levels • Balance oxygen and nutrition supply and demands	• Balance oxygen and nutrition in the systems • Maintain balance of blood flow and water	• Balance oxygen and nutrition to the heart and lungs • Maintain balance of blood flow and water in the heart, lungs, and blood vessels
Processes	• Healing • Activity • Homeostatic processes	• Healing • Workload • Oxygen exchange • Carbon dioxide exchange • Nutrient and waste exchange • Flow of blood and water	• Healing • Metabolism • Oxygen exchange • Carbon dioxide exchange • Nutrient and waste exchange • Flow of blood and water
Physiology and objects	• Body as a whole • Medications	• Cardiac and respiratory systems • Active ingredients of medications	• Heart • Lungs • Blood vessels • Active ingredients of medications
Anatomical details and attributes of objects	• Age, weight, and gender of patient • Type, dose, and concentration of prescribed medications	• Composition of medications and concentration in the body	• Degree of disease • Key physical dimensions (e.g., minimum internal diameter of blood vessels) • Chemical influence of medications

Source: Data from Burns, C.M., Enomoto, Y., and Momtahan, K., *Applications of Cognitive Work Analysis*, CRC Press, Boca Raton, FL, 2008a.

a work domain analysis constructed to illustrate relationships that are important for patient healing; Figure 4.9 illustrates the decision ladder response to an incoming patient call; Figure 4.10 illustrates an alternate way of handling the call using a strategy analysis. Other examples of CWA can be found in Bisantz and Burns' (2008) *Applications of Cognitive Work Analysis.*

4.2.10 COMPARING ANALYTICAL TECHNIQUES

Overall, the analytical techniques described in Section 4.2 are more similar than they are different. All of the techniques focus on understanding the objectives and the tasks that are needed for operators to function successfully with these systems. When comparing HTA with CWA, Salmon et al. (2011) argues that there are benefits in applying two or more techniques to inform the design of the same system. While Table 4.6 identifies the key requirements of each technique, it should be noted that, when done by an experienced analyst, similar information can be captured with each technique. Table 4.6 shows only the fundamental requirements that the technique is designed to capture.

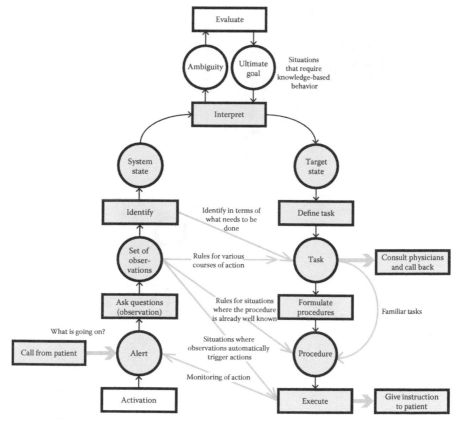

FIGURE 4.9 Decision ladder for cardiac triage. (Redrawn from Burns, C.M., Enomoto, Y., and Momtahan, K., *Applications of Cognitive Work Analysis*, CRC Press, Boca Raton, FL, 2008a.)

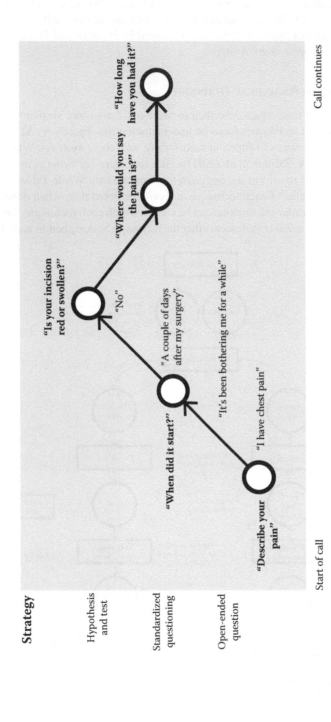

FIGURE 4.10 Strategy analysis for cardiac triage. (Redrawn from Burns, C.M., Enomoto, Y., and Momtahan, K., *Applications of Cognitive Work Analysis*, CRC Press, Boca Raton, FL, 2008a.)

TABLE 4.6

Intelligent Adaptive System (IAS) Design Requirements Captured by Different Analytical Techniques

Technique	Objectives	Task Description	Function Description	Novice and Expert Differences	Performance Control	Mental Model Description
Mission, function, task analysis (MFTA)	X	X	X	—	—	—
Hierarchical task analysis (HTA)	X	X	—	—	—	—
Hierarchical goal analysis (HGA)	X	—	—	—	X	—
Goal-directed task analysis (GDTA)	X	X	—	—	—	—
Concept mapping	X	—	—	X	—	X
Critical decision method	X	—	—	X	—	—
Applied cognitive task analysis (ACTA)	X	X	—	—	—	—
Team cognitive task analysis (Team CTA)	X	X	—	—	—	—
Cognitive work analysis (CWA)	X	X	X	X	—	—

TABLE 4.7

Knowledge Elicitation Methods Common to Different Analytical Techniques

Technique	Subject Matter Experts (SMEs)	System Documents	Observation	Interviews	Simulation	Focus Groups
Mission, function, task analysis (MFTA)	X	X	X	X	X	—
Hierarchical task analysis (HTA)	X	X	X	X	—	—
Hierarchical goal analysis (HGA)	X	—	X	X	X	X
Goal-directed task analysis	X	—	X	X	—	—
Concept mapping	X	—	—	—	—	X
Critical decision method	X	—	—	X	—	—
Applied cognitive task analysis (ACTA)	X	—	X	—	—	—
Team cognitive task analysis (Team CTA)	X	—	X	X	—	X
Cognitive work analysis (CWA)	X	X	X	X	—	—

Table 4.7 shows the primary information sources for analytical techniques, although most analysts would take advantage of any information sources available to inform their analysis.

4.3 REVIEW OF COMMON DESIGN TECHNIQUES

Three design techniques are discussed in this section. JAD takes a participatory design technique that involves operators in generating the new design. The US Department of Defense Architecture Framework (DoDAF: US Department of Defense, 2010) is a highly structured, formal design technique. EID (Vicente and Rasmussen, 1992) is a general design technique with a strong focus on visualization. Other techniques can be considered for user-interface design, but these three are representative of the major classes. For further information, readers may want to consult a text specific to user-interface design, such as Burns and Hajdukiewicz's (2004) *Ecological Interface Design* or Bennett and Flach's (2011) *Display and Interface Design: Subtle Science, Exact Art.*

4.3.1 PARTICIPATORY DESIGN

Participatory design was developed in Scandinavia in the 1960s under the name cooperative design. Currently, participatory design refers to a philosophy for user interface design that encourages deep user involvement. JAD was developed by IBM in the late 1970s and is a more technical form of participatory design. JAD integrates operators into the development cycle as frequently and as effectively as possible. The proponents of participatory design argue that it should reduce development times and improve client satisfaction. Through frequent client interaction, the development team should be able to stay on track and meet client needs. Keeping the client involved in the process creates buy-in (i.e., support for the design), largely as a result of the client feeling involved in the formation of key ideas.

Participatory design evolves through workshop sessions with clients and developers. In these sessions, key issues are discussed and problems are resolved. Visual concepts may be presented, discussed, and modified on the fly, giving the clients the sense that they are part of the design process. This is in contrast to the traditional iterative design cycle, where clients are engaged for feedback and testing but are not invited to contribute to the design cycle.

Participatory design relies heavily on rapid prototyping techniques for quickly conveying design ideas to workshop participants. It is a good technique for getting early client feedback, ensuring the team is on the right track, and understanding client preferences. Further, participatory design is useful for understanding less tangible requirements that may not be easily expressed in a statement of work document. This technique can also be used to establish consensus and reasonable expectations regarding the functionality and the look and feel of the final product. Workshops should be well planned by experienced facilitators who know how to elicit participation and mediate cases where there are differences in opinion. Users of participatory design should be cautious in situations where the client is not the end user, as the client and end user's understanding of requirements may differ. For example, a manager client contracting the development of a tool to be used by employees, or a company developing a consumer product. In these cases, participatory design should be supplemented with a technique that elicits requirements from actual end users, or includes end users in the workshops.

Gao et al. (2007) reported their experiences using participatory design in the design of a health information system that could accommodate an emergency response unit. They built handheld displays for medics, and web portals for the emergency department, transport officer, and incident commander. All parties were deeply involved in the design of the new system. Their evaluation and implementation created a systems design that reduced triage times, increased the number of patients triaged, and reduced the number of calls between team members.

4.3.2 UNITED STATES DEPARTMENT OF DEFENSE ARCHITECTURE FRAMEWORK

The DoDAF is intended to promote process and architecture consistency across joint forces and multinational development efforts.

Figure 4.11 illustrates the three architecture views used in the DoDAF: (a) operational view (OV), (b) systems view (SV), and (c) technical standards view (TV). OV and SV create models of how operation and systems constraints interact, and by building these views, can generate requirements. Table 4.8 shows the views and descriptions added to modified later versions of the DoDAF. For example, version 1.8 of the Department of National Defence/Canadian Armed Forces (CAF) Architecture Framework (DNDAF) incorporates 8 core architecture views, including the human view (HV), and 37 subviews (Wang et al., in preparation). The intention is to provide a comprehensive requirements gathering process and architecture framework.

- *Operational view*: A description of the tasks and activities, operational elements, and information exchanges required to accomplish various objectives. It contains graphical and textual products that comprise an

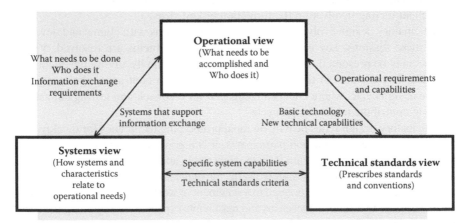

FIGURE 4.11 Fundamental linkages among Department of Defense Architecture Framework (DoDAF) views. (Redrawn from Darvill, D., Kumagai, J.K., and Youngson, G., Requirements analysis methods for complex systems (Report No. DRDC Valcartier CR 2005-076). Defence Research and Development Canada, North Val-Belair, Canada, 2006.)

TABLE 4.8

Additional Architectural Views of the Department of National Defence/ Canadian Armed Forces Architecture Framework (DNDAF) Version 1.8

DNDAF Views	Description
Common view	• Captures overarching aspects of an architecture that relate to all views • Defines the scope, context, and definitions within the architecture
Strategic view	• Identifies strategic issues (goals, outcomes, risks, and opportunities) through assessments of the business context
Capability view	• Articulates a desired effect that is in line with DND/CAF strategic intent through combinations of means and ways to perform a set of tasks or activities
Operational view	• Provides an operational-level description of the tasks, activities, business processes, and information exchange requirements to accomplish DND/CAF objectives • Defines stakeholder needs
System view	• Describes systems and families of systems and their interconnections providing for, or supporting, DND/CAF functions • Associates systems resources with the OV to demonstrate how these systems support the operational activities and facilitate the exchange of information or material among active components
Technical view	• Describes a profile of the minimal set of standards governing the implementation, arrangement, interaction, and interdependence of systems
Information view	• Defines the overall pattern or structure that is imposed on the information design and an information plan that defines information units and how they are to be completed
Security view	• Provides visibility on the attributes of the DND/CAF architecture that deal with the protection of assets

Source: Data from Wang, W., Coates, C., Dyck, W., and Au, L., Principles and approaches for developing the Canadian human view (Report No. TR 2013-022), Defence Research and Development Canada, Toronto, Canada, 2013.

identification of the operational nodes and elements, assigned tasks and activities, and information flows required between nodes. It defines the types of information exchanged, the frequency of the exchange, what tasks and activities are supported by information exchange, and the nature of the information exchange itself.

- *Systems view*: A set of graphical and textual products that describes systems and interconnections supporting various functions. It associates systems resources with the OV. These systems resources support the operational activities and facilitate the exchange of information among operational nodes.
- *Technical standards view*: The minimal set of rules governing the arrangement, interaction, and interdependence of system parts or elements. Its purpose is to ensure that a system satisfies its specified operational requirements. The TV provides the technical systems implementation guidelines

to create engineering specifications, establish common building blocks, and develop product lines. It includes a collection of the technical standards, implementation conventions, standard options, rules, and criteria organized into profiles that govern systems and system elements for a given architecture.

A number of products are associated with each DoDAF view, such as relational networks showing entities and information flow. Table 4.9 shows products for the OV. Not all of these products need to be used for a specific system architecture analysis. The information collected within the DoDAF can also be used as source data for inputs to other human factors and cognitive engineering analyses, such as MFTA, HGA, and CWA.

Some authors have noted that the representation of the operator in the DNDAF is largely implicit, which has led some to suggest that a HV should be considered

TABLE 4.9
Products Associated with the Department of Defense Architecture Framework (DoDAF) Operational View (OV)

View	Product Name	Description
ALL	Overview and summary information	Scope, purpose, intended users, environment, analytical findings
ALL	Integrated dictionary	Architecture data repository with definitions for all terms
OV	High-level operational graphic	High-level description of operational context
OV	Operational node connectivity description	Operational nodes, connectivity, and information exchange needs between nodes
OV	Operational information exchange matrix	Information exchanged between nodes and the relevant attributes of that exchange
OV	Organizational relationships chart	Organizational role, or other relationships among organizations
OV	Operational activity model	Capabilities, operational activities, relationships among activities, inputs, and outputs; may show cost or other information
OV	Operational rules model	Describes operational activity, identifies business rules that constrain activity
OV	Operational state transition description	Describes operational activity, identifies business process responses to events
OV	Operational event trace description	Describes operational activity, traces actions in a scenario or a sequence of events
OV	Logical data model	Documentation of the system data requirements and rules of the operational view

Source: Data from Darvill, D., Kumagai, J.K., and Youngson, G., Requirements analysis methods for complex systems (Report No. DRDC Valcartier CR 2005-076). Defence Research and Development Canada, North Val-Belair, Canada, 2006.

TABLE 4.10

Description of the North Atlantic Treaty Organization (NATO) Human View (HV)

HV Label	Description
HV-A concept	High-level representation of the human component
HV-B constraints	Repository for different classes of human limitations
HV-C functions	Description of the human-specific activities
HV-D roles	Definitions of the job functions for the human component
HV-E human network	Human-to-human communication patterns that occur
HV-F training	Accounting of training requirements, strategy, and implementation
HV-G metrics	Repository for human-related values and performance criteria
HV-H human dynamics	Dynamic aspects of the human system components

Source: Data from Wang, W., Coates, C., Dyck, W., and Au, L., Principles and approaches for developing the Canadian human view (Report No. TR 2013-022), Defence Research and Development Canada, Toronto, Canada, 2013.

(Wang et al., 2013). As a concept, the HV is still under development in several locations. Table 4.10 shows one example, the NATO HV.

4.3.2.1 Tools and Resources Required

Completing DoDAF views requires input from SMEs, system documents, operator observations, and operator interviews. System engineering tools are often used to make DoDAF products. The following is a list of the resources that are commonly used for the DoDAF:

- Documents
- Interviews
- Observations
- SME sessions

4.3.2.2 Strengths and Limitations

The DoDAF is a good technique if a design must integrate with a larger systems engineering process, as DoDAF views and outputs normally work well with the design processes of other units. However, the technique can be complex and time consuming, and may be unwieldy for smaller projects, or projects with less architecture planning. Also, the operator models that were developed by the DoDAF may lack richness when compared with some of the other design techniques presented.

4.3.3 Ecological Interface Design

EID uses CWA and visualization techniques to develop operator interface designs. The intent of EID is to support operator work that is skill-based, rule-based, and knowledge-based using the skills–rules–knowledge framework of Rasmussen (1983).

In particular, EID seeks to move complex work to easier cognitive levels (knowledge-based to rule-based or rule-based to skill-based). Evaluations have shown that the EID technique improves the detection and diagnosis of system flaws (Pawlak and Vicente, 1996), as well as improving operator SA (Burns et al., 2008b).

EID begins with a work domain analysis that is analyzed for information requirements. Requirements are extracted at each level of the work domain model. Requirements from the lowest levels show the capabilities of components. Requirements from higher levels show the objectives and fundamental relationships of the system; higher-level requirements typically need more advanced visualization support. Examples of potential visualization graphics for higher-level ecological support are described by Burns and Hajdukiewicz (2004) in *Ecological Interface Design*. A proper EID implements a design that meets the requirements of all levels of the work domain model.

EID is often supplemented with other CWA techniques to provide additional strength to the design. In particular, control task analysis and CTA can help determine layout and navigation issues. Persona-based design as described in *Designing for the Digital Age: How to Create Human-Centered Products and Services* by Goodwin (2009) can help determine usability requirements for different operator groups. The EID technique is further discussed in Chapter 5 in the context of facilitating operator–agent interaction in OMI designs.

4.3.3.1 Tools and Resources Required

EID uses a standard iterative design process of prototyping and evaluation with operators.

4.3.3.2 Strengths and Limitations

EID has been shown to generate novel and effective designs in many different environments (Pawlak and Vicente, 1996; Burns et al., 2008b). EID can lower cognitive load and improve expert behavior. However, unless EID is combined with other techniques, designers must be careful to also support task-based and heuristic behavior.

4.3.3.3 EID Applied to Cardiac Triage

An EID for cardiac triage was implemented on a Palm Tungsten series handheld. Figure 4.12 illustrates a screen for pain assessment. The work domain analysis identified pain assessment as a challenging cognitive function. The strategy analysis showed that expert nurses had a key strategy for handling pain assessment. They asked patients to compare the pain with known referents, such as before or after surgery, or a previous cardiac event. The work domain analysis also identified the size and the severity of pain as critical information. The comparison was aided in two key ways: (a) through the like/unlike table, which provides a quick summary; and (b) through the balance graphic, which quickly shows if the pain is growing, staying the same, or decreasing.

Figure 4.13 illustrates part of a question-asking protocol for triage. It takes the knowledge-based work of triage and moves it to a rule-based level to aid assessment. It is fully integrated with the other tools of the system to connect with other strategies and ecological views.

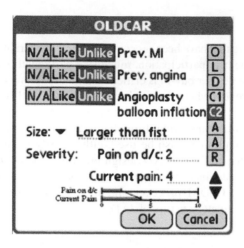

FIGURE 4.12 Screenshot of an ecological design for pain assessment. (Reprinted from Momtahan, K.L., Burns, C.M., Labinaz, M., Mesana, T., and Sherrard, H., *Studies in Health Technology and Informatics Series: Vol. 129 MEDINFO 2007: Proceedings of the 12th World Congress on Health (Medical) Informatics*, pp. 117–121, IOS Press, Amsterdam, 2007. With permission.)

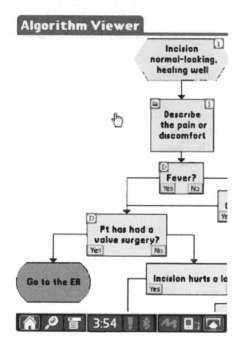

FIGURE 4.13 Screenshot of rule-based support that was developed from a knowledge-based process for triage. (Reprinted from Momtahan, K.L., Burns, C.M., Labinaz, M., Mesana, T., and Sherrard, H., *Studies in Health Technology and Informatics Series: Vol. 129 MEDINFO 2007: Proceedings of the 12th World Congress on Health (Medical) Informatics*, pp. 117–121, IOS Press, Amsterdam, 2007. With permission.)

4.3.4 COMPARING DESIGN TECHNIQUES

The design techniques that have been presented differ considerably in their intent and style. JAD focuses on user participation, much like participatory design, and is best used in situations where user buy-in is critical. The DoDAF is intended to promote a systematic design process where the results are traceable, well documented, and can be integrated with systems engineering techniques. EID, in contrast, is focused on promoting effective problem solving through design and visualization.

These three techniques can also be used, either partly or fully, in conjunction. There is no reason that an EID or JAD design process could not be systematized by following the systems engineering stages of the DoDAF. Similarly, there are situations when an EID or a DoDAF process could integrate operator consultation, taking advantage of the participatory nature of JAD.

4.4 CREATING A HYBRID TECHNIQUE

The final section of this chapter presents an example of how three analytical techniques were selected and then modified into a hybrid technique. The hybrid technique was developed to support the specific goals of a project researching IAS concepts for supporting future uninhabited aerial vehicle (UAV) operations within the CAF. The project is described in terms of its objectives, constraints, and analysis team, and a rationale is provided for the selection of the hybrid technique, which is described in more detail.

4.4.1 IDENTIFYING PROJECT REQUIREMENTS AND CONSTRAINTS

This project supports the CAF's goal of establishing UAV cells across Canada. Specifically, the project supported the implementation of IAS technologies within a UAV ground control station (GCS). A study conducted by Defence Research and Development Canada (DRDC)-Toronto identified a number of UAV-related incidents and accidents that could be attributed to a range of cognitive risks, such as operator fatigue, workload, and loss of SA. Both the complexity of future UAV missions and the capability of UAV technologies were anticipated to place significant cognitive demands on operators. Thus, the main objective of the analysis and definition phase of the project was to identify how IAS concepts could be implemented within the GCS to mitigate the impact of these cognitive risks during future CAF UAV missions.

After further exploring the aims and objectives of the project, certain project requirements and constraints were identified to guide the selection of an analytical technique:

- A preliminary analysis was conducted prior to the commencement of the project. This analysis comprised a mission analysis and top-level functional decomposition using the MFTA analytical technique described in this chapter. This initial MFTA provided the starting point for the analysis.
- Given that the future UAV operator roles were anticipated to be heavily reliant on cognitive capabilities, the analysis focused on identifying cognitive

risks as candidates for IAS support, particularly during mission phases that are heavily reliant on decision making and the acquisition and maintenance of SA. Both GDTA and CTA were identified as suitable analytical techniques.

- The analysis team was experienced at using MFTA and GDTA analytical techniques. Additionally, all members of the analysis team had backgrounds in cognitive psychology.
- Good access to relevant SMEs was possible through DRDC Toronto.
- The project included performance prediction analyses to examine IAS support options.

4.4.2 SELECTING ANALYTICAL TECHNIQUES

The project team reviewed the suitability of several of the analytical techniques described in this chapter for supporting the UAV project. The techniques were shortlisted to reflect approaches that have been used in complex decision-making environments typical of where IASs may be used, and were applicable to the analysis of CAF UAV operations.

Although the techniques differ in detail and origin, they share many of the same goals. The selection of particular techniques needed to fit with the context of CAF UAV operations and the types of IAS solutions that the project was seeking to develop and implement. The selected analysis techniques were used to decompose high-level goals into more detailed IAS implementation strategies, and elucidate the information necessary to populate the various models with the IAS, such as the expert and operator models, which were discussed in Chapter 3.

The context of CAF UAV operations and the emphasis to be placed on cognition and SA led to the development of the following criteria for selecting the most appropriate analytical techniques from those described earlier in this section:

- *Focuses on SA*: The output of the analysis has to be specific to SA requirements.
- *Effort*: The level of effort, time, and expert resources required for conducting the analysis.
- *New system design*: The analysis aims at developing a new system (i.e., initial requirements analysis) and not simply improving an existing system.
- *Provides data for task modeling*: The analysis provides output that can be used for the project's task network modeling activities.
- *Facilitates the identification of cognitive risk*: The analysis provides a model from which cognitive limitations that could potentially impair the process or actions can be readily identified.
- *Military*: The analysis is supported by military standards and has a track record of being used in the military domain.

For each of the analysis techniques described, a judgment was made based on published reports and open source literature. Table 4.11 shows that the SA-focused criterion received a yes or no judgment, while the other five criteria were judged on

TABLE 4.11

Evaluation of the Suitability of a Range of Analytical Techniques for Situation Awareness–Oriented Analysis of Canadian Armed Forces (CAF) Uninhabited Aerial Vehicle (UAV) Operations

Criterion	Analytical Techniques				
	MFTA	CTA	CWA	HGA	GDTA
SA–focused	No	No	No	No	Yes
Effort	Mid	Mid	Mid	Mid	Mid
New system design	Low	Mid	High	Mid	High
Task modeling data	Mid	Mid	Low	Low	Low
Cognitive risk	Low	High	Low	Low	Mid
Military	High	Mid	Mid	Mid	Mid

Note: Data from Banbury, S., Baker, K., Proulx, R., and Tremblay, S., Human technology interaction standing offer: TA1: Unmanned air system operator information flow and cognitive task analyses, [Contract Report for Defence Research and Development Canada, Contract No. w7719-145238/001/TOR], Thales, Ottawa, 2014.

a three-point rating scale (i.e., high–mid–low). The shaded cells indicate the most appropriate technique(s) for each of the six criteria.

The MFTA, CTA, CWA, and GDTA methods are all based on a functional modeling of the work domain rather than specific tasks for which the procedural sequence is well known. This is an essential contribution to new system design. HGA is based on a hierarchical goal structure that does not differentiate functions and tasks. When these criteria, except SA focused and, to some extent, new system design, are considered, MFTA appears to be the preferred method. MFTA is also best suited for the military given its direct connection with US Department of Defense standard MIL–HDBK–46855A. However, MFTA is considered less suitable for new system design as it is, in some cases, organized on a time base and not on an operator goal base. While CWA has been applied to military settings, it requires access to operational procedures or manuals or training material for performing work domain analysis. However, these resources were not available to the project team when the team started analysis phase.

The ability of a system to support decision makers in successfully accomplishing their mission depends on how well the system supports their goals and SA requirements; however, only the GDTA method is SA focused. Therefore, GDTA is best suited for extracting information for SA requirements in the context of UAV operations. GDTA is also appropriate for informing the design of new systems. Unfortunately, the GDTA method is not very suitable with regard to identifying cognitive risk and supporting task modeling activities. For these cases, the preferred method is the CTA technique as it focuses on the cognitive demands that are imposed on operators working in dynamic and complex environments.

TABLE 4.12

Strengths and Limitations of Three Different Analytical Techniques for the UAV Project

Technique	Strengths	Limitations
Mission, function, task analysis (MFTA)	• Supports military design process • Can be used to analyze a new system • Uses a rigorous process (systematic and consistent) • Includes analysis of task flows • Supports performance prediction studies	• Requires experienced analysts • Requires access to suitable SMEs with sufficient expertise and experience • Can be time consuming to conduct analysis on large systems
Goal-directed task analysis (GDTA)	• Identifies information requirements that support situation awareness • Identifies requirements independent of any existing interface or information system	• Does not include analysis of task flows • Does not support performance prediction studies
Cognitive task analysis (CTA)	• Develops a rich list of cognitive requirements to help define IAS support	• Does not include analysis of task flows • Does not support performance prediction studies

Source: Data from Banbury, S., Baker, K., Proulx, R., and Tremblay, S., Human technology interaction standing offer: TA1: Unmanned air system operator information flow and cognitive task analyses, [Contract Report for Defence Research and Development Canada, Contract No. w7719-145238/001/TOR], Thales, Ottawa, 2014.

In summary, three types of analysis techniques can be identified as the most appropriate for the UAV project: MFTA, GDTA, and CTA. Table 4.12 shows the strengths and limitations of the three identified analytical techniques as applicable to the UAV project.

Although each technique has its own strengths and limitations, taken as a whole the strengths of one technique make up for the limitations of another:

- MFTA provided the level of rigor required for the design of a new military system and captured task flow and performance prediction information.
- GDTA provided the decision and SA requirements information required to identify opportunities for supporting UAV operators in their highly SA-dependent role.
- CTA provided a high level of description about the cognitive risks experienced by operators to support the implementation of IAS technologies.

In addition, the analysis team was experienced with all three analytical techniques, SMEs with sufficient expertise and experience were made available to the project, and enough time and resources were available to conduct a large-scale analysis.

4.4.3   USING A MFTA/GDTA/CTA HYBRID ANALYTICAL TECHNIQUE

The analytical technique was based on a combination of the MFTA, GDTA, and CTA approaches to requirements analysis. The hybrid technique is divided into three discrete steps:

4.4.3.1   Step 1: Mission and Function Analysis

- *Mission analysis*: Mission objectives, environment, and operational constraints were identified. The mission themes were chosen to cover a broad range of functions involved in the UAV operator work domain. The output of this analysis was, in part, a composite mission scenario that was used to provide the basis for all subsequent analysis activities.
- *Function analysis*: High-level functions that were required to be performed by the operator and the system to achieve the mission objectives were identified. Function analysis was hierarchical in nature, and proceeded in a top-down fashion by decomposing abstract high-level functions into more concrete subfunctions.

4.4.3.2   Step 2: Goal, Decision, and Situation
Awareness Requirements Analysis

- *Goal analysis*: Goals that had to be reached in order to complete each high-level function were identified. Similar to a task flow analysis in MFTA, the sequence and timing of the identified goals were also noted in the goal analysis. Specifically, interest lay in whether the goals were accomplished sequentially or concurrently to support a specific function.
- *Decision analysis*: The decisions that must be made in order to achieve each goal were identified, as well as the sequence and timing of those decisions. Specifically, interest lay in whether decisions were made sequentially or concurrently to support a specific goal.
- *SA requirements analysis*: The knowledge (i.e., SA elements) that operators were required to possess in order to successfully make each decision was identified. Figure 4.14 illustrates (a) a mission and function analysis; and (b) a goal, decision, and SA requirements analysis applied to the goal of understanding the mission tasking.

4.4.3.3   Step 3: Cognitive Risk Analysis

- *Identify critical decisions*: Critical decisions were identified by determining if one or more of the following decision criticality criteria was an important concern with respect to a specific decision: (a) safety, (b) mission completion or effectiveness, (c) cost, (d) efficiency, and (e) system reliability.
- *Identify external (situation) stressors*: Based in part on DRDC Toronto's review of UAV-related incidents and accidents, it was determined if one or more of the following external stressors was significant to each critical decision:
  - *Information uncertainty*: The temporal, positional, or precisional uncertainty of information used to make the decision

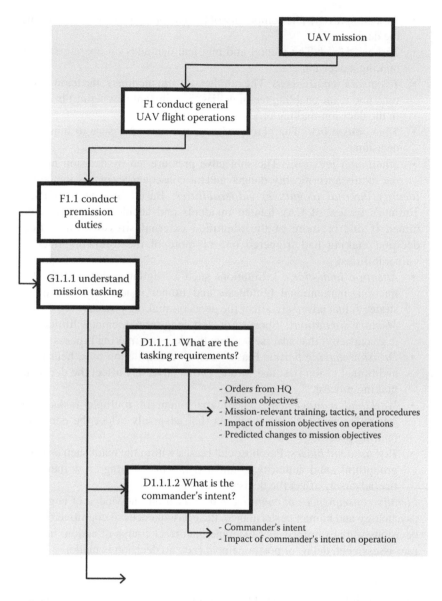

FIGURE 4.14 The first two steps of a hybrid analytical technique: (a) a mission and function analysis, and (b) a goal, decision, and situation awareness requirements analysis applied to the goal of understanding the mission tasking. To understand the mission tasking, two decisions are identified, which are supported by several situation awareness elements. (Redrawn from Banbury, S., Baker, K., Proulx, R., and Tremblay, S., Human technology interaction standing offer: TA1: Unmanned air system operator information flow and cognitive task analyses, [Contract Report for Defence Research and Development Canada, Contract No. w7719-145238/001/TOR], Thales, Ottawa, 2014.)

- *Task complexity*: The complexity of the reasoning processes underlying the decision
- *Task workload*: The mental and physical demands on operators when making a decision
- *Teamwork requirements*: The number of team members, the team structure, and team-enabling technologies (e.g., instant messaging) involved in the decision-making process
- *Time constraints*: The perceived or actual time pressure to make the decision
- *Situational pressures*: The subjective pressure felt by decision makers due to mission criticality, danger, and the consequences of mission failure
- *Identify internal (cognitive) vulnerabilities*: Based in part on DRDC Toronto's review of UAV-related incidents and accidents, it was determined if one or more of the identified external stressors for critical decision making had triggered one or more of the following internal vulnerabilities:
  - *Attention limitations*: Limitations such as vigilance decrements, distraction, inattentional blindness, and tunnel vision (i.e., attentional strategy) that adversely affect the decision-making process
  - *Memory limitations*: Short-term and long-term memory limitations (e.g., capacity) that adversely affect the decision-making process
  - *Decision-making biases*: Biases such as confirmation bias, heuristics, and tunnel vision (i.e., task strategy) that adversely affect the decision-making process
  - *Workload management*: The management of multiple tasks, task schedules, and task interruptions that adversely affect the decision-making process
  - *Psychosocial biases*: Psychosocial biases within the team such as trust, groupthink, and authority gradient (i.e., rank) among crew members that adversely affect the decision-making process
- *Identify consequences of cognitive risk*: Based on knowledge of cognitive psychology and human performance, the consequences of cognitive risk on decisions were examined. For example, incorrect course of action, inaccurate assessment, delay, or postponement (i.e., no decision is made).

Figure 4.15 illustrates a worked example of the cognitive risk analysis. It is based on the decision for "what are the tasking requirements?," which was considered critical (i.e., required cognitive risk analysis) based on its impact on safety, mission completion, and mission effectiveness. For this decision, task complexity, task workload, time constraints, and situational pressures were determined to be the most dominant external (e.g., task) stressors present. It was understood that these stressors would trigger internal (i.e., cognitive) vulnerabilities relating to memory limitations, decision-making biases, and workload management. The consequences of external stressors triggering these internal vulnerabilities were considered an inaccurate assessment of tasking requirements, which could potentially lead to incorrect courses of action included in mission planning.

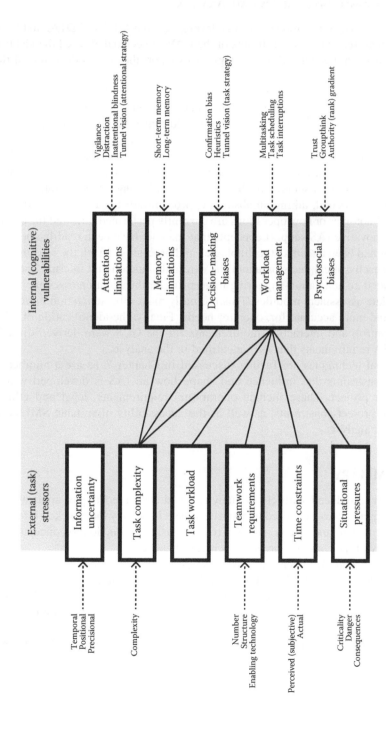

FIGURE 4.15 The third step of a hybrid analytical technique—a cognitive risk analysis to decide "What are the tasking requirements?"

4.4.4 UNDERSTANDING THE OVERALL ANALYSIS

The hybrid analytical technique that was developed from MFTA, GDTA, and CTA suited the requirements and constraints of the UAV project and allowed the identification of opportunities for IAS technologies to support the critical decisions of the UAV crew.

The analysis focused on the decision level. Specifically, the goal was to gain an understanding of what information the operator needed to make decisions and what cognitive risks might be present that would make the decision-making process more difficult. This approach had a number of advantages. First, the initial analysis of UAV incidents and accidents indicated that workload, fatigue, and loss of SA had a significant impact on operator decision making and course of action, and that providing IAS support in these cases would bring many operational and safety benefits. Second, taking an abstract approach (rather than concentrating on specific technologies or how operations are currently performed) allowed the exploration of innovative IAS support concepts that may not have been readily apparent if the focus had been on current technologies and processes. Third, the information gained for each critical decision about the external situation and internal operator state provided an adaptation-triggering fingerprint for the situation assessment and operator state assessment modules. This normally makes the adaptation more sensitive to, and more accurate for, operator needs. Finally, the identification of OMI display, control, and interaction requirements for the GCS were derived directly from the SA requirements that were identified in the analysis.

Analytical techniques are further discussed in Chapter 7, as are a number of practical constraints that influence and shape how an IAS is developed within a particular project. These include operational requirements, legal and ethical issues, and project constraints, as well as the accessibility of suitable SMEs and experienced analysts.

4.5 SUMMARY

This chapter introduced a variety of analytical and design techniques applicable to systems design. Tables were used to assist designers in quickly locating key information. The provided guidance is intended to be a basic introduction to concepts, and further study of the chosen approaches is recommended.

The role of analytical techniques in IAS design was explained. Each analytical technique—MFTA, HTA, HGA, GDTA, CTA, concept mapping, the critical decision method, Team CTA, and CWA—was introduced, and lists of expected artifacts were provided. Tools and resources were noted, strengths and limitations were detailed, and examples were included wherever possible. In the hands of experienced practitioners, these techniques can be extremely effective for extracting the information needed to create good systems design.

Three design approaches relating to interface design, each representing a major class of approach, were examined: participatory design, the DoDAF, and EID.

An example of a hybrid analytical technique using MFTA, GDTA, and CTA was offered. This example included an explanation of why the decision was made to use the specific techniques, the three discrete steps required when using a hybrid approach, and an understanding of the completed analysis. Solid analytical results are the starting point for defining and designing agents, which are closely examined in Chapter 5.

REFERENCES

Alion Micro Analysis and Design. (2009). IPME—Micro analysis and design, inc. [online]. Retrieved July 31, 2014 from http://www.maad.com/index.pl/ipme.

Annett, J. and Duncan, K. D. (1967). Task analysis and training design. *Journal of Occupational Psychology, 41*, 211–221.

Annett, J. and Stanton, N. A. (1998). Editorial. Ergonomics, 41(11), 1529–1536.

Banbury, S., Baker, K., Proulx, R., and Tremblay, S. (2014). Human technology interaction standing offer: TA1: Unmanned air system operator information flow and cognitive task analyses [Contract Report for Defence Research and Development Canada, Contract No. w7719-145238/001/TOR]. Thales, Ottawa.

Bennett, K. B. and Flach, J. M. (2011). *Display and Interface Design: Subtle Science, Exact Art.* Boca Raton, FL: CRC Press.

Bisantz, A. M. and Burns, C. M. (2008). *Applications of Cognitive Work Analysis.* Boca Raton, FL: CRC Press.

Bolstad, C. A., Riley, J. M., Jones, D. G., and Endsley, M. R. (2002). Using goal directed task analysis with army brigade officer teams. In *Proceedings of the 46th Annual Meeting of the Human Factors and Ergonomics Society*, pp. 472–476. September 30 – October 4, New York, NY: SAGE Publications.

Burns, C. M., Enomoto, Y., and Momtahan, K. (2008a). A cognitive work analysis of cardiac care nurses performing teletriage. In A. Bisantz and C. M. Burns (eds), *Applications of Cognitive Work Analysis*, pp. 149–174. Boca Raton, FL: CRC Press.

Burns, C. M. and Hajdukiewicz, J. R. (2004). *Ecological Interface Design.* Boca Raton, FL: CRC Press.

Burns, C. M., Skraaning, G., Lau, N., Jamieson, G. A., Kwok, J., Welch, R., and Andresen, G. (2008b). Evaluation of ecological interface design for nuclear process control: Situation awareness effects. *Human Factors: The Journal of the Human Factors and Ergonomics Society, 50*(4), 663–679.

Chow, R., Crebolder, J., Kobierski, R. D., and Coates, C. E. (2007). Application of hierarchical goal analysis to the Halifax class frigate operations room (Report No. TR 2007-161). Toronto: Defence Research and Development Canada.

Chow, R., Kobierski, R. D., Coates, C. E., and Crebolder, J. (2006). Applied comparison between hierarchical goal analysis and mission function task analysis. In *Proceedings of the 50th Annual Meeting of the Human Factors and Ergonomics Society*, pp. 520–524. October 16–20, New York, NY: SAGE Publications.

Cummings, M. L. and Bruni, S. (2010). Human-automation collaboration in complex multivariate resource allocation decision support systems. *International Journal of Intelligent Decision Technologies, 4*, 101–114.

Darvill, D., Kumagai, J. K., and Youngson, G. (2006). Requirements analysis methods for complex systems (Report No. DRDC Valcartier CR 2005-076). North Val-Belair: Defence Research and Development Canada.

Elm, W. C., Roth, E. M., Potter, S. S., Gualtieri, J. W., and Easter, J. R. (2004). Applied cognitive work analysis (ACWA). In N. Stanton, A. Hedge, K. Brookhuis, E. Salas, and H. Hendrick (eds), *Handbook of Human Factors and Ergonomics Methods*, pp. 357-367. Boca Raton, FL: CRC Press.

Embrey, D. (1986). SHERPA: A systematic human error reduction and prediction approach. In *Proceedings of the International Topical Meeting on Advances in Human Factors in Nuclear Power Systems*, pp.184–193. April 21–24, La Grange Park, IL: American Nuclear Society.

Endsley, M. R., Bolstad, C. A., Jones, D. G., and Riley, J. M., (2003). Situation awareness oriented design: From user's cognitive requirements to creating effective supporting technologies. In *Proceedings of the 47th Annual Meeting of the Human Factors and Ergonomics Society*, pp. 268–272. October 13–17, New York, NY: SAGE Publications.

Endsley, M. R., and Jones, D. G. (2011). Determining SA requirements. In M. R. Endsley and D. G. Jones, *Designing for Situation Awareness: An Approach to User-Centered Design*, 2nd edn., pp. 63–78. Boca Raton, FL: CRC Press.

Farrell, P. S., Hollands, J. G., Taylor, M. M., and Gamble, H. D. (1999). Perceptual control and layered protocols in interface design: I. Fundamental concepts. *International Journal of Human-Computer Studies*, 50(6), 489-520.

Flanagan, J. C. (1954). The critical incident approach. *Psychological Bulletin*, 51(4), 327–359.

Furukawa, H. and Parasuraman, R., (2003). Supporting system-centered view of operators through ecological interface design: Two experiments on human-centered automation. In *Proceedings of the 47th Annual Meeting of the Human Factors and Ergonomics Society*, pp. 567–571. October 13–17, New York, NY: SAGE Publications.

Furukawa, H., Nakatani, H., and Inagaki, T. (2004). Intention-represented ecological interface design for supporting collaboration with automation: Situation awareness and control in inexperienced scenarios. In D. A. Vincenzi, M. Mouloua, and P. A. Hancock (eds), *Human Performance, Situation Awareness, and Automation: Current Research and Trends HPSAA II, Volumes I and II*, pp. 49–55. Boca Raton, FL: Taylor & Francis.

Gao, T., Massey, T., Sarrafzadeh, M., Selavo, L., and Welsh, M., (2007). June. Participatory user centered design approaches for a large scale ad-hoc health information system. In *Proceedings of the 1st ACM Sigmobile International Workshop on Systems and Networking Support for Healthcare and Assisted Living Environments*, pp. 43–48. June 11–14, New York, NY: ACM.

Goodwin, K. (2009). *Designing for the Digital Age: How to Create Human-Centered Products and Services*. New York, NY: Wiley.

Harris, D., Stanton, N. A., Marshall, A., Young, M. S., Demagalski, J., and Salmon, P. (2005). Using SHERPA to predict design-induced error on the flight deck. *Aerospace Science and Technology*, 9(6), 525–532.

Hendy, K., Beevis, D., Lichacz, F., and Edwards, J. L. (2001). Analyzing the cognitive system from a perceptual control theory point of view (Report No. DCIEM-SL-2001-143). Toronto: Defence and Civil Institute of Environmental Medicine.

Klein, D. E., Klein, H. A., and Klein, G. A., (2000). Macrocognition: Linking cognitive psychology and cognitive ergonomics. In *Proceedings of the 5th International Conference on Human Interactions with Complex Systems*, pp. 173–177. April 30–May 2, Urbana, IL: US Army Research Laboratory.

Klein, G. A. (2000). Cognitive task analysis of teams. In J. M. C. Schraagen, S. F. Chipman, and V. J. Shalin (eds), *Cognitive Task Analysis*, pp. 417–429. Mahwah, NJ: Lawrence Erlbaum Associates.

Klein, G. A., Calderwood, R., and Macgregor, D. (1989). Critical decision method for eliciting knowledge. *IEEE Transactions on Systems, Man and Cybernetics*, 19(3), 462–472.

Lane, R., Stanton, N., and Harrison, D. (2006). Applying hierarchical task analysis to medication administration errors. *Applied Ergonomics*, 37(5), 669–679.

Lee, J. D. and Kirlik, A. (2013). *The Oxford Handbook of Cognitive Engineering*. Kettering, England: Oxford University Press.

Militello, L. and Hutton, R. (1998). Applied cognitive task analysis (ACTA): A practitioner's toolkit for understanding cognitive task demands. *Ergonomics*, 41(11), 1618–1641.

Momtahan, K. L., Burns, C. M., Labinaz, M., Mesana, T., and Sherrard, H., (2007). Using personal digital assistants and patient care algorithms to improve access to cardiac care best practices. In K. Kuhn, J. R. Warren, and T.-Y. Leong (eds), *Studies in Health Technology and Informatics Series: Vol. 129 MEDINFO 2007: Proceedings of the 12th World Congress on Health (Medical) Informatics*, pp. 117–121. August 20–24, Amsterdam, NL: IOS Press.

Novak, J. D. and Canas, A. J. (2008). The theory underlying concept maps and how to construct and use them (Report No. IHMC Technical Report Cmap Tools 2006-01 Rev 01-2008). Pensacola, FL: Florida Institute for Human and Machine Cognition.

Pawlak, W. S. and Vicente, K. J. (1996). Inducing effective operator control through ecological interface design. *International Journal of Human-Computer Studies*, *44*(5), 653–688.

Powers, W. T. (1973). *Behavior: The Control of Perception*. Chicago, IL: Aldine.

Rasmussen, J. (1983). Skills, rules, and knowledge; signals, signs, and symbols, and other distinctions in human performance models. *IEEE Transactions on Systems, Man, and Cybernetics*, *13*(3), 257–266.

Salmon, P., Jenkins, D. Stanton, N., and Walker, G. (2010). Hierarchical task analysis vs. cognitive work analysis: comparison of theory, methodology and contribution to system design. *Theoretical Issues in Ergonomics Science*, *11*(6), 504–531.

Seppelt, B. D. and Lee, J. D. (2007). Making adaptive cruise control (ACC) limits visible. *International Journal of Human-Computer Studies*, *65*(3), 192–205.

Taylor, M. M., Farrell, P. S., and Hollands, J. G. (1999). Perceptual control and layered protocols in interface design: II. The general protocol grammar. *International Journal of Human-Computer Studies*, *50*(6), 521–555.

US Department of Defense. (2010). DoDAF—DoD architecture framework version 2.02—DoD deputy chief information officer [online]. Retrieved July 31, 2014 from http://dodcio.defense.gov/dodaf20.aspx.

US Department of Defense. (2011). Department of Defense standard practice: Human engineering requirements for military systems, equipment, and facilities (Military Specification No. MIL-STD-46855A). Arsenal, AL: Army Research Development and Engineering Command Aviation and Missile Research and Engineering Center.

Vicente, K. J. (1999). *Cognitive Work Analysis: Toward Safe, Productive and Healthy Computer-Based Work*. Mahwah, NJ: Lawrence Erlbaum Associates.

Vicente, K. J. and Rasmussen, J. (1992). Ecological interface design: Theoretical foundations. *IEEE Transactions on Systems, Man and Cybernetics*, *22*(4), 589–606.

Wang, W., Coates, C. E., Dyck, W., and Au, L. (2013). Principles and approaches for developing the Canadian human view (Report No. TR 2013-022). Toronto: Defence Research and Development Canada.

5 Agent-Based, Interaction-Centered IAS Design*

5.1 OBJECTIVES

- Define agent, agent intelligence, and agent adaptivity in the context of IAS design
- Identify the benefits and advantages of an agent-based design method for OMIs
- Discuss existing agent-based design methods important to IAS design
- Discuss interaction-centered design in the context of operator–agent interaction
- Introduce a generic agent-based, interaction-centered conceptual framework for IAI design

IASs must be proactive and collaborate with operators in order to achieve overall system goals and operate efficiently and effectively in complex and dynamic environments. To better understand how the operator relates to an overall IAS, the system can be seen as a corporation, and the operator as the board of directors that governs it. The concepts outlined in Chapter 2, the conceptual architecture discussed in Chapter 3, and the analytical techniques noted in Chapter 4 are practical tools that should help systems designers as they design IASs.

A systems designer cannot create an effective IAS based on a system concept, a conceptual architecture, and analytical techniques alone. Effective IASs, with true potential to achieve system goals, are created through proper design and consideration of all HMS components, including the operator, the machine (e.g., hardware, software, and associated OMIs), and the external working environment. Once the high-level goals and structure of an IAS and its functional components and associated working mechanisms are determined, using techniques such as those outlined in Chapter 4, the components must be developed.

This chapter presents methodologies and associated principles for producing IAS software components using agent-based design methods, specifically focusing on software agent design for effective IASs. Although various system components, such as machine hardware or software entities, can be agents, the term "agent" used in this chapter specifically refers to software agents unless another type of agent is specified. As entities capable of autonomous action in an environment in order to meet their design objectives (Wooldridge and Jennings, 1995), software agents offer the potential to aid human decision making and facilitate operator interaction with OMIs.

Agent-based design methods focusing on IAI design are used to illustrate the use of an agent-based approach to IAS design. Interaction-centered design approaches in the IAS context are discussed, as well as why systems designers need to build a model of operator–agent interaction that facilitates IAS design. A generic, agent-based, interaction-centered framework for an IAI design consisting of multiple agents as a subsystem of a complex IAS is also described. The overall goal of this chapter is to provide guidance to systems designers interested in the development of IASs, and to illustrate the advantages of adopting an agent-based, interaction-centered design method.

5.2 DEFINING AGENTS IN THE CONTEXT OF IAS DESIGN

At their most generic level, agents are entities authorized to take action on behalf of humans while humans are busy doing other tasks. Machines and their components can be agents, and, in the case of IASs, the machine component can be an agent that works collaboratively with its human partner to achieve system goals. The functional components of IASs (e.g., various pieces of hardware and software) can also be agents. Note that agents in this context serve similar roles to humans who pursue careers as agents. For example, travel agents maintain close relationships with vendors and tour guides and are authorized by clients to organize vacation itineraries and book flights and hotels. By focusing on one particular area of expertise (i.e., travel), they are able to provide clients with an optimized vacation experience. Likewise, stockbrokers are human agents authorized to benefit their clients' portfolio by purchasing and selling stocks at the right time.

Agents in the context of IAS design demonstrate a shift from computers as tools used to complete a specific job to entities that cooperate with users or work independently to get things done (Allen and Ferguson, 2003). To facilitate operator–agent interaction, the IAS should include a combination of algorithmic and symbolic models for assessing and predicting the activities, awareness, intention, resources, and performance of the operator. The OMI should support operators efficiently, effectively, and naturally.

Like human agents in an organization, software agents are interconnected, and cooperate to ensure the IAS functions at an optimal level based on the conditions faced. However, little research has focused on optimizing operator–agent interaction from an HF perspective because agent-based technologies have not yet been adopted in many complex and dynamic systems designs. Additionally, many existing agent design frameworks focus on individual models, such as user, task, and domain models, rather than operator–agent interaction models. Regardless, agent-based systems have been widely used in the software engineering and artificial intelligence fields, and this research continues to evolve.

5.2.1 CORPORATE INTERACTION AS AN ANALOGY OF OPERATOR–AGENT INTERACTION

Figure 5.1 illustrates an IAS as a corporation. Operators can be thought of as the board of directors, their ultimate goal being to make the corporation profitable (i.e.,

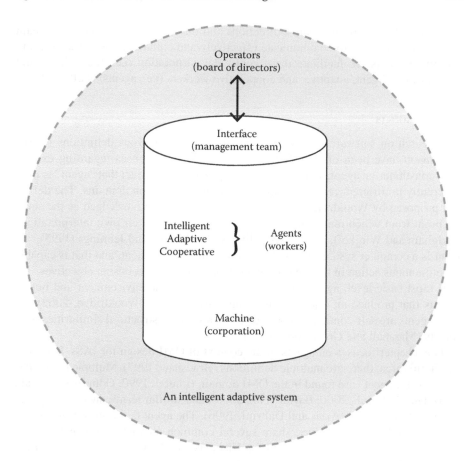

FIGURE 5.1 Corporate interaction as an analogy for operator–agent interaction within intelligent adaptive systems (IASs). Operators, like a board of directors, play a supervisory role, while agents, like workers, perform day-to-day tasks.

maximize overall system performance). The board must trust its management team (i.e., the OMI components) and communicate business matters to them. The management team acts as a bridge between the board of directors and the corporation as a whole (i.e., the system). The management team runs the business by automatically performing and delegating tasks to workers (i.e., low-level agents) and representing the best interests of the board. They keep the board of directors updated on corporation status and what the corporation's next move will be. The management team also handles any issues within the corporation, including correcting internal mistakes, without interference from the board, unless emergencies arise. In emergencies, the board must understand the corporation's position, how the management team is handling the situation, why the issue occurred, how to instruct the management team to handle the situation, and how to directly take over operations, if necessary.

Normally, the board of directors should only play a supervisory, or governance, role. This means that the management team should be able to adapt not only to changes in the business environment (i.e., task, system, and working environment), but also to changes in the board's directions and intentions. The management team also needs to be able to communicate effectively and collaborate with the board to maximize profits. To facilitate these goals, the corporation (i.e., the IAS) should consist of intelligent, adaptive, and cooperative workers (i.e., agents) at all levels.

5.2.2 AGENTS

As research on software agents continues to evolve, numerous definitions for the term *agent* have been offered, but there is a lack of consensus regarding exactly what constitutes an agent. This is further complicated by the fact that "agent" is used differently in different research communities and application domains. The definition proposed by Wooldridge and Jennings, however, is acknowledged as the starting point from which many other researchers have drawn their own interpretations (Padgham and Winikoff, 2004). According to Wooldridge and Jennings (1995), "an agent is a computer system that is situated in some environment, and that is capable of autonomous action in this environment in order to meet its design objectives." At their most basic level, agents take sensory input from the environment and output actions that produce an effect on that same environment (Wooldridge, 2002). As well, agents are self-contained blocks of code, and share structural similarities with objects (Marshall and Greenwood, 1997).

This chapter focuses on agents in the context of OMI design for IASs, but, even within this area, there are multiple definitions (Bradshaw, 1997). Multiple references to "interface agent" are found in the OMI domain (Laurel, 1990; Gómez-Sanz et al., 2000; Duvallet et al., 2000; Baylor et al., 2006), and similar terms, such as "application interface agents" (Arias and Daltrini, 1996). The agent definitions proposed by these researchers do, however, share several common characteristics that can aid OMI design for complex IASs. As Finin, Nicholas, and Mayfield (as cited in Jansen, 1999) identified, an agent's most prevalent characteristics are:

- *Adaptation*: Reactivity to changes in users and environments
- *Autonomy*: Automatically taking certain tasks from users
- *Cooperation*: Social ability among agents themselves

These three characteristics are central to the agent concepts defined in this chapter. Designers of complex IASs should always keep adaptation, autonomy, and cooperation at the forefront of their designs.

5.2.3 AGENT INTELLIGENCE

Defining intelligence in the context of software agents is not straightforward, and the definition of intelligence can vary depending on domain. In the artificial intelligence domain, an intelligent agent is defined as "an autonomous entity which observes and acts upon an environment and directs its activity toward achieving goals" (Russell and Norvig, 2009). From a knowledge engineering perspective, Wooldridge and Jennings (1995) describe three characteristics for intelligent agents to exhibit when satisfying their design objectives:

- *Reactivity*: Intelligent agents must perceive their environment and respond in a timely fashion to changes.
- *Proactiveness*: Intelligent agents must persistently take the initiative to achieve goals.
- *Social ability*: Intelligent agents must interact effectively with other agents and humans.

Padgham and Winikoff (2004) expand upon these characteristics and suggest that intelligent agents must also be

- *Situated*: Intelligent agents must exist in an environment.
- *Autonomous*: Intelligent agents must not be dependent on external control.
- *Flexible*: Intelligent agents must have multiple ways to achieve goals.
- *Robust*: Intelligent agents must be able to recover from failure.

Wooldridge and Jennings (1995) also note two types of intelligent agents: weak and strong. Weak agents exhibit characteristics of autonomous decision making, demonstrate social interaction with other agents, and are reactive and proactive to their environment. Strong agents exhibit these characteristics, but also exhibit psychological notions, such as beliefs, desires, intentions, and other proactive behavior. These proactive behaviors include (a) rationality, which is performing actions to further their goals while not performing actions that may conflict with their goals; (b) accuracy, which is performing the correct actions at the correct time; and (c) adaptability, which is learning from both the operator and the environment. Rationality can be considered a key characteristic, as intelligent agents should be rational in their pursuit of goals to optimize operator–agent interaction (Padgham and Winikoff, 2004). Strong intelligent agents are capable of automated reasoning and can choose between alternatives based on existing knowledge and beliefs in order to achieve stated goals (Kendall et al., 1996).

GPS navigation systems are an example of weak intelligent agents. They improve daily life by providing voice-assisted, turn-by-turn guidance based on operator destination input and preloaded maps. Newer models are also able to reactively and proactively modify routes based on real-time traffic and road closure data. Route guidance, however, is generally dependent on the parameters set by the engineer who programmed the navigation algorithms. Consequently, accidents may occur if too much trust is placed in technology that is neither adaptive to its surroundings nor rational enough to optimize its interaction with the operator. For example, on a dark night in 2010 a Senegalese man in Spain was guided down a dark road by his GPS; the road ended in a new reservoir, which was not on the GPS map. Consequently, the man drowned, although his passenger was able to swim to safety (Govan, 2010).

Advances in mobile technology have also led to the use of intelligent agents as personal assistants, such as Google Now and Siri. Personal assistants are examples of strong intelligent agents and offer functions tied to leveraging real-time user data, being location aware, and knowing operator interests. They are reactive, proactive, rational, accurate, and capable of learning and reasoning. For example, Google Now assists operators by automatically delivering accurate information such as weather, sports scores, meeting times, stock alerts, and public transit schedules. It proactively leverages real-time location data and personal information, such as calendars, contacts, e-mail, and public data feeds, to create "cards" of relevant information that users can access. Google Now also independently presents cards based on observation of current user context and location, as well as observation of previous user choice and behavior. For example, if an operator is a known music enthusiast, Google Now will present cards regarding upcoming concerts when the operator is within walking distance of host venues.

Although strong agents exhibit many characteristics of human intelligence, there is still a need to address the social, flexible, and adaptive abilities of true agent intelligence so that human–agent interaction behavior can fully emulate human–human interaction behavior.

5.2.4 AGENT ADAPTIVITY

Researchers in multiple domains have defined and created frameworks for intelligent agents with adaptive properties, but there has been little effort made to provide a distinct conceptual framework for software agents that exhibit the defining characteristics of adaptive intelligent agents—adaptation, autonomy, and cooperation (Jansen, 1999)— in accomplishing tasks for operators. An adaptive intelligent agent can be defined as a personification of computer code and algorithms that mimics human behavior, perception, and cognition; that can cooperate with other agents; that automatically takes action either autonomously or on behalf of the operator; and that can adapt to changes in the operator, system, or environment (Hou et al., 2011). This definition applies to agents in the context of IAS design throughout this book.

The ability to automatically take actions on behalf of the operator, and the modeling of human behavior, perception, and cognition, are what truly distinguish an IAS agent from other agents. To facilitate collaboration between operators and the IAS through an OMI, this type of agent should be able to adapt autonomously, and should exhibit descriptive, prescriptive, intelligent, adaptive, and cooperative characteristics. Agents should be able to inform operators about what is happening, and what will or should be done next; they should be able to learn operator intentions, monitor operator cognitive workload and performance, and guard operator resources and time; they should learn from past experiences and, in response, change how they behave in any given situation; and they should enable themselves to communicate and cooperate with each other and act in accordance with the results of their communication. For example, if the in-car GPS noted earlier had been able to learn from its satellite agent partner that the car was approaching the reservoir (i.e., being cooperative), it could have informed the driver of the situation (i.e., being descriptive) and advised the driver to stop immediately (i.e., being intelligent and prescriptive). If the GPS had the ability to work with other in-car speed agents to detect that the car was traveling in the wrong direction (i.e., not consistent with previously learned driver intentions) or was going too fast for the driver to stop, it could then work with an agent controlling the brakes to stop the car automatically without asking the driver for authorization (i.e., being intelligent, adaptive, and cooperative).

Lack of consensus regarding agent definitions and categorizations across fields and among practitioners has led to the creation of new terms for concepts similar to agent adaptivity. Many agent definitions proposed in the context of belief-desire-intention (BDI) frameworks inherit the characteristics of agent adaptivity, regardless of the specific terms the researchers have employed (Georgeff and Lansky, 1986; Rao and Georgeff, 1995; Busetta et al., 1999; Huber, 1999; D'Inverno et al., 2004; van Doesburg et al., 2005). For example, Barber et al. (2002) present a framework based on "sensible agents" that support the human operator by forming decision-making teams, making plans within those teams, and conducting "situation assessment to best perceive and respond to changing, complex environments." Agent-based BDI frameworks are further discussed later in this chapter.

Note that, while a broad range of applications can be classified as IAS agents, "normal" computer applications cannot. For example, a standard macro would not be classified as an adaptive intelligent agent. Macros automate tasks for the operator, but are usually dependent on the working environment and operator input. Any deviation from the initial input or changes to the environment can cause the macro to fail.

Alternatively, the Mercedes-Benz Active Parking Assist system (Mercedes-benz.ca, 2014) demonstrates agents cooperating with human operators as active partners to simplify one of the most challenging aspects of driving for new and experienced drivers alike: parallel parking. Ten ultrasonic sensors automatically scan the area to the left and right of the vehicle at speeds <36 km/h, and measure the length and depth of potential parking spaces. A "P" symbol on the dashboard describes to the driver that the search is in progress and an arrow appears to indicate a suitable parking space has been found. The driver stops, engages reverse gear, pushes a button to confirm acceptance of the recommended parking space, and removes their hands from the steering wheel.

The parking assist system continually measures the distance from the vehicles in front and behind, interprets data from the sensors, calculates the speed from the accelerator, and intelligently works out the most effective path to the parking space. The driver is required only to regulate speed with the accelerator and brake pedals. When the speed changes, the agent partners with the human to reactively adjust the control of the electromechanical steering system and change the path to safely maneuver the vehicle into the parking spot. If anything is noted as unsafe, the agent informs the driver through auditory and visual warnings and prescribes further actions, guiding the driver to a safe path.

Agents that exhibit adaptive and intelligent characteristics can act as partners to optimize operator–agent interaction in an intuitive fashion, maximizing overall IAS performance. Given the differences between human–machine interaction and operator–agent interaction, the most effective way to design an IAS interface is to incorporate a variety of intelligent adaptive agents that understand operators, act for operators, and explain and specify adaptations to operators.

5.3   BENEFITING FROM AGENT-BASED APPROACHES IN IAS DESIGN

As machine capabilities continue to increase, and humans and machines become far more closely coupled, operators will see improved human–machine interaction, advances in OMI design and technology, and direct augmentation of human performance. Agent-based concepts offer a significant benefit to human–machine interaction, which is arguably the most important single characteristic of complex software (Wooldridge, 2002). Agents provide assistance, educate, facilitate collaboration, and can independently monitor events and procedures. They adapt to the environment, to new inputs received either externally or from other agents, and to explicit inputs from the operator. They can act autonomously without operator intervention or they can cooperate with the operator, other programs, or other agents. These characteristics allow agents to communicate, assist, advise, and automatically perform actions on the operator's behalf.

According to Alan Kay (1984), the concept of software agents originated with John McCarthy and Oliver G. Selfridge in the late 1950s. They envisioned a system that, when given a goal, could carry out the appropriate functions and ask for and receive advice in human terms. Nicholas Negroponte (1969, 1970) was among the first to recognize the potential value of agents as a means to create an IAS that would offer "a better chance of making its computational and informational abilities relevant." However, research on software agents did not take off until the early 1990s, when the technology showed promise as an effective computing paradigm to implement user delegation and to make the machine and the operator active partners (Shneiderman and Maes, 1997; Grossklags and Schmidt, 2006). Notably, agent-based approaches are not the only approaches taken by IAS designers, and there is a need to consider other common design methods when addressing the OMI.

5.3.1 COMMON INTERFACE DESIGN METHODS

Originally, command-based syntax was dominant in human–machine and computer interface design. Issuing commands based on carefully designed syntax is argued to be both an efficient and an elegant way of performing repetitive operations, such as sorting and renaming files, finding and replacing text in large documents, and installing multiple applications (Preece et al., 2002). Command-based systems, however, are often isolated entities that communicate only with their human operators, and are not relevant in the realm of interconnected, networked, and distributed computing (Wooldridge, 2002).

Since the mid-1980s, most interfaces have been designed based on the concept of direct manipulation, which uses visual and auditory cues to emulate what happens to physical objects in the real world. Like command-based systems, direct manipulation requires users to explicitly initiate tasks and monitor events until the job is done (Maes, 1994; Allen and Ferguson, 2003); many users like to know about, be involved with, and have a sense of power over the software they use (Preece et al., 2002). More negatively, direct manipulation runs the risk of reducing accuracy, as users may be overwhelmed when working on multiple tasks simultaneously. Additionally, users may be made responsible for precise control that could be better handled by system intelligence and then communicated after the fact (Hutchins et al., 1985).

Object-oriented toolkits and programming languages are central to direct manipulation systems, and are arguably a dominant approach to software design because of the ubiquity of Java, C++, Visual Basic, OpenGL, and Unified Modeling Language (UML). Objects themselves are defined by Wooldridge (2002) as computational entities that encapsulate some state (i.e., status), are able to perform actions or methods on this state, and can communicate with other objects. Note that using object-oriented toolkits often results in the mixing of interface and application code.

Object-oriented systems consist of modules that are able to communicate with each other and have explicit, individual means to respond to incoming messages (Shoham, 1993). Integrating software agents within object-oriented methodologies may be advantageous, as these methodologies have been tested and offer familiarity to developers, who may be reluctant to learn and use new approaches (Iglesias et al., 1998). However, the utility of representing agents as objects is debatable because their level of abstraction as a set of attributes and methods may be too fine grained; agents are coarse-grained entities that require the use of significant system resources compared with objects (Wooldridge et al., 2000). For example, it is difficult to use objects to model concepts such as proactively generated actions, dynamic reaction to changes in the environment, and cooperating and negotiating with other self-interested entities (Wooldridge, 2002).

According to Wooldridge et al. (2000), there is "a fundamental mismatch between the concepts used by object-oriented developers and the agent oriented view." Objects have some degree of control over their internal status, but lack autonomy over their own behavior and do not have the ability to adapt and react to operator state changes or changes in other objects. Objects are also limited in that they cannot distinguish between different types of messages received from various sources, and cannot negotiate, analyze, and make decisions based on the action requested by the sender (Iglesias et al., 1998). Agents can, however, effectively perform two-way communication with the operator and other agents, and provide systems designers with the means to overcome the limitations of object-oriented design.

5.3.2 Advantages of an Agent-Based Approach

Agent-based systems have the potential to move beyond direct manipulation and object-oriented methodologies toward the indirect management of information (Benyon and Murray, 2000). For example, agents are able to determine what needs to be done to satisfy their design objectives, and do not need explicit instruction to act (Wooldridge, 2002). Wooldridge (2002) identifies the following situations where an agent-based approach may be beneficial:

- The environment is open, highly dynamic, uncertain, or complex.
- The environment (e.g., commercial or competitive) is naturally suited to models of societies of agents cooperating and competing with each other to solve complex problems.
- The distribution of data, control, or expertise is decentralized.
- The existing system contains components that are technologically obsolete but functionally essential.

Agents offer a multitude of advantages for effective IAS design. However, the key advantages are their ability to facilitate operator–agent interaction and their ability to support an interaction-centered design method. Specifically, agents promise the following benefits for interaction in the context of IASs (Hou et al., 2011):

- *More effective interaction*: Doing the right thing at the right time, and tailoring the content and the form of interaction to operator, task, system, communication, domain, and environmental contexts
- *More efficient interaction*: Enabling more rapid task completion with less work
- *More natural interaction*: Supporting spoken, written, and gestural interaction, ideally resembling human–human interaction behavior

Agents are the key to realizing indirect information management and improving the efficiency, effectiveness, and intuitiveness of IASs. However, it is crucial for systems designers to clearly understand where agents may usefully be applied, and to avoid unrealistic expectations of what agents can achieve (Wooldridge, 2002).

5.4 DESIGNING AGENTS

There is a notable lack of research focused on operator–agent interaction, as agent-based technology has not yet been adopted in many complex and dynamic system designs. From an HF perspective, the lack of theoretical development and empirical studies makes many designs costly and ineffective. Many existing frameworks focus on individual models (e.g., operator, task, and domain) rather than operator interaction with the agent system as a whole. A theory and associated design methods are needed to facilitate operator–agent interaction in any IAS. Understanding agent concepts and benefits in the context of IAS design is the first step; the next step is to apply appropriate design and implementation methods to situate individual agents within a complex IAS.

From a software engineering perspective, there are at least four major methods for implementing agents (Maes, 1994; Lieberman and Maulsby, 1996). First, the agent can be explicitly coded and integrated as part of an end-program or application. This approach ensures that agent actions are well defined, but it prevents agent reuse, requires building new systems for each change, and requires operator insight on how to effectively employ the agent. Second, the agent can be instructed by the operator using explicit directions or by being shown what to do. This allows the agent to adapt and "grow" as operator goals change. As with the first method, however, it requires a great deal of insight from the operator on how to effectively employ the agent.

Third, the agent can be designed to use the "learning approach" (Dent et al., 1992; Maes, 1994). The agent begins with some knowledge of the domain and the application, but determines what the operator would like it to do based on operator actions. This approach allows the agent to more easily adapt to the operator over time and become customized to a particular operator. The learning approach does not require the operator to have extensive domain knowledge, but the operator must know how to accomplish the task, or that the task is possible within the system.

The common major drawback to these first three methods is that the agents are implemented as an internal part of an entire system, which may require replacing existing systems that are arguably still effective without the implementation of agents. These methods seem increasingly unbeneficial and impractical when extensive research, development, and implementation costs, coupled with organizational resistance to change, must be addressed at the beginning of the interface and system design cycle.

Lieberman (1998) advocates for a fourth option: integrating agents with existing applications. Integration is usually accomplished with a standard application programming interface (API). The agent gathers information about the operator's actions via the API and then uses the information to deduce operator goals, desires, and characteristics. Of the four outlined design methods, this option most readily facilitates the integration of agents within existing systems (Riecken, 1994).

Integration also has limitations, most notably that, even within the same domain, different applications will provide different information via the API. Consequently, the benefit of agents varies from application to application. The concept of integrating agents with existing applications is worth pursuing because (a) the system does not need to be rebuilt, and (b) access to operator actions and application objects is increasingly available within a wide range of programs.

Regardless of which implementation method is used, Maes (1994) suggests that the following six agent design principles should be considered by systems designers:

- The operator must trust the agent enough to delegate action to it.
- The agent should not distract the operator from what they are doing unless it is necessary.
- The operator must understand what the agent is doing and how the agent is doing it.
- The operator must be able to exert control over the agent, both to take action and to stop it from taking action.
- The agent must be personalized for the operator in a manner consistent with the task and application.
- Ease of use issues must be addressed.

As agents are integral components of IAS automation, human–automation interaction issues, such as technological issues, human performance issues, and communication issues, should be considered when optimizing operator–agent interaction. Chen and Barnes (2014) summarize these issues into six agent design guidelines:

- Flexible operator–agent interaction
- Maintaining the operator's ultimate decision authority
- Supporting the operator's multitasking performance
- Ensuring automation transparency
- Enhancing operator–agent collaboration through visualization and training techniques
- Considering individual operator differences (e.g., spatial and attention abilities) during the design process

With these guidelines in mind (and despite the overlapping areas that exist in the design principles discussed above), systems designers must consider which design method will best facilitate operator–agent interaction within a complex IAS.

5.4.1 Agent-Based Design Methods

The relationships among agents, system components (e.g., operator and machine), and the working environment, as well as how these entities work together to help operators achieve system goals and maximize system performance, are extremely important for agent-based design methods. Agent-based design methods such as common knowledge acquisition and documentation structuring (CommonKADS), integration DEFinition (IDEF), explicit models design (EMD), ecological interface design (EID), and BDI, along with the implementation of a centralized database such as a blackboard, can help guide agent-based IAS design, and have been used in numerous contexts across multiple domains. These methods are intended to guide the analysis and design process from tentative and abstract beginnings to a more concrete and detailed design with the potential for implementation (Wooldridge, 2002).

It should be noted that the review of these agent-based design methods is not exhaustive, and each of the methods is significant and substantial in itself. The intent is not to comprehensively summarize the agent-based design methods applied to date, but to make systems designers aware of these methods, and to help them target a few key approaches for IAS design. Systems designers can study these methods in detail on their own, and references are provided to help facilitate this.

5.4.1.1 Common Knowledge Acquisition and Documentation Structuring

CommonKADS is a knowledge management and engineering approach proposed by Schreiber et al. (1999) for developing computer programs that reason and use knowledge to solve complex problems. CommonKADS has its origins in the artificial intelligence domain, and provides guidelines to design an expert system using a formalized representation of knowledge and an inference mechanism that models the capabilities of human activities such as diagnosis, planning, and design (Yoon and Hammer, 1988). A key advantage of CommonKADS is that it systematically identifies all issues related to the design of the system before any code is written, reducing the likelihood that the design will need to be revised during implementation (Edwards, 2004). However, applying CommonKADS to real-world problems can be challenging, as it only addresses specific phases of the software development life cycle and requires expert knowledge that might not be possessed by the systems designers tasked with implementing the methodology (Pavón and Gómez-Sanz, 2003). Furthermore, this design method deals mostly with human–computer interaction, and provides challenges to systems designers seeking to model interaction between agents (Iglesias et al., 1997).

CommonKADS system development is divided into phases, each encompassing the construction of one of the major CommonKADS models, which include the organization model, task model, agent model, knowledge model, communication model, and design model (Edwards, 2004). By specifying these models, CommonKADS can be applied to the design of IAS knowledge-based adaptive interface subsystems, such as control systems for UAVs. The CommonKADS design method also includes a four-step project management process that can be tailored to the needs of the systems designer: (a) objectives for the current phase are established; (b) potential risks are identified and assessed; (c) a detailed plan for the current phase is produced that takes into account previously identified objectives and risks; and (d) development work is monitored and evaluated, and the resulting conclusions are taken into account at the beginning of the next design cycle phase (Edwards, 2004).

A multiagent system extension of the CommonKADS methodology (MAS-CommonKADS), proposed by Iglesias et al. (1997), adds specific agent-related constructs to the system that relate to (a) interagent communication; (b) the division of tasks among individual agents; and (c) the implementation of multiagent systems. Organization, task, agent, knowledge, communication, and design models are used in the first four phases of system development specified by MAS-CommonKADS (Iglesias et al., 1997), which include (a) conceptualization, (b) analysis, (c) design, (d) coding and testing, (e) integration, and (f) operation and maintenance.

In addition, a coordination model is implemented that provides additional constructs for modeling agent–agent interaction (Iglesias et al., 1997). The methodology may be easy for IAS designers familiar with object-oriented modeling techniques to grasp (Iglesias et al., 1997). For example, MAS-CommonKADS was used to guide the design of an agent-based flight reservation system that provides the cheapest available fares to a given destination (Arenas and Barrera-Sanabria, 2002).

Figure 5.2 illustrates the MAS-CommonKADS architecture, which consists of models of entities (e.g., agent and operator) and the relationships between those

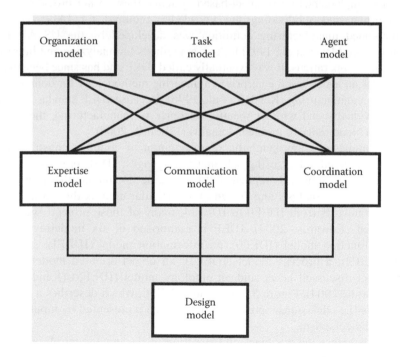

FIGURE 5.2 The multiagent system extension of the common knowledge acquisition and documentation structuring (MAS-CommonKADS) architecture, which consists of specific models and the relationships between them. The design model refines the six basic models, determines the most suitable architecture for each agent, and defines the overall requirements for the agent network.

entities (Iglesias et al., 1997; Henesey, 2006). The following models are included in the MAS-CommonKADS architecture:

- *Agent model*: Describes the characteristics of each agent
- *Task model*: Describes the agent's tasks and determines the agent's goals
- *Expertise model*: Describes the knowledge required by the agent to achieve its goals
- *Organization model*: Describes the structural relationships between the operator and agents, and between agents
- *Coordination model*: Describes the interactions, protocols, and dynamic relationships between agents
- *Communication model*: Models the dialogue between agents and describes the dynamic relationships between the operator and agents
- *Design model*: Refines the other six models and determines the most suitable architecture for each agent, as well as the overall requirements of the agent network

5.4.1.2 Integrated Computer-Aided Manufacturing Definition

Although CommonKADS methodologies offer a framework for approaching the design and implementation of agent-based systems, they do not provide a means of identifying and subdividing the knowledge required for IAIs. Integrated computer-aided manufacturing definition was developed by the US Air Force (Lydiard, 1995; Mayer et al., 1995) to look at methods for analyzing and improving manufacturing operations. It was eventually called IDEF and has since been used as the basis for an agent-based enterprise engineering methodology for domains such as discrete manufacturing (Kendall et al., 1996). Because IDEF standards address issues individually and without specific reference to manufacturing, they can be applied to a broad range of problem domains (Edwards, 2004).

IDEF languages allow symbolic representation of complex constraints that cannot be depicted solely using schematic languages. IDEF presents a collection of numbered standards represented by specific documents that provide formal guidelines for analysis and design in a particular area. Although there were plans for standards from IDEF0 to IDEF14, many of those projects were never implemented (Edwards, 2004). IDEF is comprised of six modeling methodologies: a function model (IDEF0); an information model (IDEF1); a dynamic model (IDEF2); a process model (IDEF3); an object-oriented model (IDEF4, which is not discussed here); and an ontology model (IDEF5) (Kendall et al., 1996; Edwards, 2004). Figure 5.3 illustrates IDEF0, which describes a hierarchy of functions (i.e., decisions, activities, and actions) represented by input, control, output, and mechanism.

IDEF1 describes the attributes and structure of the input, control, output, and mechanism; and IDEF2 models the time-dependent characteristics associated with the functions described in IDEF0 and IDEF1 (Kendall et al., 1996). IDEF3 allows flexible modeling of temporal concepts using a UML-based elaboration language (Edwards, 2004). IDEF facilitates formal logical representations of process constraints, allows precise specification of event timing and duration, and,

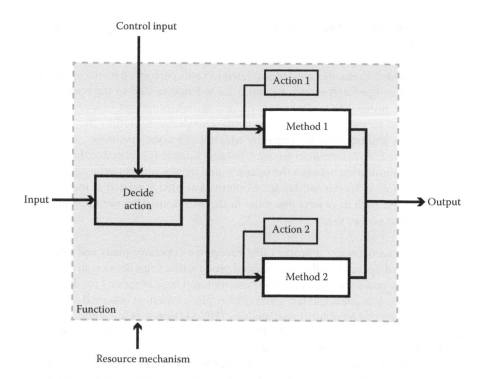

FIGURE 5.3   A detailed, generic IDEF0 function model. IDEF0 describes a hierarchy of functions (i.e., decisions, activities, and actions) represented by input, control, output, and mechanism. (Redrawn from Kendall, E.A., Malkoun, M.T., and Jiang, C.H., *Modelling and Methodologies for Enterprise Integration: Proceedings of the IFIP TC5 Working Conference on Models and Methodologies for Enterprise Integration, Queensland, Australia, November 1995*, pp. 333–344, Chapman & Hall, London, UK, 1996.)

unlike CommonKADS, provides specifications for ontology modeling (IDEF5). Consequently, integrating IDEF and CommonKADS and combining the strengths of the individual IDEF components provides a comprehensive approach to IAS design that may solve the inability of CommonKADS to identify and subdivide knowledge. For example, CommonKADS and IDEF design methods were combined to design the LOCATE workspace layout design tool (Edwards and Hendy, 2000). Principles derived from the development of LOCATE are demonstrated by a generalized help system that identifies and extends intelligent aiding principles across a wide range of software applications, and provide insight into the EMD concept.

5.4.1.3   Explicit Models Design

EMD (Edwards, 2004) subdivides the knowledge content of CommonKADS into five distinct, interacting models: task, user, system, dialogue, and world (Edwards and Hendy, 2000). A description of the models can be found in Chapter 3. The goal of EMD is to guide the design of agent-based interfaces by explicitly detailing their required knowledge (Edwards, 2004). Different types of knowledge are contained within each model:

- *Task model*: Contains knowledge related to tasks performed by the operator
- *User (i.e., operator) model*: Contains knowledge related to the operator's abilities, needs, and preferences
- *System (i.e., machine) model*: Contains knowledge related to the system itself, its abilities, and the means by which it can assist operators
- *Dialogue (i.e., interaction) model*: Contains knowledge related to the mode of communication between the operator and agents, and between agents
- *World (i.e., environment) model*: Contains knowledge related to the external world, such as objects that exist in the world, their properties, and the rules that govern them

EMD also incorporates a process that recognizes operator plans and keeps the system aware of operator goals, and a second process that helps develop strategies for accomplishing goals specific to operator assistance. These processes for decomposing user and system goals and recognizing user plans based on observed actions are best represented by hierarchical goal analysis (HGA) (Hendy et al., 2002), and HGA techniques are integrated into EMD. Figure 5.4 illustrates a generic version of the EMD architecture developed by Edwards (2004).

By making the knowledge required by agent-based interfaces explicit, EMD can determine the operator goals, the plans for achieving those goals, and how the system can most effectively assist the operator in the pursuit of those goals. EMD challenges include how to decompose knowledge into various models and how to coordinate

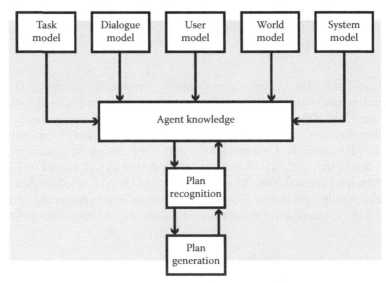

FIGURE 5.4 A generic explicit models design (EMD) architecture. EMD divides the knowledge content of common knowledge acquisition and documentation structuring (CommonKADS) into five distinct, interacting models.

knowledge among the models to build effective supporting systems. Edwards (2004) has integrated EMD with CommonKADS and IDEF to create a generic framework, which was described in Chapter 3. As part of Edwards' framework, EMD offers enhanced methods for user modeling and provides a complete snapshot of the operator, the machine, and the world they interact with, which is crucial to the design of complex IASs.

5.4.1.4 Ecological Interface Design

EID was introduced in Chapter 4 as a means to facilitate OMI design; it also shows promise as an agent-based design method. The intention of EID is to support the operator by moving complex work to more manageable cognitive levels (e.g., knowledge-based to rule-based, or rule-based to skill-based); as a result, it is well suited to address problems that occur when controlling and monitoring complex IASs. EID was developed at the Risø National Laboratory in Denmark by Vicente and Rasmussen (1992) as a framework for problem domain analysis and OMI design for complex work environments. The name derives from the incorporation of ecological psychology elements, and the framework emphasizes the importance of human interaction with working environments. While most traditional interface design methods confine their attention to human characteristics, EID also examines how humans interact with their surroundings and takes into account both physical and cognitive factors in the context of a complex system under control. To achieve this, EID offers concrete guidelines for interface design; it presents a set of techniques for constructing safe and reliable user interfaces based on cognitive control levels. EID improves the speed and accuracy of fault diagnosis and creates more flexible and robust control strategies (Vicente, 2002; Burns and Hajdukiewicz, 2004). This can be particularly helpful for operator–agent interaction support in complex systems under unusual (i.e., nonroutine) conditions that require immediate comprehension of agent or machine status and rapid action (Sheridan and Parasuraman, 2006).

EID has been applied in a variety of areas, including medicine, industrial process control, and nuclear power plants (Furukawa and Parasuraman, 2003; Burns and Hajdukiewicz, 2004; Burns et al., 2008; Davies and Burns, 2008; Lau et al., 2008; Morita et al., 2009); it recognizes that system reliability depends not only on how the system is engineered, but also on how humans interact with the system. As a result, the interface structure and content are designed to facilitate operator understanding of the system. Traditional ergonomic design techniques govern the location, clarity, and intuitiveness of controls, and an ergonomic approach can reduce the likelihood of operator errors, such as adjusting the incorrect control knob. However, without considering operator cognitive processes, ergonomic techniques cannot prevent errors of intention, such as when an operator deliberately adjusts a control for the wrong reason. EID aims to minimize errors of intention by taking into account how humans make decisions and analyze problems.

Figure 5.5 illustrates the deductive path of reasoning central to EID. Interface design should not force cognitive control to a higher level than the demands of the task, but should still provide support for (a) skill-based behavior, (b) rule-based behavior, and (c) knowledge-based behavior (Vicente and Rasmussen, 1992).

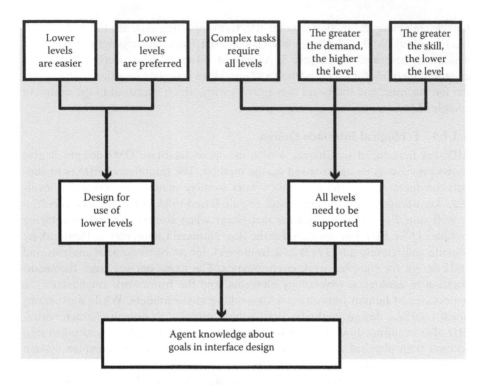

FIGURE 5.5 The deductive path of reasoning in ecological interface design (EID). Cognitive control should provide support for all levels without forcing cognitive control to a higher level than the task requires. (Redrawn from Vicente, K.J. and Rasmussen, J., *IEEE Transactions on Systems, Man and Cybernetics*, 22, 589–606, 1992.)

Knowledge-based behavior correlates with analytical problem solving based on symbolic representation (i.e., high level), while rule-based and skill-based behavior correlate with perception and action (i.e., low level).

Several studies have investigated the effectiveness of EID in the context of operator–agent interaction (Furukawa and Parasuraman, 2003; Furukawa et al., 2004; Seppelt and Lee, 2007; Cummings and Bruni, 2010). In these studies, the EID design method was used to graphically represent the capabilities and limitations of agents, the intentions of agents, and the quality of plan revisions proposed by agents. These studies consistently found that EID was effective in supporting operator understanding of agent behavior, and enhanced the operator's ability to predict future agent behavior. EID can promote operator–agent trust and contribute to effective IAS design by improving SA in abnormal and complex situations.

5.4.1.5 Belief-Desire-Intention

BDI is a well-researched agent-based design method (Bratman, 1987; Bratman et al., 1988; Brazier et al., 1997; Georgeff et al., 1998; Jarvis et al., 2005; Sudeikat et al., 2007; Briggs and Scheutz, 2012) that models the cognitive and adaptation processes of human behavior, and has its basis in philosophy (Bratman, 1987). BDI is rooted in the notion that software agents should be "rational" in their pursuit of goals (Padgham and Winikoff, 2004), and should be capable of behavior and decisions as appropriate as those of their human counterparts (van Doesburg et al., 2005).

Figure 5.6 illustrates an example BDI agent model. *Beliefs* encompass the knowledge that an agent has about itself, the operator, the environment, or other agents, and its inference mechanisms, which can be incomplete or incorrect. *Desires* correspond to the tasks allocated to an agent. Note that system goals are required to be logically consistent, but an agent will not be able to achieve all system goals even if they are consistent. Agents must instead fixate on subsets of available desires. The chosen desires are defined as *intentions* to which the agent commits resources for achieving. An agent typically pursues achieving an intention until it is either satisfied or no longer available (Cohen and Levesque, 1990).

The key data structure in the BDI model is a plan library; a plan defines a means to realize a goal (Padgham and Winikoff, 2004). An agent's plan library represents its procedural knowledge about producing system states, and specifies what courses of action the agent may take to achieve its intentions (Doane and Sohn, 2000).

Figure 5.7 illustrates Barber et al. (2002)'s sensible agents architecture, which resembles BDI but uses different terminology to describe similar concepts. The architecture is divided into four modules: (a) a perspective modeler that contains the agent's explicit model, which fits its local subjective viewpoint of the world (e.g., itself, other agents, and the environment); (b) an autonomy reasoner that determines the appropriate decision-making framework for agent goals (e.g., command-driven, consensus, locally autonomous/master); (c) an action planner that interprets domain-specific goals, generates plans to achieve those goals, and executes the plans, and includes "a communication actuator" for sending messages to other agents; and (d) a conflict resolution advisor that identifies and proposes resolution strategies and belief conflicts between agents.

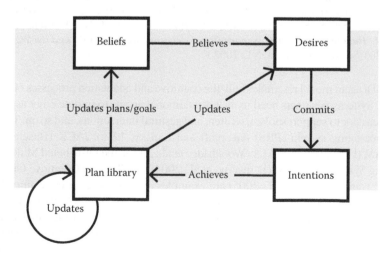

FIGURE 5.6 Belief-desire-intention (BDI) agent model. Beliefs are the agent's knowledge and inference mechanisms; desires are the tasks allocated to the agent; and intentions are subsets of available desires. The plan library provides the agent with ways to achieve its goals.

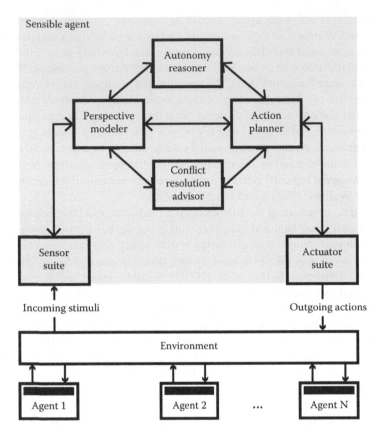

FIGURE 5.7 Sensible agents architecture, which uses different terminology to describe concepts similar to those found in belief-desire-intention (BDI). (Redrawn from Barber, K.S., Goel, A., Kim, J.N., Martin, C.E., Han, D.C., Lam, D.N., MacMahon, M., and McKay, R., *Human Factors in Transportation, Communication, Health, and the Workplace*, pp. 273–286, Shaker, Maastricht, NL, 2002.)

The BDI agent model resembles only the cognitive and adaptation processes of human behavior. Systems designers need to include additional agents to imitate other aspects of human behavior to support spoken, written, and gestural interactions, and so on. The procedural reasoning system (PRS) (Georgeff and Lansky, 1986), JACK (Busetta et al., 1999), JAM (Huber, 1999), GAIA (Wooldridge et al., 2000), the Distributed Multi-Agent Reasoning System (dMARS) (D'Inverno et al., 2004), and Tactical Cognitive Opponent (TACOP) (van Doesburg et al., 2005) are examples of agent-based architectures based on BDI that have been applied to complex, real-world systems. Of these, PRS and dMARS are perhaps the best recognized, and they have many similarities, including the relative simplicity of transitioning from the design phase to real-world implementation (Wooldridge, 2002). PRS and dMARS systems have been used in numerous industrial applications, including air traffic management, space shuttle fault diagnosis, business process control software for call centers and internet service providers, and a mission-based simulation system for the Australian military (D'Inverno et al., 2004).

5.4.1.6 Blackboard

The idea of a centralized data and knowledge repository, or blackboard, is central to many agent-based system frameworks (Marshall and Greenwood, 1997; Berger, 2007; Hou et al., 2011) and is evident in the generic agent-based design framework described later in this chapter. The blackboard model, first proposed by Newell in 1962, describes cooperative goal-directed action (Kempe, 1994), and was first implemented as a design method in the Hearsay-II speech understanding system developed in the mid-1970s (Nii, 1986). It has since been expanded upon by Nii (1986) and others (Marshall and Greenwood, 1997; Berger, 2007) and integrated into various agent-based systems.

Figure 5.8 illustrates a blackboard-based system, which has three main components: (a) a database, or blackboard, that contains information and hypotheses; (b) knowledge sources, which are usually sets of rules, that monitor the status of the blackboard and create new data or hypotheses; and (c) a control mechanism/inference engine that monitors and adjusts the hypotheses and subsequent testing. The database and knowledge sources are deliberately kept apart, and the knowledge sources are further divided into independent experts (i.e., agents) that update the blackboard when internal constraints match the blackboard status. Each expert contributes to a problem resolution with its knowledge of how to search for a partial solution to parts of the problem faced (Kempe, 1994). The control mechanism mediates the interaction of multiple knowledge sources and implements a feedback loop between the operator and blackboard interface that adapts over time (Marshall and Greenwood, 1997). As an expert system model, the blackboard exhibits some characteristics seen in IAS components, such as the situation state module and adaptation module, which were discussed in Chapter 3. Various agents collect and update data to build expert models based on sets of rules from a variety of knowledge sources. Inference engines then deduce the situation state and create new hypotheses for the expert model based on the updated data and blackboard internal constraints. The control mechanism (one or a group of agents) controls the adaptation module to adjust the hypotheses and blackboard status while providing feedback to the operator.

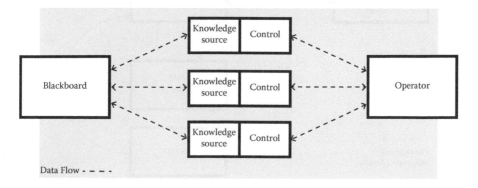

FIGURE 5.8 A blackboard model, containing the blackboard (i.e., a database), knowledge sources, and controls. The operator receives filtered information, and the agents update the blackboard when internal constraints match the blackboard status. (Redrawn from Nii, H.P., *AI Magazine*, 7, 38–53, 1986.)

The blackboard concept has been applied in numerous design contexts. In the military aviation domain, Berger (2007) demonstrated a knowledge-based adaptive resource management agent (KARMA) that evolves the basic blackboard design method by addressing its intrinsic deficiencies, and implements the concurrent execution of multiple knowledge sources to support real-time tactical mission planning in dynamic and uncertain environments. The goal of KARMA is to facilitate the collaboration of multiple autonomous decision makers (agents and operators) to meet a variety of common and individual goals.

Marshall and Greenwood (1997) designed a blackboard-centered generic architecture that has been applied in the medical domain to a handheld insulin dosage advice system and data log for diabetes outpatients. The system adapts to changes in operator state by modifying its interface, while analyzing data to automatically identify and alert the physician about patterns and anomalies in operator blood sugar levels. In this architecture, the blackboard acts as a central data repository for patient data that is then analyzed by various agents based on domain, task, and user models. Figure 5.9 illustrates agents integrated into a control layer comprised of overseer and scheduler agents that moderate agent interaction with the system as a whole, and an interaction layer comprised of interface, data, and reasoning agents that place

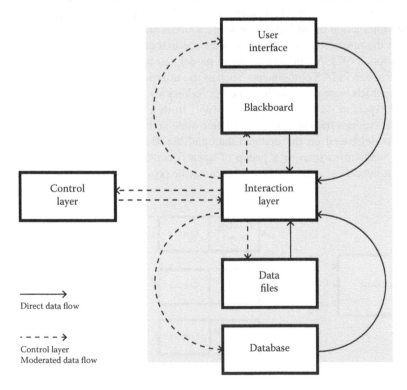

FIGURE 5.9 Blackboard agent architecture for an insulin dosage system. (Redrawn from Marshall, P. and Greenwood, S., *PAAM 97: Proceedings of the 2nd International Conference on the Practical Application of Intelligent Agents and Multi-Agent Technology*, pp. 319–332, 1997.)

data, interact with data, and share data with other agents in both the control and interaction layers of the blackboard. Agents in both layers access a database of diabetes events, such as health details, blood sugar level, and physician advice, which are further abstracted into a data file structure. The end result is an advice system that enables physicians to accurately advise patients, based on their medical history, on how to maintain a consistent blood sugar level and improve their overall health.

The blackboard's centralized data and knowledge repository enables human operators to interact with an interface that has the flexibility of a well-trained human subordinate, and, when combined with a control system, offers IAS designers adaptive automation potential (Miller et al., 2005). One notable evolution of the blackboard concept, developed to enhance operator–agent interaction by facilitating the effective, flexible tasking of multiple agents while keeping the operator in the decision-making loop, is Playbook (Miller et al., 2005). Playbook consists of a user interface, an intelligent analysis and planning component, and a library of shared task models (Miller et al., 2004). Shared task models are hierarchically and sequentially organized, and facilitate communication between operator and agent about plans, goals, methods, and resource usage (Miller et al., 2005).

Table 5.1 shows the benefits and limitations of specific agent-based design methods. By integrating descriptive, prescriptive, intelligent, adaptive, and cooperative adaptive intelligent agents, systems designers have the potential of achieving optimal IAS performance.

5.4.2 Interaction-Centered Design

Analytical techniques and design methodologies usually consist of a collection of models and a set of guidelines associated with those models (Wooldridge, 2002). The models used widely in agent-based design methods and in the intelligent interface technology community (Edwards, 2004; Zhu and Zhou, 2006, 2008) explicitly represent the domains of applications, task constraints, and the flexibility inherent in human interaction with complex IASs (Hou et al., 2011). However, they are not sufficient to describe the interaction between an operator and agents, nor do they reflect the work environment (with the exception of the EID design method) and the dynamic and interaction-centered nature of an IAS.

IAS design should follow an interaction-centered design method that emphasizes augmenting human judgment and responsibilities and exhibits an understanding of how activities are conducted by operators and how operators act and react to events. Note that interaction-centered design differs from interaction design, which focuses on designing systems to satisfy user needs and desires, and making the interface useful through goal-oriented design and the use of personas, cognitive dimensions, and other techniques (Preece et al., 2002; Cooper, 2004; Cooper et al., 2012).

Interaction-centered design views interaction as an important characteristic for agents that exhibit adaptation, autonomy, and cooperation. The extent of this importance was only clearly emphasized after Benyon and Murray (1993) defined the interaction model as similar to a dialogue model. Beaudouin-Lafon (2000), defines an interaction model as "a set of principles, rules and properties that guide the design of an interface. It describes how to combine interaction techniques in a meaningful

TABLE 5.1

Benefits and Limitations of Agent-Based Design Methods

Design Method	Benefits	Limitations
Common knowledge acquisition and documentation structuring (CommonKADS)	• Identifies issues related to the design of the system before code is written, reducing the need for code revision • Includes a four-step project management process that can be tailored to the needs of the designer • Multiagent system common knowledge acquisition and documentation structuring (MAS-CommonKADS) facilitates: (a) interagent communication; (b) the division of tasks among individual agents; (c) the implementation of multiagent systems; and (d) a coordination model for modeling agent–agent interaction	• Only addresses specific phases of the software development life cycle and requires expert knowledge often not possessed by engineers • Primary focus is human–computer interaction, not agent–agent interaction • Does not provide a means of identifying and subdividing the knowledge required for an IAS
Integrated computer-aided manufacturing definition (IDEF)	• Sufficiently generic to be applied to a broad range of problem domains • Allows symbolic representation of complex constraints that cannot be depicted solely using schematic language (e.g., UML) • Facilitates formal logical representations of process constraints • Allows for precise specification of event timing and duration • Provides specifications for ontology modeling	• Does not address the dynamic nature of operator-agent interaction • Does not reflect the IAS working environment
Explicit models design (EMD)	• Explicitly defines agent knowledge through task, user, system, dialogue, and world models • Recognizes user plans and keeps the system aware of what an operator is trying to accomplish • Develops strategies for accomplishing goals specific to assisting the operator • Includes enhanced methods for user modeling • Provides a complete snapshot of the operator, system, and world	• Must determine how to decompose knowledge into the various models • Must determine how to coordinate knowledge among the models

TABLE 5.1   (Continued)
Benefits and Limitations of Agent-Based Design Methods

Design Method	Benefits	Limitations
Ecological interface design (EID)	• Supports the operator by moving complex work to more manageable cognitive levels • Offers concrete guidelines on interface design to optimize usability and safety • Improves fault diagnosis speed and accuracy • Creates more flexible and robust control strategies • Supports operator–agent interaction under nonroutine conditions when quick comprehension of machine or agent status and rapid action are needed • Minimizes errors of intention by incorporating cognitive factors into the interface that take into account how people make decisions and analyze problems • Supports the operator's understanding of agent behavior • Enhances the operator's ability to predict future agent behavior • Promotes operator–agent trust • Improves operator situation awareness	• Not enough empirical evidence to demonstrate the benefits of using EID to support the interaction between operators and agents in both normal and abnormal situations, especially for the real-world IAS applications
Belief-desire-intention (BDI)	• Models the cognitive and adaptation processes of human behavior • Helps software agents be rational in pursuit of their goals • Well-established design method and has been applied to multiple real-world and theoretical systems • BDI-based procedural reasoning system (PRS) and distributed multiagent reasoning system (dMARS) architectures are relatively simple to transition from the design phase to real-world implementation	• Resembles only cognitive and adaptation processes of human behavior • Systems designers need to implement human like agents to imitate other aspects of human behavior not covered by the BDI design method
Blackboard	• Well established and has been applied to numerous design contexts • Enables human operators to interact with an interface with the same flexibility as with well-trained human subordinates • Offers IAS designers the potential of adaptive automation when combined with a control system	• Does not reflect the working environment • Is insufficient to address the dynamic and interaction-centered nature of an IAS

and consistent way and defines the 'look and feel' of the interaction from the user's perspective."

Interaction models have their basis in perceptual control theory (PCT), which asserts that all behavior results from controlling perception (Powers, 1973). In the context of systems design, perception is a subconscious signal value that can be measured by sensors and occurs at many different levels of abstraction, and "control" refers to operators (or machines) taking action to bring their partners to a reference state and maintaining the reference state by reacting to external influences that would alter it (Taylor et al., 1999). PCT is controversial in that it contradicts other common rationales for human behavior, such as (a) stimulus–response, in which the operator takes in stimuli and emits responses that are functions of the incoming stimuli, and (b) cognitive psychology, in which operators plan and decide actions to accomplish a desired result and then execute those actions (Farrell et al., 1999). However, it is crucial for systems designers to consider all potential rationales for behavior. Farrell et al. (1999) and others (Hou et al., 2007) have investigated the design implications of PCT for multiple agent interaction in military applications in the HF context, and PCT and interaction models have been implemented in agent-based systems to aid the construction of control stations for unmanned vehicle operators (Hou et al., 2007).

Cognitive and stimulus–response approaches emphasize consistent and correct operator actions through learning and repetition, while PCT emphasizes understanding the perception of possible actions undertaken by operators and their machine partners (Farrell et al., 1999). The goal of an operator in the PCT context is to reduce the difference between the operator's perception of his or her current state and his or her desired condition for that state (Farrell et al., 1999; Taylor et al., 1999).

Layered protocol theory (LPT: Farrell et al., 1999; Taylor et al., 1999) is an example of an IAS design application of PCT that relates specifically to communication issues. In contrast to PCT's assertion that "all behavior is the control of perception" (Powers, 1973), LPT suggests that "all communication is the control of belief" (Farrell et al., 1999). Unlike BDI, in which beliefs are determined by an agent's inference mechanisms and its knowledge about internal and external worlds, beliefs in LPT are determined by groups of several perceptual signals. Operator–agent interaction in the context of LPT is guided by the perceived differences between the reference beliefs of both partners and their current beliefs.

Properties of the interaction model can be used to evaluate specific interaction designs. Thus, an interaction model should represent at least three properties of the control, controlled systems, and operator: (a) what system changes the operator wants to make; (b) why the changes should be made, with respect to system goals and current status; and (c) what operator activities need to be undertaken to achieve the desired system status. Additionally, the model should present the nature of any concurrent activities, and provide choices to the operator based on the current system status. To be useful to IAS design, an interaction model must be both descriptive and prescriptive; it should aid decision making by describing what an operator actually does and specify what the operator should do next. Keeping this in mind, Hou et al. (2011) proposed an operator–agent interaction model that adopts interaction-centered design through the use of agents that interact in an IAI as a subsystem of an IAS. Interaction-centered design is also central to the IAS design framework introduced later in this chapter.

5.4.3 Operator–Agent Interaction

An IAS must contain many agents to understand operators, act for operators, and explain and specify adaptations to operators. Human–machine interaction is essentially operator–agent interaction. With assistance from various software agents, an operator's tasks move from direct information manipulation to indirect information management; operator focus becomes overall system operation and performance, instead of more detailed system functions (i.e., tasks). Detailed system functions should be delegated to various embedded agents, and these agents should inherit the adaptation, autonomy, and cooperation characteristics of adaptive intelligent agents. Optimized IASs should provide flexible task allocation between operator and agents while maintaining overall operator authority.

To optimize operator–agent interaction, agents should be able to tell operators what is happening (i.e., descriptive) and what should or will be done next (i.e., prescriptive). Agents should be able to learn operators' intentions, monitor their cognitive workload and performance, and guard their resources and time (i.e., intelligent). They should also be able to learn from past experiences and change how they behave in any given situation (i.e., adaptive). Additionally, these agents should enable themselves to communicate and cooperate with each other and act according to the results of their communication (i.e., cooperative). An IAS is a system that gathers agents that are not only descriptive and prescriptive, but also intelligent, adaptive, and cooperative.

5.5 AGENT-BASED, INTERACTION-CENTERED IAS DESIGN FRAMEWORK

This section presents a generic conceptual agent-based framework for the design of an IAI, an IAS subsystem. It also includes a review of agent structure and adaptation processes from an interaction-centered design perspective. Figure 5.10 illustrates that this framework is hierarchical, can be generalized, and provides systems designers with a cooperative multiagent architecture that keeps track of system resources and coordinates agent activities.

The IAI conceptual framework consists of four different function groups of hierarchically managed agents that communicate with each other and collaborate with the operator to increase SA and maximize IAS performance. The IAI architecture is a cooperative multiagent structure that can keep track of all system resources and coordinate agent activities. As such, this architecture plays a crucial role in enabling operator–agent interaction.

Typically, the four function groups of an IAI represent a hierarchy of managing, senior, working, and junior agents. The managing agent is singular, while other groups have multiple agents communicating with each other and working collaboratively. Senior agents oversee the internal system, external system, and environmental status; manage the flow of information and operational routines; and communicate with internal and external sensors, the operator, and other embedded agents. Working agents gather data from senior and junior agents, which is then used to reason and model operator state and machine, task, and environmental status, and provide feedback and decision-making aids to the senior agents, if necessary. Junior agents act on

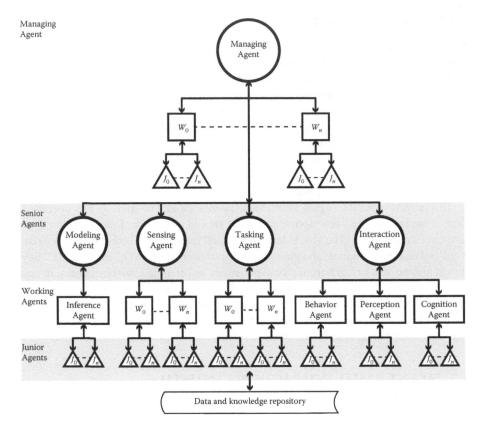

FIGURE 5.10 A generic, agent-based conceptual framework for an IAI. This framework is hierarchical and consists of four different function groups.

direct inputs from the operator, machine, task, and environment to provide information to working agents for further analysis and decision making.

Due to the conceptual nature of this framework, agent contributions to operator–agent interaction are examined at the managing, senior, and working levels only. Junior agents, labeled as J_0 to J_n in Figure 5.10, handle the details of individual tasks. They communicate with other agents in the same group and report to their working agents. For example, an eye gaze-tracking agent is a junior agent that monitors operator eye movement and reports eye gaze-tracking information to its working agent, the perception agent. Junior agents also communicate with other agents through their superior working and senior agents.

5.5.1 Managing Agent

The managing agent is the only agent in its group. It is responsible to, and communicates directly with, the operator. Figure 5.10 illustrates how the managing agent acts like the CEO of an organization on a management team. It receives instructions from the operator and communicates and coordinates with four senior agents to collect information, or knowledge, about operator state, machine status, task status, and environmental status. The managing agent makes decisions and alters the interface through its own working and junior-level agents. It manages information flow and data output, shares knowledge with the other senior agents, controls display characteristics, prioritizes system reactions to emergencies, decides adaptation, assigns tasks, and governs the automation level of working and junior agents.

5.5.2 SENIOR AGENTS

Four senior agents manage high-level tasks in order to achieve high-level system goals. The senior agent group consists of a modeling agent, a sensing agent, a tasking agent, and an interaction agent. These senior agents manage and coordinate tasks, information flow, communications, and feedback among the operator, agents, tasks, and environment. They communicate with each other and with their own working agents. They update the managing agent about changes in operator state and machine, task, and environmental status. They also give advice about adaptation strategies.

- *Modeling agent*: Gathers information from the sensing agent, tasking agent, and interaction agent about current operator state and machine, task, and environmental status and compares newly collected data with embedded models. It then provides updated models to various agents and the database, or data and knowledge repository. The modeling agent also has a working agent (i.e., inference agent) that collects information from the sensing agent, tasking agent, and interaction agent for the processes of comparison and inference.
- *Sensing agent*: Gathers information from system hardware (i.e., sensors) and a data link to keep the managing agent and modeling agent updated on the status of internal and external assets, environments, and so on. It also provides feedback on the effect of system adaptation elicited in response to system changes. The sensing agent has a number of working agents.
- *Tasking agent*: Gathers information to keep the managing agent and modeling agent updated on the current status of internal and external tasks. It also provides feedback on the effects of system adaptation on internal and external tasks. The tasking agent has a number of working agents.
- *Interaction agent*: Gathers and integrates information from its three working agents (i.e., behavior agent, perception agent, and cognition agent) about the state of the operator when interacting with other system components. It provides feedback on the operator's physiological attributes, cognitive characteristics, and psychological states. It coordinates with the modeling agent to process information and communicates with the managing agent. It also makes recommendations to the managing agent to manage interface output. These recommendations facilitate operator–agent interaction by deciding optimal modality, controlling display characteristics, assigning tasks and automation levels to different agents, prioritizing the emergency subsystem, and so on.

5.5.3 WORKING AGENTS

Working agents perform individual aspects of major tasks and assist senior agents in achieving their goals. This function group includes an inference agent that works for the modeling agent, and a behavior agent, perception agent, and cognition agent that work for the interaction agent. They communicate within the group, report to their superior agent, and are shown as W_0 to W_n in Figure 5.10. They also communicate with working or junior agents in other groups through their superior agents.

- *Behavior agent*: Collects operator input data from a keyboard, mouse, joystick, touchscreen, and so on. Data are fed to the interaction agent and integrated with data from other working agents. The interaction agent coordinates with the modeling agent to process the integrated data set and inform the managing agent about the operator's current behavioral state.
- *Perception agent*: Collects data regarding an operator's arm movements, facial expressions, eye gaze characteristics, gestural changes, and so on. Data are fed to the interaction agent and integrated with data from other working agents. The interaction agent coordinates with the modeling agent to process the integrated data set and inform the managing agent about the operator's attention, fatigue, frustration, fear, excitement, and so on.
- *Cognition agent*: Collects operator physiological data through psychophysiological monitoring devices, such as electroencephalography, electrocardiography, functional near-infrared imaging, and so on (operator monitoring approaches are further discussed in Chapter 6). Data are fed to the interaction agent and integrated with data from other working agents. The interaction agent coordinates with the modeling agent to process the integrated data set and inform the managing agent about the operator's workload, situation awareness, complacency, skill degradation, etc.
- *Inference agent*: Collects information from the sensing agent, tasking agent, and interaction agent through the modeling agent. It systematically organizes and categorizes information about operator state and machine, task, and environmental status. Through reasoning, information is processed and transformed into knowledge about system components. The knowledge is then disseminated to aid the decision making of relevant agents through the modeling agent and managing agent.

The managing agent and each senior agent have their own working-level agents, and each working-level agent has its own junior-level agents. All agents constantly update their own status and send the collected information to their superiors and the data and knowledge repository.

In the conceptual IAI hierarchy, working agents take responsibility for detailed work. They can work independently or cooperate with other agents under the supervision of a senior agent. For example, the behavior agent, perception agent, and cognition agent work independently to collect data about operator behavioral, perceptive, and cognitive states, and then send that data to the interaction agent. They also coordinate with each other through the interaction agent to reflect the current operator state. Other descriptive information comes from the sensing agent and tasking agent about current machine status, tasks, and goals. Based on the information described by these three senior agents, the modeling agent communicates with its working agent, the inference agent, to reason and model the operator, machine, tasks, and environment. The inference agent classifies, compares, and evaluates all the data sets in the system database. The results are sent back to the modeling agent to inform it of possible changes in operator state and machine, task, and environmental status. The modeling agent updates embedded models and informs the other senior agents about current status and possible changes. Using built-in criteria and algorithms, the managing agent and interaction agent decide how to make necessary adjustments (i.e., adaptation) in response to changes in operator goals and tasks, machine status, or environmental status. These changes include decisions regarding adaptation methods, modality, automation levels, and future actions.

5.5.4 FUNCTION ALLOCATION AND ADAPTATION PROCESSES

Agents use adaptation processes to optimize operator–agent interaction and maximize IAS performance, exhibiting the characteristics of the interaction-centered design method. Adaptation is accomplished by keeping the IAI aware of activities, resources, timelines, and operator intentions, plans, and performance. Agents are able to interact with different aspects of operator behavior, including physiological attributes (e.g., eye, mouth, body, etc.), cognitive characteristics (e.g., capacity, recognition, learning, reasoning, decision, trust, etc.), knowledge basis (e.g., the environment, machine, operator, tasks, etc.), and psychological states (e.g., concentration, SA, vigilance, fatigue, patience, etc.). In the case of the IAI framework, junior agents gather basic information about the operator, machine, environment, and the processes associated with various tasks. Then, working agents interpret this information and process it into knowledge of system components. Finally, senior agents build adaptation strategies into the interface and react to the changes through the managing agent.

From a management perspective, an agent is authorized to allocate appropriate tasks to the operator and other agents. Figure 5.11 illustrates how this function allocation follows adaptation philosophies that answer a four-step process for function allocation and adaptation based on the concept of W5+ (Duric et al., 2002; Maat and Pantic, 2007). W5+ has four steps:

1. *Knowledge acquisition*: Information is gathered from various junior and working agents through the sensing, tasking, and interaction agents about operator state and machine, task, and environmental status. The purpose is to inform the system and the operator of what is currently happening.
2. *Attention*: Involves various junior and working agents through the managing agent and senior-level modeling agents and the working-level inference agent. The managing and modeling agents enable operators to be aware of what should or will happen next by indicating what changes (i.e., adaptations) are needed, and where and when the changes will occur. They also provide the operator with feedback and alerts.

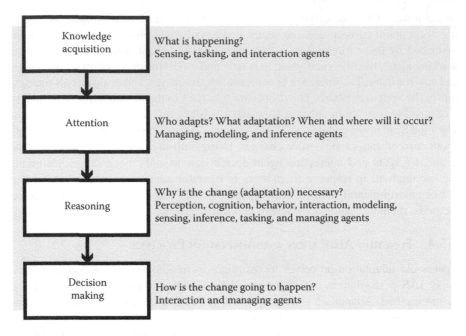

FIGURE 5.11 Agent adaptation process. Function allocation follows a four-step process that involves gathering information, alerting the operator, determining necessary changes, and applying decision-making criteria to determine the method of adaptation.

3. *Reasoning*: Involves various junior and working agents through the managing agent, the senior-level sensing agent, tasking agent, interaction agent, and modeling agent, and the working-level behavior agent, perception agent, cognition agent, and inference agent. Junior agents collect information about operator state and machine, task, and environmental status. Working agents classify, interpret, reason, and compare information. The inference agent compares current variable states with existing database models. It judges whether changes are necessary and suggests through the modeling agent what models should be used for adaptation. The generated knowledge is then communicated to the managing agent and senior agents to ensure that they understand why the changes are necessary.

4. *Decision making*: After completing the knowledge acquisition, attention, and reasoning phases, the managing agent, senior-level interaction agent, and tasking agent evaluate proposed methods of changes to the reasoning process by applying two decision-making criteria. The primary criterion looks at human performance consequences, such as mental workload, SA, complacency, and skill degradation. The secondary criterion looks at automation reliability and the costs of the identified consequences (Parasuraman et al., 2000). Once these criteria are applied, a method of adaptation is chosen and automation levels for various agents are identified. At the same time, proper feedback or alerts are presented to the operator, including the managing agent, senior agents, working agents, and junior

agents. An example was given to illustrate the communication and collaboration among various agents at all four levels with the aim of facilitating operator-agent interaction.

5.5.5 AGENT INTERACTION

The following examples illustrate the communication and collaboration exhibited by various agents in the framework. If an input is captured by an eye-tracking junior agent, its data are sent to the perception agent for processing along with inputs from speech recognition, arm movement, facial expression, and gesture change agents. The output of the perception agent consists of the knowledge of where on the interface an operator is currently looking. This output is fed to its senior interaction agent. The interaction agent combines this output with outputs from the behavior agent and the cognition agent, and processes these data in order to understand the operator's behavior, SA, workload, and so on.

In this example, the operator is found not to be monitoring the area of the interface that should be focused on. Based on this assessment, the interaction agent describes the situation and provides recommendations, such as sending a reminder message to the operator, the modeling agent, and the managing agent. The modeling agent combines these data with data from other agents and then transmits all data through the sensing agent and tasking agent to its working-level inference agent. The inference agent then compares all inputs with the current operator state and machine, task, and environmental status stored in the data and knowledge repository. After data are processed, the inference agent sends its outputs about the operator's current attention and other related information, such as losing a data link to an asset on the scene, through its modeling agent to the interaction agent and managing agent. According to its built-in protocols, the modeling agent provides a new and enhanced action plan (e.g., augmenting visual modality with auditory modality because an asset might have been lost and the operator's attention is needed immediately). The managing agent then makes decisions based on action recommendations provided by the interaction agent and modeling agent. As a result, the managing agent may task its own working and junior agents to change the output of the interface, such as combining flashing text in a message window and an auditory alert.

Figure 5.12 illustrates the structure of a working-level perception agent. It contains six arguments (i.e., inputs) that are the return values (i.e., outputs) of six junior agents responsible for collecting facial data, haptic states, arm movement data, speech recognition data, eye gaze-tracking data, and gesture changes. Each junior agent has its own inputs from its own sensors. After decoding, the junior agent compares new data with previous knowledge based on embedded algorithms and models. The return values are then encoded and delivered to the perception agent.

The perception agent gathers all outputs (i.e., knowledge) from the junior agents and compares it with results from its own embedded models. Its outputs consist of information about workload levels, SA, frustration, fear, excitement, and so on. The outputs become inputs and are fed to the interaction agent, which interprets operator state. The interaction agent then passes this information through the modeling agent to the inference agent. These three agents work together, comparing current

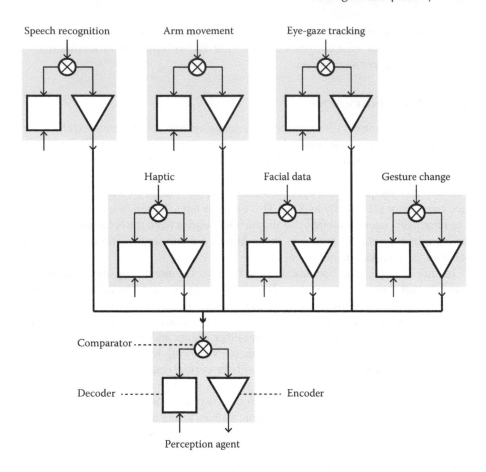

FIGURE 5.12 The working-level perception agent as a standard computer program. It is fed coded information from six junior agents, which it decodes and compares with its own embedded models. It then encodes the results and feeds them to the interaction agent.

operator state with previously stored data. Using these data, they infer future states for workload, SA, complacency, and skill degradation based on cognitive models. Through the managing agent, the system and the operator are informed of why and what events are happening and which events need to be prioritized according to predefined rules.

Similarly, the managing agent also has arguments (i.e., inputs) from the senior-level sensing agent, tasking agent, modeling agent, and interaction agent. These senior agents have knowledge of the interface; task processes; task environments; emergency situations; physical, emotional, and cognitive states of the operator; current automation levels of other agents; and so on. The returns (i.e., outputs) of the managing agent are adjustments to other actions, such as the most effective modality for operator–machine interaction, and the new models chosen by the modeling agent to adapt to the operator, machine, tasks, and environment. For the sensing agent, inputs include all data from internal and external sensors and data links. The outputs are the current and next machine status and task status. For the interaction agent, inputs are the states of an operator's behavior, perception, and cognition. Outputs are the possible next stages of the process and recommended actions that the operator and the interface will take. For the modeling agent, inputs include information from the senior sensing agent, tasking agent, and interaction agent, and its own working inference agent. Outputs are adaptation strategies and new models of the operator, machine, tasks, and environment to be used by other senior agents.

Agents require accurate and appropriate data to interact effectively with their human partner. The most effective way for a systems designer to ensure that an IAS has the knowledge it needs to work at optimal performance levels is by incorporating a variety of operator monitoring methods into the system. Chapter 6 presents an overview of operator monitoring approaches and technologies that can be implemented into an IAS to facilitate real-time operator state assessment.

5.6   SUMMARY

This chapter defined agents in the context of IAS design by examining corporate interaction as a representation of effective operator–agent interaction, which uses multiple intelligent agents in a defined hierarchy. Common agent attributes were discussed, including adaptation, autonomy, and cooperation, and agent intelligence was illustrated in the example of GPS navigation systems that can modify routes based on real-time traffic and road closure data.

Common interface design methods were reviewed, and benefits of agent-based methods for designing IASs were discussed. Agent-based design principles and guidelines were offered, and major agent-based implementation methods were outlined. Six popular design methods were examined – CommonKADS, IDEF, EMD, EID, BDI, and the blackboard model.

The benefits of an agent-based, interaction-centered approach to IAS design were highlighted, including the ability to facilitate operator–agent interaction more effectively, efficiently, and naturally to resemble human–human interaction behavior. To demonstrate the autonomous properties of current agent-based IAS technology, the Mercedes Active Parking Assist system was discussed.

A conceptual agent-based hierarchical framework was introduced for the design of an IAI subsystem that increases SA and maximizes IAS performance. The framework consists of four different function groups of hierarchically managed agents that communicate with each other and collaborate with the operator, including the managing agent, senior agents, working agents, and junior agents. An example was given to illustrate the communication and collaboration among various agents at all four levels with the aim of facilitating operator-agent interaction.

Ideally, operator–agent interaction is both descriptive and prescriptive. Well-designed agents allow operators to function at a higher level as controllers of automation teams that manage the details of the tasks required to achieve overall system goals. Human considerations, specifically approaches for monitoring operator state and performance, are discussed in Chapter 6.

REFERENCES

Allen, J. and Ferguson, G. (2003). Agent based architectures for dynamic crisis management (Report No. ARFL-IF-RS-TR-2003-146). Rome, NY: Defense Advanced Research Projects Agency.

Arenas, A. E. and Barrera-Sanabria, G., (2002). Applying the MAS-CommonKADS methodology to the flights reservation problem: Integrating coordination and expertise. In *Proceedings of the Fifth Joint Conference on Knowledge-Based Software Engineering*, pp. 3–12. September 11–13, Amsterdam, NL: IOS Press.

Arias, C. I. S. and Daltrini, B. M. (1996). A multi-agent environment for user interface design. In *Proceedings of the 22nd EUROMICRO Conference of EUROMICRO 96. Beyond 2000: Hardware and Software Design Strategies*, pp. 242–247. 2–5 September, Prague, CZ: IEEE.

Barber, K. S., Goel, A., Kim, J. N., Martin, C. E., Han, D. C., Lam, D. N., MacMahon, M., and McKay, R. (2002). Sensible agents: Augmenting and empowering decision-makers. In D. De Waard, K. Brookhuis, J. Moraal, and A. Toffetti (eds), *Human Factors in Transportation, Communication, Health, and the Workplace*, pp. 273–286. Maastricht, NL: Shaker.

Baylor, A. L., Rosenberg-Kima, R. B., and Plant, E. A., (2006). Interface agents as social models: The impact of appearance on females' attitude toward engineering. In *Proceedings of the International Conference on Human Factors in Computing Systems (CHI'06)*, pp. 526–531. April 22–27, New York, NY: ACM Press.

Beaudouin-Lafon, M., (2000). Instrumental interaction: An interaction model for designing post-wimp user interfaces. In *Proceedings of the SIGCHI Conference on Human Factors in Computing Systems (CHI '00)*, pp. 446–453. April 1–6, New York, NY: ACM Press.

Benyon, D. R. and Murray, D. M. (1993). Adaptive systems: From intelligent tutoring to autonomous agents. *Knowledge-Based Systems*, 6(4), 197–219.

Benyon, D. R. and Murray, D. M. (2000). Special issue on intelligent interface technology: Editor's introduction. *Interacting with Computers*, 12(4), 315–322.

Berger, J. (2007). KARMA: Knowledge-based adaptive resource management agent (Tech. Rep. 2003-278). Saint-Gabriel-de-Valcartier, Canada: Defence Research and Development.

Bradshaw, J. M. (ed.). (1997). *Software Agents*. Cambridge, MA: AAAI Press.

Bratman, M. E. (1987). *Intentions, Plans, and Practical Reason*. Cambridge, MA: Harvard University Press.

Bratman, M. E., Israel, D. J., and Pollack, M. E. (1988). Plans and resource-bounded practical reasoning. *Computational Intelligence*, 4(3), 349–355.

Brazier, F., Dunin-keplicz, B., Treur, J., and Verbrugge, R., (1997). Modelling internal dynamic behaviour of BDI agents. In J. J. Ch. Meyer and P. Y. Schobbens (eds), *Proceedings of the Third International Workshop on Formal Schools of Agents*, pp. 36–56. January 15–17, Berlin, Germany: Springer.

Briggs, G. and Scheutz, M., (2012). Multi-modal belief updates in multi-robot human-robot dialogue interactions. In R. Rzepka, M. Ptaszynski, and P. Dybala (eds), *Proceedings of the AISB/IACAP Symposium: Linguistic and Cognitive Approaches to Dialogue Agents*, pp. 67–72. July 2–6, Birmingham, UK: The Society for the Study of Artificial Intelligence and Simulation of Behaviour.

Burns, C. M. and Hajdukiewicz, J. R. (2004). *Ecological Interface Design*. Boca Raton, FL: CRC Press.

Burns, C. M., Skraaning, G., Lau, N., Jamieson, G., Kwok, J., Welch, R., and Andresen, G. (2008). Evaluation of ecological interface design for nuclear process control: Situation awareness effects. *Human Factors: The Journal of the Human Factors and Ergonomics Society*, 50(4), 663–679.

Busetta, P., Rönnquist, R., Hodgson, A., and Lucas, A. (1999). *JACK Intelligent Agents— Components for Intelligent Agents in Java.* Melbourne, Australia: Agent Oriented Software.

Chen, J. Y. C. and Barnes, M. J. (2014). Human-agent teaming for multi-robot control: A review of human factors issues. *IEEE Transactions on Human-Machine Systems, 44*(1), 13–29.

Cohen, P. R. and Levesque, H. J. (1990). Intention is choice with commitment. *Artificial Intelligence, 42*(2), 213–261.

Cooper, A. (2004). *The Inmates Are Running the Asylum: Why High Tech Products Drive Us Crazy and How to Restore the Sanity,* 2nd edn. Indianapolis, IN: SAMS Publishing.

Cooper, A., Reimann, R., and Cronin, D. (2012). *About Face 3: The Essentials of Interaction Design.* Hoboken, NJ: Wiley.

Cummings, M. L. and Bruni, S. (2010). Human-automation collaboration in complex multivariate resource allocation decision support systems. *International Journal of Intelligent Decision Technologies, 4*(2), 101–114.

Davies, T. C. and Burns, C. M. (2008). Advances in cognitive work analysis and the design of ecological visual and auditory displays. *Cognitive Technology, 13*(2), 17–23.

Dent, L., Boticario, J., McDermott, J., Mitchell, T., and Zabowski, D., (1992). A personal learning apprentice. In *AAAI-92: Proceedings of the 10th National Conference on Artificial Intelligence,* pp. 96–103. July 12–16, Palo Alto, CA: AAAI.

D'Inverno, M., Luck, M., Georgeff, M., Kinny, D., and Wooldridge, M. (2004). The dMARS architecture: A specification of the distributed multi-agent reasoning system. *Autonomous Agents and Multi-Agent Systems, 9*(1–2), 5–53.

Doane, S. M. and Sohn, Y. W. (2000). Adapt: A predictive cognitive model of user visual attention and action planning. *User Modeling and User-Adapted Interaction, 10*(1), 1–45.

Duric, Z., Gray, W. D., Heishman, R., Li, F., Rosenfeld, A., Schoelles, M. J., Shunn, C. and Wechsler, H. (2002). Integrating perceptual and cognitive modeling for adaptive and intelligent human-computer interaction. *Proceedings of the IEEE, 90*(7), 1272–1289.

Duvallet, C., Boukachour, H., and Cardon, A. (2000). Intelligent and self-adaptive interface. In *Proceedings of the 13th International Conference on Industrial and Engineering Applications of Artificial Intelligence and Expert Systems: Intelligent Problem Solving: Methodologies and Approaches,* pp. 711–716. 19–22 June, Berlin, Germany: Springer Berlin Heidelberg.

Edwards, J. L. (2004). A generic agent-based framework for the design and development of UAV/UCAV control systems (Report No. CR 2004-062). Toronto, Canada: Defence Research and Development.

Edwards, J. L. and Hendy, K. C. (2000). A testbed for intelligent aiding in adaptive interfaces. In *Spring Symposium on Adaptive User Interfaces* (SS-00-01). Stanford, CA: Stanford University.

Farrell, P. S., Hollands, J. G., Taylor, M. M., and Gamble, H. D. (1999). Perceptual control and layered protocols in interface design: I. Fundamental concepts. *International Journal of Human-Computer Studies, 50*(6), 489–520.

Furukawa, H. and Parasuraman, R., (2003). Supporting system-centered view of operators through ecological interface design: Two experiments on human-centered automation. In *Proceedings of the 47th Annual Meeting of the Human Factors and Ergonomics Society,* pp. 567–571. October 13–17, New York, NY: SAGE Publications.

Furukawa, H., Nakatani, H., and Inagaki, T. (2004). Intention-represented ecological interface design for supporting collaboration with automation: Situation awareness and control in inexperienced scenarios. In D. A. Vincenzi, M. Mouloua, and P. A. Hancock (eds), *Human Performance, Situation Awareness, and Automation: Current Research and Trends (HPSAA II),* vol. 2, no. 1, pp. 49–55. Hove, UK: Psychology Press.

Georgeff, M. P. and Lansky, A. L. (1986). Procedural knowledge. *Proceedings of the IEEE*, *74*(10), 1383–1398.

Georgeff, M. P., Pell, B., Pollack, M., Tambe, M., and Wooldridge, M., (1998). The belief-desire-intention model of agency. In *Proceedings of the Fifth International Workshop on Intelligent Agents V: Agents Theories, Architectures, and Languages*, pp. 1–10. July 4–7, Berlin, Germany: Springer.

Gómez-Sanz, J., Pavón, J., and Garijo, F., (2000). Intelligent interface agents behavior modeling. In *Proceedings of MICAI 2000: Advances in Artificial Intelligence*, pp. 598–609. April 11–14, Berlin, Germany: Springer.

Govan, F. (2010). Man dies after satnav sends him into a reservoir. *The Telegraph*. Retrieved October 4, 2010 from http://www.telegraph.co.uk/news/worldnews/europe/spain/8041361/Man-dies-after-satnav-sends-him-into-a-reservoir.html.

Grossklags, J. and Schmidt, C. (2006). Software agents and market (in) efficiency: A human trader experiment. *IEEE Transactions on Systems, Man, and Cybernetics—Part C: Applications and Reviews*, *36*(1), 56–67.

Hendy, K. C., Beevis, D., Lichasz, F., and Edwards, J. (2002). Analyzing the cognitive system from a perceptual control theory point of view. In M. D. McNeese and M. A. Vidulich (eds), *Cognitive Systems Engineering in Military Aviation Environments: Avoiding Cogminutia Fragmentosa!*, pp. 201–250. Dayton, OH: Wright-Patterson Air Force Base.

Henesey, L. (2006). Multi-agent systems for container terminal management (Doctoral dissertation, Blekinge Institute of Technology, Sweden). Retrieved July 28, 2014 from http://www.bth.se/fou/forskinfo.nsf/all/8db2e55374d0d79ec1257237005c4855.

Hou, M., Kobierski, R., and Brown, M. (2007). Intelligent adaptive interfaces for the control of multiple UAVs. *Journal of Cognitive Engineering and Decision Making*, *1*(3), 327–362.

Hou, M., Zhu, H., Zhou, M., and Arrabito, G. R. (2011). Optimizing operator-agent interaction in intelligent adaptive interface design: A conceptual framework. *IEEE Transactions on Systems, Man, and Cybernetics—Part C: Applications and Reviews*, *41*(2), 161–178.

Huber, M. J., (1999). JAM: A BDI-theoretic mobile agent architecture. In *Proceedings of the 3rd Annual Conference on Autonomous Agents*, pp. 236–243. May 1–5, New York, NY: ACM Press.

Hutchins, E. L., Hollan, J. D., and Norman, D. A. (1985). Direct manipulation interfaces. *Human-Computer Interaction*, *1*(4), 311–338.

Iglesias, C. A., Garijo, M., and Centeno-González, J., (1998). A survey of agent-oriented methodologies. In J. P. Müller, M. P. Singh, and A. S. Rao (eds), *Proceedings of the 5th International Workshop on Intelligent Agents V: Agent Theories, Architectures, and Languages (ATAL '98)*, pp. 313–330. July 4–7, Berlin, Germany: Springer.

Iglesias, C. A., Garijo, M., Gonzalez, J. C., and Velasco, J. R., (1997). Analysis and design of multi-agent systems using mas-commonkads. In M. P. Singh, A. S. Rao, and M. J. Wooldridge (eds), *Proceedings of the 4th International Workshop on Intelligent Agents IV: Agent Theories, Architectures, and Languages (ATAL '97)*, pp. 313–327. July 24–26, Berlin, Germany: Springer.

Jansen, B. J. (1999). A software agent for performance enhancement of an information retrieval engine (Doctoral dissertation), A & M University, Texas: UMI Dissertation Services.

Jarvis, B., Corbett, D., and Jain, L., (2005). Beyond trust: A BDI model for building confidence in an agent's intentions. In *Proceedings of the 9th International Conference on Knowledge-Based and Intelligent Information and Engineering Systems*, pp. 844–850. September 14–16, Berlin, Germany: Springer.

Kay, A. (1984). Computer software. *Scientific American*, *251*(3), 53–59.

Kempe, M. (1994). A framework for the blackboard model. In *TRI-Ada'94 Tutorial: Advanced Object-Oriented Programming with Ada 9X*, pp. 1–4. Baltimore, MD: TRI-Ada.

Kendall, E. A., Malkoun, M. T., and Jiang, C. H., (1995). A methodology for developing agent based systems for enterprise integration. In P. Bernus and L. Nemes (eds), *Modelling and Methodologies for Enterprise Integration: Proceedings of the IFIP TC5 Working Conference on Models and Methodologies for Enterprise Integration, Queensland, Australia,* pp. 333–344. November, London, UK: Chapman & Hall.

Lau, N., Skraaning, G., Jamieson, G., and Burns, C. M. (2008). Ecological interface design in the nuclear domain: An empirical evaluation of ecological displays for the secondary subsystems of a boiling water reactor plant simulator. *IEEE Transactions on Nuclear Science,* 55(6), 3597–3610.

Laurel, B. (1990). Interface agents: Metaphors with character. In B. Laurel (ed.), *The Art of Human-Computer Interface Design,* pp. 207–219. Boston, MA: Addison-Wesley.

Lieberman, H. (1998). Integrating user interface agents with conventional applications. *Knowledge-Based Systems,* 11(1), 15–23.

Lieberman, H. and Maulsby, D. (1996). Instructible agents: Software that just keeps getting better. *IBM Systems Journal,* 35(3–4), 539–556.

Lydiard, T. J. (1995). *Using IDEF3 to Capture the Air Campaign Planning Process.* Edinburgh, UK: University of Edinburgh Artificial Intelligence Applications Institute. Retrieved July 29, 2014 from: http://www.aiai.ed.ac.uk/~arpi/ACP-MODELS/ACP-OVERVIEW/96-MAR/HTML-DOC/idef3.html.

Maat, L. and Pantic, M. (2007). Gaze-x: Adaptive, affective, multimodal interface for single-user office scenarios. In T. S. Huang (ed.), *Artificial Intelligence for Human Computing, LNAI Vol. 4451 2007,* pp. 251–271. Berlin, Germany: Springer.

Maes, P. (1994). Agents that reduce work and information overload. *Communications of the ACM,* 37(7), 30–40.

Marshall, P. and Greenwood, S., (1997). A generic multi-agent adaptive interface for information management implemented in a medical domain. In *Proceedings of the 2nd International Conference on the Practical Application of Intelligent Agents and Multi-Agent Technology (PAAM 97),* pp. 319–332. April 21–23. Blackpool, United Kingdom: Practical Application Company Ltd.

Mayer, R. J., Menzel, C. P., Painter, M. K., deWitte, P. S., Blinn, T., and Perakath, B. (1995). Information integration for concurrent engineering (IICE) IDEF3 process description capture method report (Tech Rep. AL-TR-1995-XXXX). College Station, TX: Knowledge Based Systems.

Mercedes-benz.ca. (2014). Mercedes-Benz Canada—Comfort—Assistive technologies. Retrieved March 19, 2014 from http://www.mercedes-benz.ca/content/canada/mpc/mpc_canada_website/en/home_mpc/passengercars/home/new_cars/models/m-class/w166/facts/comfort/assistancesystems.html.

Miller, C., Funk, H., Wu, P., Goldman, R., Meisner, J., and Chapman, M., (2005). The playbook approach to adaptive automation. In *Proceedings of the 49th Annual Meeting of the Human Factors and Ergonomics Society,* pp. 15–19. September 26–30, New York, NY: SAGE Publications.

Miller, C., Goldman, R., Funk, H., Wu, P., and Pate, B., (2004). A playbook approach to variable autonomy control: Application for control of multiple, heterogeneous unmanned air vehicles. In *Proceedings of FORUM 60, the Annual Meeting of the American Helicopter Society,* pp. 7–10. June 7–10, Alexandria, VA: American Helicopter Society.

Morita, P. P., Burns, C. M., and Calil, S. J., (2009). The influence of strong recommendations, good incident reports and a monitoring system over an incident investigation system for healthcare facilities. In *Proceedings of the 53rd Annual Meeting of the Human Factors and Ergonomics Society,* pp. 1679–1683. October 19–23, New York, NY: SAGE Publications.

Negroponte, N. (1969). Towards a theory of architecture machines. *Journal of Architectural Education,* 23(2), 9–12.

Negroponte, N. (1970). *The Architecture Machine: Towards a More Human Environment.* Cambridge, MA: MIT Press.

Nii, H. P. (1986). The blackboard model of problem solving and the evolution of blackboard architectures. *AI Magazine, 7*(2), 38–53.

Padgham, L. and Winikoff, M. (2004). *Developing Intelligent Agent Systems: A Practical Guide.* West Sussex, UK: Wiley.

Parasuraman, R., Sheridan, T. B., and Wickens, C. D. (2000). A model for types and levels of human interaction with automation. *IEEE Transactions on Systems, Man, and Cybernetics—Part A: Systems and Humans, 30*(3), 286–297.

Pavón, J. and Gómez-Sanz, J. (2003). Agent oriented software engineering with INGENIAS. In *Multi-Agent Systems and Applications III*, pp. 394–403. Berlin, Germany: Springer.

Powers, W. T. (1973). *Behavior: The Control of Perception.* New York, NY: Hawthorne.

Preece, J., Rogers, Y., and Sharp, H. (2002). *Interaction Design: Beyond Human-Computer Interaction*, 1st edn. New York: Wiley.

Rao, A. and Georgeff, M., (1995). BDI agents: From theory to practice. In *Proceedings of the International Conference on Multi-Agent Systems (ICMAS'95)*, pp. 312–319. June 12–14, Palo Alto, CA: AAAI.

Riecken, D. (1994). Intelligent agents. *Communications of the ACM, 37*(7), 18–21.

Russell, S. and Norvig, J. P. (2009). *Artificial Intelligence: A Modern Approach*, 3rd edn. Englewood Cliffs, NJ: Prentice Hall.

Schreiber, G., Akkermans, H., Anjewierden, A., de Hoog, R., Shadbolt, N., Van de Velde, W., and Wielinga, B. (1999). *Knowledge Engineering and Management: The CommonKADS Methodology.* Cambridge, MA: MIT Press.

Seppelt, B. D. and Lee, J. D. (2007). Making adaptive cruise control (ACC) limits visible. *International Journal of Human-Computer Studies, 65*(3), 192–205.

Sheridan, T. B. and Parasuraman, R. (2006). Human-automation interaction. In R. S. Nickerson (ed.), *Reviews of Human Factors and Ergonomics*, vol. 1, pp. 89–129. Santa Monica, CA: HFES.

Shneiderman, B. and Maes, P. (1997). Direct manipulations vs. interface agents. *Interactions, 4*(6), 42–61.

Shoham, Y. (1993). Agent-oriented programming. *Artificial Intelligence, 60*(1), 51–92.

Sudeikat, J., Braubach, L., Pokahr, A., Lamersdorf, W., and Renz, W. (2007). Validation of BDI agents. In *Proceedings of the Fourth International Workshop on Programming Multi-Agent Systems (ProMAS'06)*, pp. 185–200. Berlin, Germany: Springer.

Taylor, M. M., Farrell, P. S., and Hollands, J. G. (1999). Perceptual control and layered protocols in interface design: II. The general protocol grammar. *International Journal of Human-Computer Studies, 50*(6), 521–555.

van Doesburg, W. A., Heuvelink, A., and van den Broek, E. L., (2005). TACOP: A cognitive agent for a naval training simulation environment. In *Proceedings of the Fourth International Joint Conference on Autonomous Agents and Multiagent Systems (AAMAS '05)*, pp. 34–41. July 25–29, New York, NY: ACM Press.

Vicente, K. J. (2002). Ecological interface design: Progress and challenges. *Human Factors: The Journal of the Human Factors and Ergonomics Society, 44*(1), 62–78.

Vicente, K. J. and Rasmussen, J. (1992). Ecological interface design: Theoretical foundations. *IEEE Transactions on Systems, Man and Cybernetics, 22*(4), 589–606.

Wooldridge, M. (2002). *An Introduction to Multiagent Systems.* West Sussex, UK: Wiley.

Wooldridge, M. and Jennings, N. R. (1995). Intelligent agents: Theory and practice. *Knowledge Engineering Review, 10*(2), 115–152.

Wooldridge, M., Jennings, N. R., and Kinny, D. (2000). The Gaia methodology for agent-oriented analysis and design. *Journal of Autonomous Agents and Multi-Agent Systems, 3*(3), 285–312.

Yoon, W. and Hammer, J. (1988). Deep-reasoning fault diagnosis: An aid and a model. *IEEE Transactions on Systems, Man, and Cybernetics*, *18*(4), 659–675.

Zhu, H. and Zhou, M. (2006). Role-based collaboration and its kernel mechanisms. *IEEE Transactions on Systems, Man and Cybernetics, Part C: Applications and Reviews*, *36*(4), 578–589.

Zhu, H. and Zhou, M. (2008). Roles in information systems: A survey. *IEEE Transactions on Systems, Man and Cybernetics, Part C: Applications and Reviews*, *38*(3), 377–396.

6 Operator State Monitoring Approaches*

6.1 OBJECTIVES

- Provide an overview of operator state monitoring approaches and technologies that can be implemented into IASs
- Discuss behavioral-based monitoring approaches
- Discuss psychophysiological-based monitoring approaches
- Discuss contextual-based monitoring approaches
- Discuss subjective-based monitoring approaches
- Examine the Cognitive Cockpit cognition monitor module as an example of combination-based monitoring approach

Operator state describes the general condition of a human operator interacting with a system. The concept includes behavioral activity, physiological patterns, and psychological states, and is strongly context dependent (Pleydell-Pearce et al., 1999). Well-designed IASs should monitor operator state and enable flexible task allocation between the operator and the machine to reduce operator workload and fatigue. Well-designed systems should automatically assign tasks to the machine during periods of high stress (i.e., overload), and should reengage operators during periods of underload. IASs can either return tasks to active operator control, keeping the operator in-the-loop, or provide stimuli to refocus supervisory operator attention, keeping the operator on-the-loop (Chen and Barnes, 2014). An ideal IAS design allocates tasks collaboratively while emphasizing operator needs.

Technologies that monitor real-time psychophysiological changes provide an unobtrusive method for assessing operator internal state. Without interrupting current tasks, operator state monitoring technologies enable IASs to invoke dynamic task allocation based on operator needs and adapt accordingly. These technologies have great potential to facilitate intelligent adaptation. For example, an electroencephalogram (EEG) can be used to monitor cognitive workload by measuring electrical activity through scalp-mounted sensors (Berka et al., 2004). Electrocardiograms (ECGs), skin conductivity sensors, and respiration rate sensors have all previously been employed to determine mood (McCraty et al., 1995; Schnell et al., 2007; Figner and Murphy, 2010), while eye trackers and blink monitors have been used to ascertain fatigue levels (Barr et al., 2009; Highway Safety Group, n.d.; Fraunhofer-Gesellschaft, 2010).

Although there is correlation between operator state constructs (e.g., cognitive workload, arousal, engagement) and psychophysiological measurements, there are multiple challenges to using these approaches. Psychophysiological relationships are heavily task dependent. For example, tasks relying on physical activity can cause inaccuracy in cardiovascular measurements. There is also considerable variation in how humans react to systems under different conditions, and variation increases when operator-specific attributes, such as spatial and attentional abilities, are considered (Chen and Terrence, 2009).

IAS adaptation to operator variance can reduce overall operator error and make the system easier for humans to learn and use. To facilitate this symbiotic interaction, IAS designers must consider the merits of IAS conceptual architectures (as discussed in Chapter 3) and operator state monitoring technologies that facilitate intelligent adaptation. IASs can intelligently adapt to unique situations and individual operators by obtaining behavioral, psychophysiological, contextual, and subjective data about the operator in real time. Using operator state monitoring approaches can create more robust operator–machine interaction, while increasing efficiency.

This chapter provides an overview of current operator state monitoring approaches and technologies that can potentially be implemented into IASs. Monitoring approaches discussed include behavioral-based monitoring, psychophysiological-based monitoring, contexual-based monitoring, subjective-based monitoring, and combination-based monitoring. Rule-based data fusion, a combination-based approach for drawing conclusions using strict if-then type approaches, and the use of artificial neural networks to support decision making by automating inference generation are also discussed. The cognition monitor (COGMON) module of the Cognitive Cockpit project is also examined as a real-world example of combination-based operator state monitoring technology. As designed systems transition from passive to active, and perhaps even become predictive machines that are able to adapt to new operators, environments, and scenarios, the use of operator feedback data to facilitate intelligent adaptation is increasingly required to provide a higher degree of error resistance.

6.2   USING OPERATOR STATE MONITORING APPROACHES

"Fatigued," "stressed," "contented," "fearful," and "distracted" are all words that describe a human's overall state, or mode, of being. These modes have a significant impact on human ability to efficiently complete tasks when interacting with HMSs. For example, fatigued operators are more likely to perform at lower performance levels than fully alert operators. Additionally, overworked and stressed operators are more prone to errors, and operators who are content are more likely to exhibit higher productivity than those who are fearful or distracted.

Operator state changes dynamically over time depending on individual circumstance, and includes physical, emotional, and cognitive states that can be inferred through a variety of means. Approaches used to elicit data and draw conclusions about operator state can normally be classified into one of four categories:

* *Behavioral-based monitoring:* Monitoring and making inferences from what the operator is doing

- *Psychophysiological-based monitoring*: Monitoring and making inferences from the operator's state of body and mind
- *Contexual-based monitoring*: Monitoring and making inferences from the operator's surroundings or working environment
- *Subjective-based monitoring*: Monitoring and making inferences from what the operator communicates about his or her own state

In general, no single measurement is likely to provide a full or accurate picture of an operator's true state. However, a solid snapshot of operator state can be obtained by using a combination of several indicators. For example, lie detection is typically assessed using a constellation of physiological data, such as an abnormally fast heart rate, excessive sweating, and elevated body temperature. However, these same physiological responses can also be brought about by operator context, such as a high-stress interrogation environment, an elevated ambient temperature, or an underlying medical condition.

Thus, to avoid misinterpreting operator state assessments, researchers will often use several measurements and multiple data elicitation methods simultaneously. Figure 6.1 illustrates combination-based monitoring, which is the fusion of data from multiple sources across four categories of data acquisition. Combination-based monitoring provides the basis for IAS adaptation that best supports the operator.

Recent research has set out to demonstrate the viability of systems that monitor operator workload and adjust assigned tasks based on real-time estimates of operator workload (Prinzel et al., 1999; Young et al., 2004; Berka et al., 2004; Wilson and Russell, 2007; Durkee et al., 2013; Pfeffer et al., 2013). Figure 6.2 illustrates a

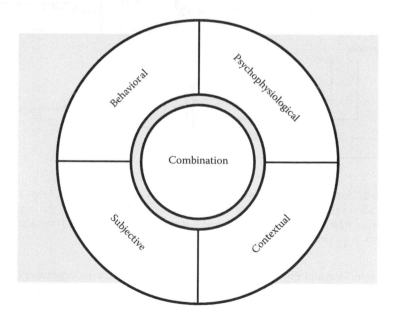

FIGURE 6.1 Visual relationship of the four primary types of operator feedback: behavioral, psychophysiological, contextual, and subjective. Combination-based monitoring draws on multiple subtypes.

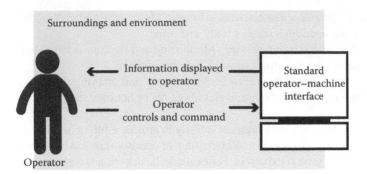

FIGURE 6.2   Standard interface without an operator feedback loop.

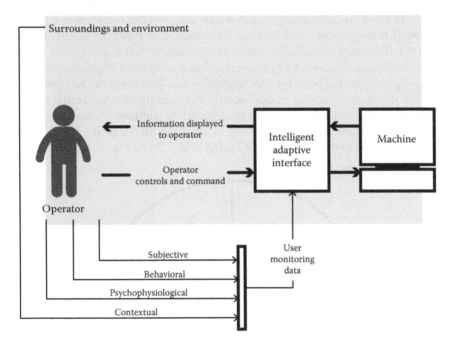

FIGURE 6.3   Intelligent adaptive system (IAS) with an operator feedback loop. The feedback loop allows the interface to adapt the information displayed and how the controls and commands are interpreted, as well as to adjust system-level automation and interaction based on information gathered about the operator and the environment.

standard interface and Figure 6.3 illustrates an interface that incorporates operator state monitoring in a feedback loop and has the ability to adapt to how information is displayed, how controls and commands are interpreted, and the level of system automation and interaction. When coupled with an interface capable of engaging adaptation, operator state feedback has the potential to increase system robustness, safety, and efficiency.

6.3   BEHAVIORAL-BASED MONITORING

Behavioral-based monitoring refers to inference of operator state by observation of operator actions in response to tasks, the working environment, or other stimuli (Wood, 2006). Operator actions can be conscious, subconscious, voluntary, or involuntary; all actions have the potential to provide meaningful data that can assist in accurately determining operator state.

Operators, as humans, rely on facial expression and posture to determine the state of others in a social context. The ability to recognize subtle changes in body language is difficult to replicate in a computer algorithm, and is, therefore, not well suited for integration into IASs. Aspects of operator behavior that are easily monitored by IASs exhibit a smaller degree of variability and allow measurements to be made under controlled parameters.

Eye tracking or gaze tracking using cameras and image analysis software is useful to gain insight into the motivation behind operator behavior. Eye-related measurements are often tied to behavior, but they actually follow psychophysiological approaches, and are discussed later in this chapter. How the operator manipulates interface controls can also provide important information on operator state, and variations can be monitored by a computer in a straightforward fashion. For example, if a standard point-and-click interface is being used, software (e.g., the perception agent discussed in Chapter 5) that monitors click rate and duty time can serve as a basis for creating hypotheses about operator attention level.

Voice recognition and auditory analysis also has significant potential as a tool for obtaining operator state information. Table 6.1 shows a summary of relevant behavioral-based monitoring approaches.

TABLE 6.1
Summary of Relevant Behavioral-Based Operator State Monitoring Approaches

Behavioral Feedback Approach	Summary	Examples of Current Uses
Eye tracking	• Monitors the operator's visual attention and cognitive activity	• Fatigue monitoring systems in vehicle operation • Advertisement design and evaluation
Operator–control interaction monitoring	• Monitors how the operator is interacting with the controls available • Key measurements include reaction time, haptic pressure, and input frequency	• E-commerce sites that monitor purchasing habits to present items that they may be interested in
Voice recognition and auditory analysis	• Monitors auditory cues that might provide information on operator state • Key measurements include frequency (pitch), tone, and timbre	• Voice recognition software

6.3.1 Operator–Control Interaction Monitoring

How an operator interacts with interface controls can provide clues to help determine operator state. Slower reaction times or a decrease in control use may indicate the onset of fatigue or boredom. On the other hand, excessive control and overresponsive reactions may indicate fear, anxiety, or nervousness. In either case, data can easily be incorporated into sensing mechanisms (e.g., the behavior agent discussed in Chapter 5) to contribute to the overall inference about operator state. Reaction time, haptic pressure, and input frequency are examples of parameters that can be directly measured.

Operator–control interaction measurements have been noted for their potential application in military contexts. For example, Wood (2006) proposed an intelligent control framework for the behavioral-based customization of warfighter interfaces. In Wood's framework, operator actions, system status information, and external events are used as inputs to the system, which then creates a model of the operator's context as the machine understands it, and a data table that outlines what subsystems can be adjusted, and the automation levels associated with current tasks. From their knowledge of the selected automation levels, the subsystems are then able to determine what tasks they should perform and to what extent they should perform them. Information gathered about automation levels is also linked back to the operator model. Although the intelligent control framework was designed specifically for military applications, it can be applied to any adaptive system.

How the operator exerts control over the system provides valuable insight into operator state and is a key point for researchers that should be studied further. For example, an intelligent adaptive interface (IAI) could monitor the pressure an operator exerts on various keys on a keyboard, the number of words or characters typed per minute, and the number of spelling or grammatical errors that result. A light keyboard touch with a decreased word per minute count and a higher degree of errors made could reasonably be interpreted as fatigue, boredom, or frustration. Well-designed HMSs can mitigate negative emotions by including a component that helps operators manage their emotional state while interacting with the system (Klein et al., 2002).

6.3.2 Voice Recognition and Auditory Analysis

Verbal communication, such as speaking through software to interact with the system and directly communicating with other operators and remote workstations, is a major source of workload for operators (Owens et al., 1984). Tone, pitch, and other characteristics of verbal speech can provide insight into operator state (Murray and Arnott, 1993), and a significant correlation has been noted regarding perceived situational stress in operator voice, reduced performance, and increased stress measured by other approaches (Schiflett and Loikith, 1980). Auditory and vocal tasks exhibit fundamentally different performance metrics from visual and manual tasks (Angell et al., 2006). Accordingly, algorithms have been developed that determine operator emotion by analyzing involuntary changes in speech patterns and cues (Hirschberg, 2010; Gravano and Hirschberg, 2011; Gravano et al., 2012). Additionally, latency of verbal responses to stimuli can indicate workload levels. For example, slow responses to communication may indicate high levels of cognitive resources being allocated to other tasks.

There are two main techniques for capturing vocalization data. The first technique uses electrodes attached to the skin around the larynx. The electrodes monitor the degree of vocal fold contact, determining through electrical means the frequency of speech generated in the throat. The second, less invasive, technique involves using microphones to capture speech, and then using analysis software to determine characteristics such as pitch, which is determined by the fundamental frequency of the voice, and the loudness and timbre or tone quality of the utterance.

Speech recognition, also known as automatic speech recognition or computer speech recognition, converts spoken words to text. Speech recognition systems normally recognize generic human speech (of a particular language), where voice recognition systems are trained to specific operators based on their unique vocal sounds. The term "voice recognition" is sometimes used to refer to speech recognition where the recognition system is trained to a particular speaker, seen in language training software such as Rosetta Stone® (2014). In these cases, there is an element of speaker recognition, which attempts to identify the person speaking in order to better recognize what is being said.

Vocalization systems also enable the system to detect and correct the pronunciation and verbal fluency of the operator, and can potentially improve operator interpersonal and communication skills by providing audible feedback. Verbal and auditory skill augmentation would be helpful as part of the intelligent tutoring system discussed in Chapter 8, as improvised explosive device disposal questioning often involves interviewing witnesses with diverse linguistic backgrounds.

6.4   PSYCHOPHYSIOLOGICAL-BASED MONITORING

Behavioral measurements are useful when determining operator state, but should not be the only approach relied on to provide a complete assessment (Parasuraman, 2003). Emotional and behavioral states are easier to identify than operator state related to task performance. Wickens (2002, 2008) developed multiple resource theory as a way to define cognitive state, or specifically mental workload, both qualitatively and quantitatively. Multiple resource theory added the concept of multitasking to existing information processing theories, which allows systems designers to better predict an operator's mental workload when interacting with a particular HMS.

Psychophysiology examines interactions between the mind and the body by recording how the body is currently functioning and relating the data to previously recorded behavior. The field is based on the premise that changes in the human body are related to changes in behavior, affect, and motivational state (Blanchard et al., 2007). Psychophysiological measurements relevant to IAS design include (a) EEG, (b) near-infrared spectroscopy (NIRS), (c) electrodermal response (EDR), (d) ECG, (e) heart rate variability (HRV), (f) eye tracking, (g) respiration measurements, (h) skin temperature measurements, and (i) electromyography (EMG). These measurements can be used alone, or in combination with one another.

Neuroergonomics and augmented cognition (AugCog) are two fields that rely on psychophysiological measurements. Operator state monitoring approaches and techniques from these fields can facilitate IAS design. Neuroergonomics is defined by Parasuraman (2012) as "the study of the human brain in relation to performance at work and in everyday settings"; it dictates that the relationship between the human brain and the operator's environment must be measured in order to fully understand the relationships among cognition, operator action, and the emergence of artifacts (Parasuraman, 2003). Neural circuitry changes both structurally and functionally when operators interact with their environment, and brain activity-monitoring technologies such as EEG and NIRS are able to detect these changes (Fidopiastis and Nicholson, 2010). AugCog uses the same type of modern neuroscientific tools and methodologies as neuroergonomics to determine real-time cognitive state in order to adapt information, technology, and the environment to meet the operator's needs (Schmorrow and Stanney, 2008). Unlike neuroergonomic systems, AugCog systems use a multimodal array of sensors to predict more than one cognitive state. The AugCog adaptations are more varied and connected not only to the on and off status of the system, but also to system content, interface, and tasking (Scerbo, 2007). Thus, AugCog technologies are more suitable for intelligent tutoring systems, which are further discussed in Chapter 8, and issues of operator cognitive state in high-workload environments such as the control of UAVs. The goal of AugCog research is to improve the information management capabilities of IASs using psychophysiological sensors to detect suboptimal operator performance, diagnose operator state issues, and modify human–machine interaction in real time to mitigate performance gaps (Hou and Fidopiastis, 2014).

Central to the use of psychophysiological measurements in IAS design is the assumption that there is an optimal cognitive and physical operator state for a given task. Though workload, arousal, and engagement can all be successfully measured by various psychophysiological-based monitoring technologies and applied to IAS design, mental workload is the key performance variable. Currently, a clearly defined and universally accepted definition of mental workload does not exist (Cain, 2007). Generally speaking, mental workload refers to the difference between the processing resources available to the operator and the resource demands required by the task (Sanders and McCormick, 1993) and can be defined as the amount of processing capacity expended during task performance (Eggemeier, 1988).

Mental workload cannot be directly observed; it must be inferred either from the observation of overt behavior or from measuring psychological and physiological indicators produced by the human body (Cain, 2007). Psychophysiological measurements are useful metrics of mental workload, as they provide immediate feedback about the operator's mental processing of the task at hand (Gale and Christie, 1987; Parasuraman, 1990; Kramer, 1991; Byrne and Parasuraman, 1996).

The human nervous system is divided hierarchically into the central nervous system, which consists of the brain and spinal cord, and the peripheral nervous system, which consists of the somatic nervous system (i.e., voluntary muscle movement and detection of the external environment through the five senses) and the autonomic nervous system (i.e., heart rate, respiration rate, eye movement, etc.). Psychophysiological measurements have unique properties that make them ideal operator state monitoring approaches for implementation into IASs; the psychophysiological measurements most relevant for IAS design are noted as either central nervous system or "brain state" measurements, such as EEG and NIRS, or peripheral nervous system measurements, such as cardiovascular measurements, eye tracking, and respiration. Table 6.2 shows relevant psychophysiological monitoring technologies.

TABLE 6.2
Summary of Relevant Psychophysiological Monitoring Technologies

	Psychophysiological Monitoring Technology	Summary	Examples of Current Uses
Central Nervous System	Electroencephalogram (EEG)	• Electrode sensors placed on the scalp monitor electrical activity in the brain	• Clinically used to diagnose and research seizures
	Near-infrared spectroscopy (NIRS)	• Near-infrared light monitors blood oxygenation levels in the brain	• Clinically used to determine brain areas that are associated with information processing
Peripheral Nervous System	Electrodermal response (EDR)	• Electrical conductivity of the skin is measured to determine sweat levels	• Polygraphs
	Cardiovascular (electrocardiogram (ECG), heart rate variation (HRV), heart rate)	• Electrode sensors monitor heart activity	• Clinical or field assessment of stress
	Eye tracking	• Monitors the operator's visual attention and cognitive activity	• Fatigue monitoring systems in vehicle operation • Advertisement design and evaluation
	Respiration measurements	• Monitors breathing rate	• Polygraphs
	Skin temperature measurements	• Thermocouples monitor skin temperature for variations	• Polygraphs
	Electromyogram (EMG)	• Electrodes placed in or on specific muscles determine muscle activity	• Physiotherapy diagnostic tool

6.4.1 ELECTROENCEPHALOGRAM

An EEG records electrical activity produced by the synchronous firing of many neurons within different areas of the brain (Niedermeyer and Lopes da Silva, 1999). Typical EEG systems consist of a network of electrodes (i.e., sensors) placed in contact with the scalp at strategic locations. The number of electrodes used to detect different brain state changes depends upon the context of use. As few as 12 electrodes are common in clinical settings, while 256 electrodes are sometimes used in research settings. Each sensor is capable of detecting small voltage changes within the brain, which can be recorded, processed, and analyzed for patterns. The system is unable to measure the activity of individual neurons, which are too small, and instead measures the activity of individual groups of neurons. These measurements are recorded as graphs of brainwaves on a time scale, which can be correlated to a specific stimulus (e.g., a specific sensory, cognitive, or motor event) to determine event-related potential (ERP) and to get a clear understanding of exactly what electrical activity took place when the stimulus was presented (Karamouzis, 2006).

A key advantage of ERP is its ability to noninvasively provide information about brain activity with a high temporal resolution (Gumenyuk et al., 2004). There is, however, a risk that ERP measurements may become contaminated by electrical activity generated by eye, neck, and body movements (Kramer et al., 1996). Note that other psychophysiological measurements can also suffer from contamination from external (e.g., electrical interference) and internal (e.g., neuronal cross talk) sources.

Certain EEG patterns have a close association with behavior and operator state (e.g., Pope et al., 1995; Dussault et al., 2005; Berka et al., 2007). For example, workload level, high-order cognition, verbal processing, and image processing all have unique patterns that can be detected by EEG brain scans. Thus, inferences about overall operator state can be made by monitoring EEG signals and noting which regions of the brain exhibit the greatest activity and what frequencies take a dominant role.

As a method for measuring brain activity during real-world task performance, EEG has a central role in neuroergonomics for adaptive automation (Pope et al., 1995; Parasuraman and Hancock, 2004; Parasuraman, 2012) and AugCog for medical and military contexts (Schmorrow and Kruse, 2004; Kruse, 2007; Wilson and Russell, 2007; Vartak et al., 2008; Sciarini et al., 2009; Fidopiastis et al., 2009; Morishima et al., 2010). Because of its high temporal resolution (i.e., ms), EEG measurements are often used to evaluate temporal changes due to brain activities across different brain regions (Nunez, 2000).

EEGs have been used for many years, mainly in the medical and research communities (St. John et al., 2003; Prinzel et al., 2003; Russell, 2005; Schnell et al., 2007; Trejo et al., 2007; Tremoulet et al., 2007; Mathan et al., 2007; Stanney et al., 2011). Medical applications include diagnosing epilepsy, testing for brain death, and monitoring the depth of anesthesia. EEGs have also played a role in the design of brain–computer interfaces. Studies have been conducted to determine methods for controlling external actuators and devices by simply "thinking" in a certain way. For example, Prinzel et al. (1999) examined the effectiveness of a closed-loop system designed to monitor operator engagement level and adapt automation levels based on operator EEG measurements. Study results showed that the experimental group's performance was significantly better than the control group's performance. Additionally, real-time analysis of EEG has been used to modify task characteristics to better match operator functional state (Wilson and Russell, 2006; Schnell et al., 2007; Lei and Roetting, 2011).

Different strategies and styles are associated with distinct patterns of EEG activity. This sometimes creates challenges for interpreting EEG measurements accurately, as individual brains can differ in organization. EEG also produces a large number of data points, which must then be meticulously interpreted by researchers. Given that McCraty et al. (1995) successfully linked EEG to heart rate measurements such as HRV, it may be appropriate to use less complex approaches when implementing psychophysiological-based monitoring technologies into IASs. Table 6.3 shows the benefits and limitations of EEG measurements.

6.4.2 NEAR-INFRARED SPECTROSCOPY

NIRS measurements, like EEG, are used to monitor central nervous system activity during real-world task performance. NIRS is a noninvasive brain imaging technique that monitors signals thought to be markers for brain activation; this is done by using light in the near-infrared range to penetrate the skull and measure the ratio of oxygenated hemoglobin to deoxygenated hemoglobin in the brain's active areas (Villringer and Chance, 1997).

Understanding the coordinating efforts of brain regions both spatially and temporally can be advantageous for IASs responsible for monitoring operator workload. To date, most research has been clinical in nature. For example, Izzegtoglu et al. (2007) developed a functional NIRS system that monitors hemodynamic changes in the

---

TABLE 6.3
Benefits and Limitations of Electroencephalogram (EEG) Measurements

Benefits	Limitations
• EEG measurements are noninvasive • Portable and field-ready EEG measuring technologies are currently available • EEG measurements have a high temporal resolution • Workload level, high-order cognition, verbal processing, and image processing all have unique patterns that can be detected by EEG brain scans	• EEG data is represented by complex waveforms with a vast number of data points that require sophisticated signal-processing equipment (Cain, 2007) • EEG approaches are prone to artifacts such as far-field potentials • Individual brains may differ in their organization, providing distinct patterns of EEG activity • Event-related potentials (ERPs) have low signal-to-noise ratios (Gaillard and Kramer, 2000) • ERP measurements may become contaminated by other electrical activity generated by the heart, as well as eye, neck, and body movements

frontal cortex, which are thought to mediate the brain processes related to memory and problem solving. In the military domain, Keebler et al. (2009) used a functional NIRS system to differentiate the decisions made by novice and experienced operators learning to identify military vehicles. The high temporal resolution (i.e., ms) of NIRS technology affords a unique opportunity to evaluate the temporal changes due to brain activities associated with real-world task performance.

NIRS measurements are optical, which is beneficial as the measurements are less affected by electrical interference in real-world environments. Notably, the changes captured by NIRS are similar to the blood oxygenation level-dependent signal measured during functional magnetic resonance imaging; however, NIRS does not require a superconducting magnet or radioactive contrast. The key challenge of NIRS is that measuring blood oxygenation levels using light in the near-infrared range only allows measurement of cognitive state changes reflected in the gray matter of the cortex (Villringer and Chance, 1997). As with any technology transitioning to the applied domain, more work needs to be done in understanding signals, as well as developing a tool set to process signals. Table 6.4 shows the benefits and limitations of NIRS measurements.

6.4.3 ELECTRODERMAL RESPONSE

EDR, also known as galvanic skin response (GSR), measures the skin's electrical resistance, or impedance. The skin's electrical properties are partially governed by the body's autonomic responses. Electrodermal activity is an indicator of mental effort, arousal and vigilance levels, and may also respond to affective stimuli (e.g., Sharpe et al., 1995; Fernandez, 1997; Chen and Vertegaal, 2004; Kapoor et al., 2007).

When humans are mentally, emotionally, or physically aroused, a response is triggered in the skin. During excitation, glands in the skin fill with sweat, which is salty and acts as a weak conductor. This autonomic response is similar to the elevated heart rate commonly experienced in stressful situations. While electrodermal changes do not provide a direct measurement of cognition, they can act as a rough indicator of stress and workload levels. In particular, research has shown that large fluctuations in human autonomic response are linked to states of distraction and an increased chance of reduced task performance. Notably, qualitative and emotional

---

TABLE 6.4
Benefits and Limitations of Near-Infrared Spectroscopy (NIRS)

Benefits	Limitations
• Can measure changes using only light (i.e., hemoglobin level) that previously required expensive apparatus such as a superconducting magnet and radioactive contrast • Not affected by electrical interference • High temporal resolution in ms	• NIRS is an emergent technology and analysis software is not mature • Current studies focus on only a few brain areas, limiting applied results (Villringer and Chance, 1997)

aspects of affect, such as a "positive" versus "negative" reaction, or the expression of fear versus anger, joy, disgust, and so on, are not reflected in EDR and must be inferred from other sources, such as EMG (Figner and Murphy, 2010).

EDR measurement consists of an electrode couple being placed on the surface of the skin and noting the resistance to an electrical current passed between its two points. Skin and muscle tissue response to external and internal stimuli varies resistance. Higher arousal, which often occurs with increased operator involvement, will cause a fall in skin resistance within 0.2–0.5 s; reduced arousal, which often coincides with operator withdrawal, will cause a rise in skin resistance (McCleary, 1950).

Traditional techniques for measuring EDR include strapping electrodes to the index and middle finger, while more recent research, such as that of HandWave (Strauss et al., 2005), has explored untethered and wireless approaches. Also, the GSR Temp 2X (Thought Technology Ltd, 2010) offers a model in which operators have only to lightly place their fingers on top of the device, rather than being strapped to it. In every case, the device is attached to a computer. Electrodermal monitoring is most often applied in academic and research settings (St. John et al., 2003; Schnell et al., 2007), but the most commonly known use of EDR is for polygraph devices (i.e., lie detectors).

It is important to interpret electrodermal output cautiously, as there is a relationship between sympathetic activity and emotional arousal, even if the specific emotion being elicited cannot be identified. Additionally, the signal is relatively slow and there is a notable latency between stimulus and response when compared with other psychophysiological measurements (Figner and Murphy, 2010). Systems designers seeking to make operator workload inferences should use EDR with other psychophysiological measurements as part of a combination-based monitoring approach. Table 6.5 shows the benefits and limitations of EDR.

TABLE 6.5

Benefits and Limitations of Electrodermal Response (EDR)

Benefits	Limitations
• EDR can indicate states of emotional arousal • Less sensitive to environmental noise as compared with other psychophysiological measurements	• Poor temporal resolution as compared with other psychophysiological measurements • Slow signal as compared with other psychophysiological measurements • Notable latency between stimulus and response as compared with other psychophysiological measurements • It is difficult to identify specific emotions as a single measurement can result from several different causes • Qualitative and emotional aspects of affect are not reflected in EDR and must be inferred from other sources

6.4.4 Cardiovascular Measurements

Cardiovascular activity is the most commonly used index of cognitive workload. It is relatively unobtrusive, reliable, easy to use, and easy to interpret (Fahrenberg and Wientjes, 2000), and appears to be readily accepted by those in an operational environment (Prinzel et al., 2003). Parameters of cardiac activity relevant to a psychophysiological-based approach are electrical changes measured by an ECG, such as heart rate and HRV. Blood pressure is not an ideal parameter for mental workload measurement; it correlates with cognitive demand, but does not appear to be very sensitive, and it is prone to exercise artifacts (Castor, 2003).

6.4.4.1 Electrocardiogram

An ECG measures the electrical activity of the heart over time. The heart emits the highest electrical activity of all the body's organs, providing robust psycho-physiological data about operator workload that might be more difficult to detect via EEG alone. While an EEG provides valuable information on mental load, ECG provides systems designers with information about motor-related activity (Chen and Vertegaal, 2004).

ECG measurements are derived from electrodes placed on the skin. A minimum of two electrodes are required, but the vast majority of ECG setups use three, five, or twelve-lead ECGs. Leads are placed at strategic locations on and around the chest. Depending on the equipment used, leads can also be connected to the arms and legs.

ECGs are used primarily in clinical settings to measure and diagnose abnormal rhythms detectable through electrical means; heart attacks and evidence of other past afflictions can be seen through the ECG. McCraty et al. (1995) also demonstrated that specific emotions could be distinguished based upon the power spectrum of the ECG. Examples of ECGs used in operator state monitoring can be found but are largely still in the research and development phases (St. John et al., 2003; Prinzel et al., 2003; Schnell et al., 2007; Trejo et al., 2007; Tremoulet et al., 2007).

6.4.4.2 Heart Rate and Heart Rate Variability

The main use of the ECG is to calculate the heart rate and heart rate variability (HRV) of the operator. Heart rate is the most basic time measurement of the auto-nomic nervous system's influence on the heart, and is calculated by detecting the first waveform peak in the raw ECG signal, and then converting the time between two consecutive peaks from interbeat interval (in ms) to beats per minute.

HRV is defined as the variation of the time interval between heartbeats, and thus can also be calculated based on the interbeat interval captured by the ECG. HRV data provides a powerful, objective, and noninvasive tool that can be used to explore the dynamic interaction among physiological, mental, emotional, and behavioral processes. For example, in cases of stress, there is a tendency for increased HRV in the lower-frequency ranges. Stress measurements provided by ECG low-frequency components correlate well with the mental load of operators during complex visual tasks as measured by the National Aeronautics and Space Administration Task Load Index (NASA TLX) subjective workload assessment tool (McCraty et al., 1995).

HRV generally increases and decreases as a function of cognitive workload, and research suggests that "mid-band" HRV can accurately measure changes in mental workload while retaining the properties of diagnosticity, sensitivity, reliability, and ease of use required in an effective measurement solution (Prinzel et al., 2003). Researchers such as Chen and Vertegaal (2004) have demonstrated that HRV and the ECG are directly related, and can be used to measure active or passive participation in a task. Consequently, HRV and other ECG measurements have the potential to be used as a psychophysiological trigger for invoking intelligent adaptation.

Heart rate and HRV are affected by many influences, including respiration rate and physical effort, which may make measurements inaccurate if not used in combination with other operator state monitoring approaches (De Waard, 1996; Cain, 2007). As well, heart rate measurements are prone to artifacts, are sensitive to factors other than workload, and are not diagnostic with the variety of processing demands imposed on the operator. Table 6.6 shows the benefits and limitations of cardiovascular measurements.

6.4.5 EYE TRACKING

Eye tracking is the process of measuring either the point of gaze or the motion of an eye relative to the head. In the context of psychophysiological operator state monitoring approaches, eye tracking refers to the process of measuring where the operator is looking at any given moment. This information provides insight into the motivation behind operator actions (Bednarik, 2005). In graphical user interface (GUI) displays, the components of the GUI, such as specific window panels or menu items, that receive the greatest amount of visual attention are likely to coincide with, or at least be related to, items that are most important from the operator's perspective.

TABLE 6.6
Benefits and Limitations of Cardiovascular Measurements

Benefits	Limitations
• Heart rate measurements are unobtrusive, reliable, and easy to use and interpret	• The accuracy of heart rate measurements is affected by respiration, physical work, and emotional strain, which may make measurements inaccurate if not used in combination with other measurements (Cain, 2007)
• Heart rate measurements are sensitive to cognitive demands and attention	
• Heart rate variability (HRV) can be used to assess the positive or negative valence (i.e., attractiveness) of emotion	
• HRV can provide measurements of both cognitive effort and compensatory effort, depending on the application (Byrne and Parasuraman, 1996)	• Heart rate measurements are prone to artifacts
	• Heart rate measurements are sensitive to factors other than workload (Gaillard and Kramer, 2000)
• HRV and electrocardiography (ECG) are directly related, and can be used to measure active or passive participation tasks (Chen and Vertegaal, 2004)	• Heart rate measurements are not diagnostic with the variety of processing demands imposed on the operator (Gaillard and Kramer, 2000)

IAIs (as discussed in Chapter 2) can exploit this knowledge to adapt what type of information is being displayed, how long it is displayed for, the frequency of display, and any required emphasis markers.

Current eye-tracking systems use optical sensors combined with a camera pointed at the operator's eye or face. Image-processing software determines the position of the operator's eyes and pupils. This information can then be used to calculate gaze angle and point of focus. Blink rate is also useful, and can be measured in conjunction with pupil diameter and eye gaze. In combination, blink rate and eye gaze can help the IAI discern between task demand and general fatigue. Setups vary greatly, but generally fall into two categories: those that have optical sensors mounted on operator headgear, and those that have optical sensors mounted remotely (e.g., on a desktop). Some eye trackers require the head to be stable (e.g., with a chin rest), and some function remotely and automatically to track the head during motion (e.g., goggles). Remote head and eye trackers are the least invasive, as the operator is not required to wear the device. The eye tracker can sit on a desktop or be mounted to a display, thus requiring minimal external setup aside from calibration. Eye frame trackers are worn like eye glasses and are also not invasive, though they partially obstruct the operator's view (Vidal et al., 2012; Booth et al., 2013).

The most popular eye-tracking method is the use of video images from which the eye position is extracted, while other methods are based on the electrooculogram (EOG), which measures the resting potential of the retina to record eye movements. A major benefit of the EOG lies in the minimal amount of power and computation required for signal processing. Bulling et al. (2008) propose that the EOG is a novel measurement technique for wearable eye tracking and for recognizing operator activity and attention in mobile settings. With the EOG, electrodes are typically placed around the eye so that, when the eye moves from the center position toward an electrode, this electrode "sees" the positive side of the retina and the opposite electrode "sees" the negative side of the retina. Consequently, a potential difference occurs between the electrodes. Assuming that the resting potential is constant, the recorded potential is a measure for the eye position.

Under ideal conditions, EOG measurements facilitate extremely accurate measurements of eye fixation (i.e., when eye gaze pauses in a certain position). Blink rate correlates with visual workload, and blink rate reduces when visual workload increases. Increased blink rate and longer blink duration can be used to detect fatigue and the onset of sleep. Saccade rate (i.e., the rate at which the eye moves to another position) provides an index of visual scan rate and an approximate measurement of visual shift. Finally, changes in pupil diameter can be helpful to determine changes in cognitive activity when operators perform tasks, as pupil diameter can be measured in real time without aggregating data.

Eye trackers have been successfully deployed in academic and commercial applications (St. John et al., 2004; Prendinger et al., 2007; Schnell et al., 2007; Tremoulet et al., 2007). For example, eye trackers are used in marketing applications to determine how consumers react to advertisements, product shelf distributions, websites, and package design. Furthermore, the United States National Highway Traffic Safety Administration has supported research to develop eye tracking-based in-vehicle systems to monitor driver alertness and performance (Knipling and Wierwille, 1994).

Key indicators of driver vigilance and attention include eyelid movement, pupil movement, face orientation, and gaze (Ji and Yang, 2002). If fatigue levels are determined to be too high, alarms and warning signals are activated to alert the vehicle operator to take extra precautions or terminate the excursion.

As technology evolves, systems become less intrusive and more capable of capturing robust data in real time (Bergasa et al., 2006; Singh et al., 2011). Research in the field has led to retail eye-tracking systems from Delphi Electronics and Safety, EyeAlert, Fraunhofer, Seeing Machines, Smart Eye AB, and other manufacturers for real-time assessment of drowsiness levels in automobile drivers and aircraft pilots (Barr et al., 2009; Highway Safety Group, n.d.; Fraunhofer-Gesellschaft, 2010).

Data collection regarding the actions of the operator is not complete without an understanding of which interface features are being attended to (Bednarik, 2005). Eye-movement tracking provides instant information about the location of the operator's visual attention, and can be used as a trigger for intelligent adaptation. Assuming that the object of attention is also located on the top of the cognitive processing "stack" of the operator, knowledge about gaze, blink rate, eye movement, pupil diameter, and so on can help the system to understand which interface features were of interest, and in what order and for how long the operator attended to each feature. Table 6.7 shows the benefits and limitations of eye-tracking measurements.

6.4.6 RESPIRATION MEASUREMENTS

Respiration is a key indicator of operator state (Schnell et al., 2007). Symptoms of respiratory distress include shortness of breath, increased breath rate, changes in skin color, and partial or complete loss of consciousness. High-frequency shallow breaths have been correlated to both high-stress environments and high-cognitive workload environments. Respiration also has a strong influence on heart rate, and respiration rate measurements used in conjunction with ECG measurements can be used to assess the degree to which cardiac changes are artifacts of respiratory variability.

Respiration rate is most often measured using a strain gauge band sensor worn around the chest, or the "doll's eye" seen in aircraft respiratory systems. The strain

---

TABLE 6.7
Benefits and Limitations of Eye Tracking

Benefits	Limitations
• Blinks and eye gaze can discern task demand and fatigue	• Lighting conditions may affect measurement accuracy
• Eye-tracking measurements are correlated with mental workload	• Eye-tracking technologies do not take into account the fact that fixation patterns will differ depending on the environment
• Eye-tracking measurements are especially useful when combined with electroencephalography (EEG) data	• Electrooculogram (EOG) measurements are not diagnostic regarding specific varieties of workload that may be involved in operator–task interaction (Gaillard and Kramer, 2000)

gauge band is elastic and contains a material that changes electrical properties when stretched; expansion and deflation of the lungs causes the band to stretch and contract, and the resulting signal is fed to the appropriate filters prior to signal analysis; the "doll's eye" gauge is located within an aircraft cockpit's pressure breathing system and indicates when the pilot is inhaling (i.e., demanding air from the system).

Note that respiration measurements are not diagnostic and can be difficult to capture. As such, respiration measurement is not useful on its own, and must be integrated with other psychophysiological measurements (Veltman and Gaillard, 1996; Scerbo et al., 2001). Table 6.8 shows the benefits and limitations of respiration measurement.

6.4.7 SKIN TEMPERATURE MEASUREMENTS

Skin temperature is measured with thermocouples placed at strategic locations on the body. Variation in skin temperature indicates changes in autonomic activity. Increased activity in the sympathetic nervous system (part of the autonomic nervous system) results in vasoconstriction of peripheral arteries, which lowers skin temperature at the extremities (Schnell et al., 2007). Decrements in peripheral temperature can be used to measure stress and arousal. Like respiration measurements, skin temperature measurements are not diagnostic and are most useful when they are integrated with other psychophysiological measurements. Table 6.9 shows the benefits and limitations of skin temperature measurement.

6.4.8 ELECTROMYOGRAPHY

EMG is used to measure and record the electrical activity of skeletal muscles. When groups of muscle cells voluntarily or involuntarily contract, electric impulses are detected.

TABLE 6.8
Benefits and Limitations of Respiration Measurements

Benefits	Limitations
• Respiration can provide valuable measurements of workload and stress when combined with electrocardiogram (ECG)	• Respiration measurements are not useful measurements on their own. They must be integrated with other measurements (Veltman and Gaillard, 1996; Scerbo et al., 2001) • Respiration measurements are not diagnostic • Respiration measurements are difficult to capture using sensors

---

TABLE 6.9
Benefits and Limitations of Skin Temperature Measurements

Benefits	Limitations
• Decreased skin temperature indicates increased activity in the sympathetic nervous system	• Skin temperature measurements are not useful measurements on their own. They must be integrated with other measurements

Muscle activity is predominantly associated with frequencies in the range of 10 Hz and upwards and is particularly marked between 50 and 150 Hz. Frequency is indicative of effector workload, and more physically onerous acts are associated with higher-amplitude EMG activity. Alterations in peripheral load associated with the control of devices such as joysticks can be mapped in real time, and are associated with higher-amplitude EMG activity. EMG measurements also correlate with state variables (Trejo et al., 2007), such as drowsiness and fatigue. During drowsiness, sleep onset, and sleep itself, there is a progressive decrease in muscle activity. A decrease in muscle tonus, or muscle state, may indicate dangerously low levels of alertness, such as during rapid eye movement sleep (Andreassi, 2007); higher arousal states are associated with increased muscle tonus.

EMG measurements are usually taken using electrodes on the surface of the skin, but intramuscular electrode methods, which involve inserting fine electrodes through the skin into muscle tissues, are also available. Applying intramuscular electrodes requires specialized technicians and can cause undue pain and stress to the subject; it is primarily used for clinical diagnosis and in some forms of muscle rehabilitation to determine the activity of small portions of a larger muscle (Andreassi, 2007). Surface EMGs are the most common method for clinical and academic use. Table 6.10 shows the benefits and limitations of EMG measurement.

6.4.9 CHOOSING PSYCHOPHYSIOLOGICAL MONITORING APPROACHES

Researchers have been determining operator state and mental workload through psychophysiological measurement for a long time—from the eye-tracking work conducted by Fitts et al. (1950) to the multimodal computerized airborne research platform at the University of Iowa (Schnell et al., 2007). Integrating psychophysiological monitoring technologies into research development platforms provides clear, objective measurements of operator state as it pertains to an undertaken task and the external environment. Consequently, task context is an important factor in determining the relevance of including a psychophysiological approach in an IAS design (Mathan et al., 2005; Hou and Fidopiastis, 2014).

Psychophysiological monitoring approaches subscribe to similar conventions of operator state and have clearly defined monitoring technologies and metrics, but measurement selection depends on the nature of the task, the operational definition of cognitive state, and the context of the external environment. The ideal one-to-one mapping of a single psychophysiological measurement to operator state does not exist. Instead, systems designers can apply a many-to-one mapping of psychophysiological measurements to a particular operator state (e.g., pupil dilation and EEG

TABLE 6.10
Benefits and Limitations of Electromyography (EMG)

Benefits	Limitations
• EMG measurements related to the control of devices can be mapped in real time	• Intramuscular EMGs require specialized technicians and can cause pain and undue stress to the operator
• EMG measurements correlate with state variables such as drowsiness and fatigue	

for mental workload). Note also that a single measurement may be sensitive to more than one psychological state. For example, EEG can measure both engagement and workload (Berka et al., 2007). As well, some metrics may overlap with several psychophysiological states, resulting in a many-to-many mapping of monitoring technologies and an overcomplicated monitoring approach.

Psychophysiological measurements provide important advantages for facilitating intelligent adaptation in IASs. When coupled with behavioral analysis, direct measurement of psychophysiological signals can help to increase the level of accuracy in determining adverse operator states. Mental workload has a very real and predictable effect on psychophysiological indicators such as brain electrical activity, heart rate, and respiration rate (McCraty, 2009; Wickens et al., 2012). Thus, psychophysiological measurement can provide a reliable, objective, and noninvasive index of cognitive load during operational task performance. Another advantage of psychophysiological measurements is that they can provide continual, objective, and often quantitative parameters for predicting operator state. Furthermore, psychophysiological measurements can help assess the desirability of systems in cases where performance measurements alone fail to differentiate among available choices (Cain, 2007). The following are general benefits of using psychophysiological measurements:

- *Objective outcomes*: Psychophysiological measurements produce unbiased, quantitative outcomes unaffected by self-reporting.
- *Unobtrusive sensor apparatus*: The sensors used to capture psychophysiological measurements are predominantly noninvasive and typically do not negatively impact the performance of the operator.
- *Data convergence*: Using multiple psychophysiological measurements can improve accuracy when determining the mental workload of the operator.
- *Immediate and continuous results*: Psychophysiological measurements produce results quickly enough to be advantageous to overall HMS goals and can often be obtained continuously.

Researchers in the AugCog domain have concluded that psychophysiological measurements suffer from numerous issues that may preclude their use in the field, including lack of construct and internal validity, such as seen in highly correlated independent measurements; lack of statistical validity, such as seen in low sample sizes; missing data; and inherently noisy sensor data (St. John et al., 2004; Cummings, 2010). Psychophysiological measurements also encompass the following general limitations, which may preclude their implementation in a particular IAS:

- *Specialized equipment*: Measurements of cognitive state require integration of special equipment into the system and technological expertise. Many off-the-shelf products have not been validated for use outside the medical domain or for integration into other application systems within the domain. The feasibility of providing real-time feedback to address lags in system adaption and system platform effects has not been well studied. Additionally, these equipment and systems require regular maintenance and part replacement, which increases cost.

- *Data acquisition issues*: Accurate assessment of operator state requires high-quality, artifact-free raw data. However, filters and artifact removal strategies for psychophysiological measurements are neither standardized nor easily understood. Issues such as latency and recovery time must also be addressed further.
- *Data processing issues*: The large amounts of data collected from psychophysiological monitoring technologies require computing technology capable of real-time processing to provide meaningful results for system adaptation. Additionally, some of the data collected are not easily understood by data analysis software, and the ability to write custom code to resolve the issue is often not readily available.
- *Efficacy and effectiveness issues*: Biosensor choices and their respective metrics derive from unstandardized operational definitions of cognitive constructs, which in turn suffer from the biases inherent in the brain theory chosen to support the operational definition. Because there are no standards and guidelines to sensor and metric selection, basic assumptions on how to operationally define the human state, calculate the sensor metric, apply data fusion techniques, and define appropriate methods for state classification are anyone's best guess. Efficacy of sensor selection and algorithm implementation is best determined in laboratory-based test beds, which may not provide effective field deployable IAS solutions. Linking simulation-based experiments to the actual task requires sensors and associated algorithms to be integrated within a real-world context to connect between efficacy and effectiveness. Tools for validating these translated algorithms for use within real-time dynamic IASs are still under development (Hou and Fidopiastis, 2014).

Table 6.11 shows the benefits and limitations of psychophysiological measurements in general.

6.5   CONTEXUAL-BASED MONITORING

Human behavior and associated psychophysiological readings exhibit considerable variability depending on environmental context. It is, therefore, very important to include a contextual arm in any operator state monitoring system. Context provides information that enables interpretation of an operator's behavioral, psychophysiological, and subjective changes. For example, consider the development of an operator state monitoring device for aircraft pilots. Takeoff and landing can be associated with dramatic changes in psychophysiological variables such as heart rate. If the pilot's psychophysiological data is being monitored without knowledge of environmental context, then incorrect inferences about pilot state can be made.

Contextual information is also used by the situation assessment module to track progress toward specific mission or work activity objectives, and predictable contextual factors can offer insight into the impact of context on overall operator state. To facilitate this type of insight, the IAS's operator state monitoring component must collect information on low-level contextual factors known to influence performance. Table 6.12 shows key technologies for contexual-based monitoring.

TABLE 6.11
General Benefits and Limitations of Psychophysiological-Based Monitoring Approaches

Benefits	Limitations
• Psychophysiological measurements are useful when subjective approaches or performance measurements prove insensitive due to covert changes in operator strategies (Cain, 2007) • Psychophysiological measurements are relatively unobtrusive. They only require the attachment of sensors to the body, which is usually quite acceptable to the operator (Gaillard & Kramer, 2000) • Measurements can often be obtained continuously (Gaillard & Kramer, 2000) • Psychophysiological measurements produce results in enough time to be advantageous to overall human-machine system goals • Psychophysiological measurements produce unbiased, quantitative outcomes unaffected by self-reporting • The sensors used to capture psychophysiological measurements are predominantly non-invasive and typically do not negatively impact operator performance • Using multiple psychophysiological measurements can improve accuracy when determining the mental workload of the operator	• Psychophysiological-measurement systems often need to be tailored to each individual rather than using group norms, making interpretation more involved (Cain, 2007) • Psychophysiological measurements may be disturbed by other physiological variables (e.g., respiration, eye movement) and artifacts (e.g., physical activity) that cannot always be easily controlled. They are also sensitive to interference from other response systems (Gaillard & Kramer, 2000) • Psychophysiological measurements of workload are difficult to interpret when applied to issues of underload (Byrne and Parasuraman, 1996) • An ideal one-to-one mapping of a single psychophysiological measure to operator state does not exist • Most psychophysiological measurements require integrating special equipment into the system and require technology expertise • Filters and artifact removal strategies for psychophysiological measurements are neither standardized nor easily understood • Latency and recovery time issues must be addressed • Data collected from psychophysiological monitoring technologies are large and require computing technology capable of real-time processing to provide meaningful results for system adaptation • Some of the data collected are not easily understood by data analysis software and the ability to write custom code to resolve the issue is often not readily available • Tools for validating efficacy algorithms for effective use within real-time dynamic HMSs are still under development (Hou and Fidopiastis, 2014)

TABLE 6.12

Summary of Relevant Contextual Feedback Approaches

Contextual Feedback Approach	Summary	Examples of Current Uses
Ambient sensors	• Monitors environment such as ambient lighting, temperature, humidity, and background noise	• Photosensitive sensors, microphones, thermocouples
System state sensors	• Monitors status and operation of system the operator is controlling or supervising	• Varies depending on the system in question

6.5.1 Ambient Sensors

The operator state monitoring component of IASs can make use of a number of sensors to gather information about ambient factors, such as ambient noise, ambient temperature, and luminance. Increments in noise are known to influence performance, and extremes in ambient temperature and luminance are both well documented to have deleterious effects on performance (Banbury et al., 2001; Boyce et al., 2000; Pilcher et al., 2002). Noise, ambient light, and temperature all have commercial, off-the-shelf monitoring applications. Ambient sensors contribute to situation assessment, although they do not contribute to system state assessment.

6.5.2 System State Sensors

System state sensors are also important to incorporate into the operator state monitoring component of IASs. In this context, "system" refers to the machine that the operator is working with, and "state" is the machine's status (as discussed in Chapter 3). For example, if the operator is a pilot, the system being operated is the aircraft, and important system state sensors would monitor G-loading, airspeed, acceleration, pitch, roll, and engine status. Though information on system status can potentially identify faulty equipment, it can also assist in predicting operator reaction. For example, emergency situations will likely induce emotions of fear and higher degrees of cognitive workload, which can lead to increased operator stress. Alternatively, prolonged periods of nominal system operation may lead to boredom and fatigue. It is also useful to record and monitor operator actions, as these may indicate the various procedures that were undertaken and how they affected both operator and system. This data can provide an operator model that can be used in future systems. By gaining insight into the operation of the system through the use of system state sensors, the type of operator state that the individual operator is likely to experience (e.g., workload, arousal) can be narrowed down before any behavioral or psychophysiological measurements are made.

6.6   SUBJECTIVE-BASED MONITORING

Subjective-based monitoring refers to approaches that elicit data about operator state by asking the operator. Subjective techniques can only be based on what the operator remembers and their interpretation of their experience (Cain, 2007). Explicit input can be valuable both before and during the actual task. Operator input obtained before the actual task is seen in integrated design approaches, and can be used to set up or preemptively adjust an interface to be customized to the individual. Input collected in real time can also be used as a feedback loop for IASs in the same way as behavioral or psychophysiological data is used to determine operator state and update the IAS operator model. Note that subjective measurements are highly suited to assessing technologies that aid judgment and decision making, but are less suited to assessing physical or mechanical aids for repetitive or highly learned tasks (Cain, 2007).

In conventional settings, subjective-based monitoring is often accomplished through paper and pencil tasks. Data collection, however, can easily be automated. For example, provisions can be made for the operator to signal subjective states of concern (i.e., the operator's perception of their current level of workload). Very high-workload states and potential performance deterioration can be signaled using speech or a single button press. Similarly, the recognition of chronic underarousal and the onset of sleep can also be communicated. Under more manageable workload conditions, the operator can also signal overly demanding task elements. Research has shown that these forms of subjective data are invaluable indices of operator state (Bonner et al., 2000; Alpert et al., 2003). Incorporating subjective measurements within a closed-loop system directly links the operator with onboard monitoring systems, and, as a result, keeps the operator in-the-loop.

6.7   COMBINATION-BASED MONITORING

Combination-based monitoring moves from several data streams to an overall conclusion regarding operator state. Systems designers use the following monitoring criteria to determine when to adapt the level of automation:

- *Performance measurements*: Monitoring criteria that reflect operator performance, such as speed, accuracy, and so on
- *Critical events*: Monitoring factors that indicate system failures or critical changes in the external environment, such as loss of power
- *System models*: Monitoring IAS performance based on an a previously established pattern of performance
- *Operator state assessments:* Monitoring operator state, or mental workload, by taking measurements, including psychophysiological measurements
- *Hybrid approaches*: Combining one or more of the above strategies (Byrne and Parasuraman, 1996; Scerbo, 1996; Scerbo et al., 2001; Kaber and Endsley, 2004)

Though researchers have suggested that IASs should include real-time operator state assessments in their design (Parasuraman et al., 2007; Scerbo, 2007), a single measurement or feedback loop is unlikely to provide a full and accurate picture of the cognitive, physical, and emotional states of the operator. Multiple feedback channels are required for the greatest accuracy, each channel providing a small piece of the puzzle (Pleydell-Pearce et al., 1999; Wilson and Russell, 2006).

Hybrid (i.e., combination-based) approaches recognize the importance of parallel processing, which is the ability to simultaneously process multiple incoming stimuli or variables of differing quality, such as simultaneous processing of the color, motion, shape, and depth components of vision. Hybrid approaches are also capable of performing complex multivariate analyses to improve inferences about operator state. Although various forms of data can be treated as separate variables, the relationships between different data sources also contain valuable information—trends in one data set can be better explained in the context of other data sets. For example, the absence of reaction to a mild threat, such as a low-altitude warning, could indicate that the operator is confident and in control; it could also indicate a loss of SA caused by dangerously low levels of arousal.

Another example of hidden trends revealed in the relationships between data sets can be seen in combining EEG and NIRS with eye tracking and EDR, which has the potential to provide real-time data for investigating operator state such as intention, attention, and cognitive load. As well, combining EEG with medical imaging technologies such as magnetic resonance imaging allows the same operator state variable to be monitored over time in specific areas of the brain. By considering data from multiple streams, the effects of anomalous variations in behavioral and psychophysiological responses between individuals can be avoided. Two of the more common methods for combination-based monitoring are rule-based data fusion and artificial neural networks.

6.7.1 RULE-BASED DATA FUSION

Rule-based data fusion can be thought of as a structured decision tree. It uses a series of if-then statements that can be queried to determine a conclusion on operator state. Rule-based data fusion is very easy to implement into software, but it can only function on very simple and predictable systems. Systems designers using rule-based data fusion must understand every case and develop a rule for all cases. This can be problematic when dealing with operators, whose actions are not always predictable. To be effective in an operator state monitoring context, rule-based data fusion must use strictly controlled scope and parameters.

6.7.2 ARTIFICIAL NEURAL NETWORKS

An artificial neural network (ANN) is a decision-making algorithm based on the biological neural system. Inputs are presented to the ANN and each is assigned an output through an interwoven cascade of simulated software neurons. Like any function, an input is mapped to a desired output. ANNs have been successfully used to accurately classify cognitive workload in a variety of environments (Russell, 2005). Figure 6.4 illustrates an ANN (Fausett, 1994).

Although a much simpler and abstract version, the ANN simulates what happens in the biological world. Inputs to the neuron are compared, the neuron

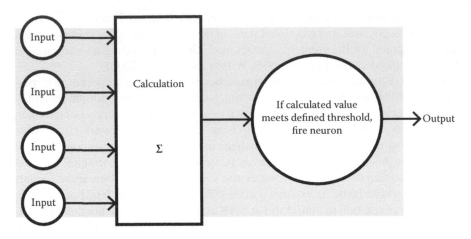

FIGURE 6.4   Simplified model of a basic artificial neural network (ANN).

performs a calculation, which can be as simple as a sum of all inputs, and, if the resulting value exceeds a predetermined threshold value, the neuron fires. ANNs can be arranged in single or multiple layers. Figure 6.5 illustrates a multilayered ANN, in which the outputs of some layers act as inputs to other layers. The overall purpose of the ANN is as a structure that can assign outputs to given inputs.

Neural networks have proven benefits in visual processing, audio processing, signal conditioning, and other applications. ANNs are robust, can be applied to almost any problem, and are well suited both for unstructured or unpredictable problems and for data that exhibit a high degree of noise and variability, such as data collected

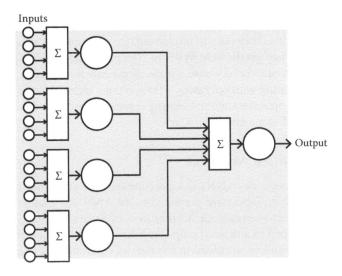

FIGURE 6.5   Simplified model of a multilayered artificial neural network consisting of five individual artificial neurons.

from operators. ANNs can also be used to search for hidden patterns within data (Pleydell-Pearce et al., 1999).

One of the major drawbacks of ANNs is that, like the biological mind, they need to be trained. This means that extensive data sets with assigned desired outputs need to be available. Also, computation time can become an obstacle for larger ANNs. ANNs are also not considered a good strategy in most applications if they lack transparency (Mooshage et al., 2002).

Despite disadvantages, several researchers have successfully applied ANNs to operator state monitoring (Nicholson et al., 2000; Russell, 2005; Wilson and Russell, 2006, 2007; Pleydell-Pearce et al., 1999). For example, Wilson and Russell (2007) used an ANN to detect periods of high and low workload in real time while operators monitored and determined priorities for a group of four autonomous UAVs. Data collected from the operator via ECG, EOG, and EEG were used to train the ANN, and were then used by the ANN to provide real-time estimates of operator state during subsequent task performance. Nicholson et al. (2000) trained single-layer ANNs to determine emotions elicited through non-verbal spoken communication by referencing a database of phoneme-balanced words.

6.8   AN EXAMPLE COMBINATION-BASED MONITORING SYSTEM

The United Kingdom's Cognitive Cockpit program, which was discussed in Chapter 3, is an example of an IAS that enhances context-sensitive communication with its operator. It uses an interaction-centered approach to develop an integrated system that keeps the operator in charge.

Function allocation between operator and system (i.e., machine) is very flexible in the Cognitive Cockpit: either the operator (i.e., pilot) or the system can perform information-acquisition, information-analysis, decision-making, and action roles. The system is only able to perform with the consent of the operator, although there are some exceptions that can take place if the operator is incapacitated, such as the aircraft taking control if the pilot loses consciousness due to a high-gravity maneuver.

Figure 6.6 illustrates the Cognitive Cockpit's cognition monitor module (COGMON), which uses combination-based monitoring to facilitate intelligent adaptation from a variety of inputs and outputs (Willis, 2005). COGMON facilitates intelligent adaptation by providing real-time analysis of the operator's psychological, physiological, and behavioral states. Primary functions include continuous monitoring of workload and producing inferences about current attentional focus, ongoing cognition, and intentions. COGMON is also capable of detecting dangerously high and low levels of arousal and provides information about the operator's objective and subjective states within a mission context. This information is used to optimize operator performance and safety, and provides a basis for the implementation of intelligent and adaptive pilot aiding.

COGMON has 32 analog-to-digital converters capable of recording behavioral, psychological, physiological, and situational data in real time. They facilitate analysis of both low and high-frequency EEG measurements. The system can perform near real-time analysis of all 32 channels and can examine coherence functions across 10

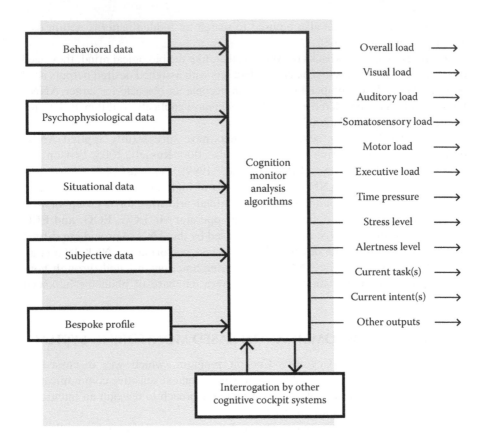

FIGURE 6.6 Overview of cognition monitor (COGMON) inputs and outputs. (Redrawn from Willis, A.R., *Presented at the 2nd Situation Awareness and Team Cognition Workshop*, Laval University, Laval, Canada, 2005.)

different frequency bandwidths for each channel. These data provide information about alertness, but also allow inferences to be made about current cognitive activity.

COGMON examines blink rate and eye movement in detail. These provide information about visual workload and can be used to determine the current locus of fixation. Heart rate and HRV are determined and respiration rate is measured and analyzed. Muscle tonus, a useful measurement of alertness and limb activity, is recorded and analyzed using EMG biosensors. Central and peripheral cutaneous temperatures and electrodermal activities are continuously measured. Combined, these measurements provide information about activity levels in the autonomic nervous system.

COGMON logs auditory input to the operator via headphones, as well as ambient noise in the external environment. Operator vocalizations are monitored using microphones and larynx EMG measurements. Other ambient factors, including luminance and temperature, are also monitored.

COGMON contains a large amount of statistical and analytical software; these algorithms are designed to perform analysis of COGMON's inputs and process data from a variety of sources. These combined analyses are a considerable improvement in inferences about operator state compared with single-source analyses. The analytical routines performed by COGMON are also aimed at identifying pilot bespoke profiles, as the philosophy underlying COGMON assumes that the suitability of workload measurements may vary from context to context, and from pilot to pilot.

Although COGMON does not claim to be the definitive workload assessment system, it is one of the most comprehensive attempts to produce such a system. COGMON is, like other systems, constrained by the maturity and availability of additional hardware and software. For example, reliable optical tracking of visual fixation could considerably enhance the monitor's ability to assess visual workload and determine the locus of visual attention.

COGMON is also partially dependent upon the rate of progress made in other IAS components. For example, software aimed at monitoring operator interaction with system controls cannot be fully implemented until the simulation environment is reasonably mature. Software aimed at the receipt and analysis of general situational data also requires that the situation assessment subsystem reach a sufficient level of maturity. Additionally, the complexity of systems such as COGMON means that its inner functions may be hard to visualize, making it difficult for operators to examine COGMON's current internal status.

6.9 SUMMARY

This chapter looked closely at operator state monitoring approaches, and offered guidance on determining appropriate options. The main goal of operator monitoring—to optimize human–machine interaction by continually updating the system on operator state—is achieved only if the most applicable technologies are integrated into the system. To facilitate this, an overview of common monitoring approaches was provided.

Behavioral-based monitoring approaches, specifically operator–control interaction monitoring and auditory analysis, were discussed briefly, while psychophysiological-based monitoring approaches were examined in detail. Psychophysiological-based monitoring technologies, namely EEG, NIRS, EDR, cardiovascular measurements, eye tracking, respiration measurements, skin temperature measurements, and EMG, were introduced. Benefits and limitations for each monitoring approach were also discussed.

Contexual-based monitoring approaches, which look at environmental factors, and subjective-based monitoring approaches, which ask direct questions of the operator, were noted. Further study of monitoring approaches is necessary for systems designers considering implementing them within a particular project.

The use of a combination-based approach was suggested to moderate the limitations associated with individual monitoring approaches; combining multiple monitoring approaches provides a clearer, more objective understanding of operator state. Rule-based data fusion, a combination-based approach for drawing conclusions based on strict if-then type methods, and ANNs, which support decision making by automating inference generation, were also examined. A detailed example of combination-based operator monitoring technology was provided using COGMON.

Interface adaptation to operators through built-in feedback loops was discussed as a method for more efficient human–automation interaction with the potential for minimizing errors. Individual operator psychophysiological measurements were also identified as possible adaptation triggers; adaptation triggers are one of the IAS design considerations discussed in Chapter 7.

REFERENCES

Alpert, S. R., Karat, J., Karat, C. M., Brodie, C., and Vergo, J. G. (2003). User attitudes regarding a user-adaptive e-commerce web site. *User Modeling and User-Adapted Interaction*, *13*(4), 373–396.

Andreassi, J. L. (2007). *Psychophysiology: Human Behavior and Physiological Response*, 5th edn. New York, NY: Psychology Press.

Angell, L. S., Auflick, J., Austria, P. A., Kochhar, D. S., Tijerina, L., Biever, W., Diptiman, T., Hogsett, J., and Kiger, S. (2006). Driver workload metrics task 2 final report (Report No. HS-810 635). Washington, DC: US Department of Transportation.

Banbury, S. P., Macken, W. J., Tremblay, S., and Jones, D. M. (2001). Auditory distraction and short-term memory: Phenomena and practical implications. *Human Factors: The Journal of the Human Factors and Ergonomics Society*, *43*(1), 12–29.

Barr, L., Popkin, S., and Howarth, H. (2009). An evaluation of emerging driver fatigue detection measures and technologies (Report No. FMCSA-RRR-09-005). Washington, DC: US Department of Transportation.

Bednarik, R. (2005). Potentials of eye-movement tracking in adaptive systems. In S. Weibelzahl, A. Paramythis, and J. Masthoff (eds), *Fourth Workshop on the Evaluation of Adaptive Systems,* pp. 1–8. July 24–30, Edinburgh, UK. Retrieved July 31, 2014 from http://homepages.abdn.ac.uk/j.masthoff/pages/Publications/UCDEAS05.pdf.

Bergasa, L. M., Nuevo, J., Sotelo, M. A., Barea, R., and Lopez, M. E. (2006). Real-time system for monitoring driver vigilance. *IEEE Transactions on Intelligent Transportation Systems*, *7*(1), 63–77.

Berka, C., Levendowski, D. J., Cvetinovic, M. M., Petrovic, M. M., Davis, G., Limicao, M. N., Zivkovic, V. T., Popovic, M. V., and Olmstead, R. (2004). Real-time analysis of EEG indexes of alertness, cognition, and memory acquired with a wireless EEG headset. *International Journal of Human-Computer Interaction*, *17*(2), 151–170.

Berka, C., Levendowski, D. J., Lumicao, M. N., Yau, A., Davis, G., Zivkovic, V. T., Olmstead, R. E., Tremoulet, P. D., and Craven, P. L. (2007). EEG correlates of task engagement and mental workload in vigilance, learning, and memory tasks. *Aviation, Space, and Environmental Medicine*, *78*(Supplement 1), B231–B244.

Blanchard, E., Calhoun, P., and Frasson, C. (2007). Towards advanced learner modeling: Discussions on quasi real-time adaptation with physiological data. In *Proceedings of the Seventh IEEE International Conference on Advanced Learning Technologies*, pp. 809–813. July 18–20, Niigata, Japan: IEEE.

Bonner, M. C., Taylor, R. M., Fletcher, K., and Miller, C. (2000). Adaptive automation and decision aiding in the military fast-jet domain. In *Proceedings of the Conference on Human Performance, Situation Awareness and Automation: User Centred Design for the New Millennium*, pp. 154–159. Savannah, GA: SA Technologies, Inc.

Booth, T., Sridharan, S., McNamara, A., Grimm, C., and Bailey, R. (2013). Guiding attention in controlled real-world environments. In *Proceedings of the 2013 ACM Symposium on Applied Perception*, pp. 75–82. August 22–23, New York, NY: ACM.

Boyce, P. R., Eklund, N. H., and Simpson, S. N. (2000). Individual lighting control: Task performance, mood, and illuminance. *Journal of the Illuminating Engineering Society*, *29*(1), 131–142.

Bulling, A., Roggen, D., and Troster, G. (2008). It's in your eyes—Towards context-awareness and mobile HCI using wearable EOG goggles. In *Proceedings of the Tenth International Conference on Ubiquitous Computing*, pp. 84–93. September 21–24, New York, NY: ACM.

Byrne, E. A. and Parasuraman, R. (1996). Psychophysiology and adaptive automation. *Biological Psychology*, 42(3), 249–268.

Cain, B. (2007). A review of the mental workload literature (Report No. RTO-TR-HFM-121-Part-II). Toronto, Canada: Defense Research and Development Canada.

Castor, M. C. (2003). Final report for GARTEUR flight mechanics action group FM AG13: GARTEUR handbook of mental workload measurement (Report No. GARTEUR TP145). Paris, France: Group for Aeronautical Research and Technology in Europe, Flight Mechanics Action Group.

Chen, D. and Vertegaal, R. (2004). Using mental load for managing interruptions in physiologically attentive user interfaces. In *Proceedings of the CHI '04 Extended Abstracts on Human Factors in Computing Systems*, pp. 1513–1516. April 24–29, New York, NY: ACM.

Chen, J. Y. C. and Barnes, M. J. (2014). Human-agent teaming for multi-robot control: A review of human factors issues. *IEEE Transactions on Human-Machine Systems*, 44(1), 13–29.

Chen, J. Y. C. and Terrence, P. I. (2009). Effects of imperfect automation and individual differences on concurrent performance of military and robotics tasks in a simulated multitasking environment. *Ergonomics*, 52(8), 907–920.

Cummings, M. L. (2010). Technology impedances to augmented cognition. *Ergonomics in Design*, 18(2), 25–27.

De Waard, D. (1996). The measurement of drivers' mental workload (Unpublished doctoral dissertation). Groningen, Netherlands: Groningen University Traffic Research Center.

Durkee, K., Geyer, A., Pappada, S., Ortiz, A., and Galster, S. (2013). Real-time workload assessment as a foundation for human performance augmentation. In D. D. Schmorrow and C. M. Fidopiastis (eds), *Foundations of Augmented Cognition*, vol. 8027, pp. 279–288. Berlin, Germany: Springer.

Dussault, C., Jouanin, J. C., Philippe, M., and Guezennec, C. Y. (2005). EEG and ECG changes during simulator operation reflect mental workload and vigilance. *Aviation, Space, and Environmental Medicine*, 76(4), 344–351.

Eggemeier, F. T. (1988). Properties of workload assessment techniques. *Advances in Psychology*, 52, 41–62.

Fahrenberg, J. and Wientjes, C. J. E. (2000). Recording methods in applied environments. In R. W. Backs and W. Boucsein (eds), *Engineering Psychology: Issues and Applications*, pp. 111–136. London, UK: Lawrence Erlbaum Associates.

Fausett, L. (1994). *Fundamentals of Neural Networks: Architectures, Algorithms, and Applications*. Englewood Cliffs, NJ: Prentice Hall.

Fernandez, R. (1997). Stochastic modeling of physiological signals with hidden Markov models: A step toward frustration detection in human-computer interfaces (Master's thesis). Cambridge, MA: Massachusetts Institute of Technology Department of Electrical Engineering and Computer Science. Retrieved July 31, 2014 from http://citeseerx.ist.psu.edu.

Fidopiastis, C. M., Drexler, J., Barber, D., Cosenzo, K., Barnes, M., Chen, J. Y. C., and Nicholson, D. (2009). Impact of automation and task load on unmanned system operator's eye movement patterns. In D. D. Schmorrow, I. V. Estabrooke, and M. Grootjen (eds), *Foundations of Augmented Cognition. Neuroergonomics and Operational Neuroscience Lecture Notes in Computer Science*, vol. 5638, pp. 229–238. Berlin, Germany: Springer.

Fidopiastis, C. M. and Nicholson, D. M. (2010). Neuroergonomics: From theory to practice. In T. Marek, W. Karwowski, and V. Rice (Eds.) *Advancing the Understanding of Human Performance: Neuroergonomics, Human Factors Design, and Special Populations* (pp. 354–359), Boca Raton, FL: CRC Press.

Figner, B. and Murphy, R. O. (2010). Using skin conductance in judgment and decision making research. In M. Schulte-Mecklenbeck, A. Kuhberger, and R. Ranyard (eds), *A Handbook of Process Tracing Methods for Decision Research: A Critical Review and User's Guide*, pp. 163–184. New York, NY: Psychology Press.

Fitts, P. M., Jones, R. E., and Milton, J. L. (1950). Eye movements of aircraft pilots during instrument-landing approaches. *Aeronautical Engineering Review*, 9, 24–29.

Fraunhofer-Gesellschaft. (2010). Eyetracker warns against momentary driver drowsiness. Retrieved July 31, 2014 from http://www.fraunhofer.de/en/press/research-news/2010/10/eye-tracker-driver-drowsiness.html.

Gaillard, A. W. and Kramer, A. F. (2000). Theoretical and methodological issues in psychophysiological research. In R. W. Backs and W. Boucsein (eds), *Engineering Psychophysiology: Issues and Applications*, pp. 31–58. Mahwah, NJ: Lawrence Erlbaum Associates.

Gale, A. and Christie, B. (1987). Psychophysiology and the electronic workplace: The future. In A. Gale and B. Christie (eds), *Psychophysiology and the Electronic Workplace*, pp. 315–333. Chichester, UK: Wiley.

Gravano, A. and Hirschberg, J. (2011). Turn-taking cues in task-oriented dialogue. *Computer Speech and Language*, 25(3), 601–634.

Gravano, A., Hirschberg, J., and Beňuš, Š. (2012). Affirmative cue words in task-oriented dialogue. *Computational Linguistics*, 38(1), 1–39.

Gumenyuk, V., Korzyukov, O., Alho, K., Escera, C., and Näätänen, R. (2004). Effects of auditory distraction on electrophysiological brain activity and performance in children aged 8–13 years. *Psychophysiology*, 41(1), 30–36.

Highway Safety Group (n.d.). DD 850 fatigue warning system: Making good drivers safer! Retrieved July 31, 2014 from http://www.driverfatiguemonitor.com/dfm/dfm.html.

Hirschberg, J. (2010). Deceptive speech: Clues from spoken language. In F. Chen (ed.), *Speech Technology*, pp. 79–88. New York, NY: Springer US.

Hou, M. and Fidopiastis, C. M. (2014). Untangling operator monitoring approaches when designing intelligent adaptive systems for operational environments. In *Proceedings of the Human Computer Interaction International Conference 2014*, (25)26–34. June 22–27, Crete, Greece.

Izzegtoglu, M., Bunce, S. C., Izzetoglu, K., Onaral, B., and Pourrezaei, K. (2007). Functional brain imaging using near-infrared technology. *IEEE Engineering in Medicine and Biology Magazine*, 26(4), 38–46.

Ji, Q. and Yang, X. (2002). Real-time eye, gaze, and face pose tracking for monitoring driver vigilance. *Real-Time Imaging*, 8(5), 357–377.

Kaber, D. B. and Endsley, M. R. (2004). The effects of level of automation and adaptive automation on human performance, situation awareness and workload in a dynamic control task. *Theoretical Issues in Ergonomics Science*, 5(2), 113–153.

Kapoor, A., Burleson, W., and Picard, R. W. (2007). Automatic prediction of frustration. *International Journal of Human-Computer Studies*, 65(8), 724–736.

Karamouzis, S. T. (2006). The use of psychophysiological measures for designing adaptive learning systems. In I. Maglogiannis, K. Karpouzis, and M. Bramer (eds), *Artificial Intelligence Applications and Innovations*, 3rd edn., pp. 417–424. New York, NY: Springer US.

Keebler, J. R., Sciarini, W. L., Fidopiastis, C., Jentsch, F., and Nicholson, D. (2009). Use of functional near infrared imaging to investigate neural correlates of expertise in military target identification. In *Proceedings of the 53rd Annual Meeting of the Human Factors and Ergonomics Society*, pp. 151–154. October 19–23, New York: SAGE Publications.

Klein, J., Moon, Y., and Picard, R. W. (2002). This computer responds to user frustration: Theory, design, and results. *Interacting with Computers*, 14(2), 119–140,

Knipling, R. R. and Wierwille, W. W. (1994). Vehicle-based drowsy driver detection: Current status and future prospects. In *Proceedings of the IVHS America Fourth Annual Meeting*, pp. 245–256. April 17–20, Washington, DC: National Highway Traffic Safety Administration, Office of Crash Avoidance Research.

Kramer, A. F. (1991). Physiological metrics of mental workload: A review of recent progress. In D. L. Damos (ed.), *Multiple-Task Performance*, pp. 279–328. London, UK: Taylor and Francis.

Kramer, A. F., Trejo, L. J., and Humphrey, D. G. (1996). Psychophysiological measures of workload: Potential applications to adaptively automated systems. In R. Parasuraman and M. Mouloua (eds), *Automation and Human Performance: Theory and Applications*, pp. 137–162. Mahwah, NJ: Lawrence Erlbaum Associates.

Kruse, A. A. (2007). Operational neuroscience: Neurophysiological measures in applied environments. *Aviation, Space, and Environmental Medicine*, 78(5), B191–B194.

Lei, S. and Roetting, M. (2011). Influence of task combination on EEG spectrum modulation for driver workload estimation. *Human Factors: The Journal of the Human Factors and Ergonomics Society*, 53(2), 168–179.

Mathan, S., Dorneich, M., and Whitlow, S. (2005). Automation etiquette in the augmented cognition context. In *Proceedings of the 11th International Conference on Human-Computer Interaction (1st Annual Augmented Cognition International)*, pp. 560–569. July 22–27, Mahwah, NJ: Lawrence Erlbaum Associates.

Mathan, S., Whitlow, S., Dorneich, M., Ververs, P., and Davis, G. (2007). Neurophysiological estimation of interruptibility: Demonstrating feasibility in a field context. In D. D. Schmorrow, D. M. Nicholson, J. M. Drexler, and L. M. Reeves (eds), *Foundations of Augmented Cognition*, 4th edn., pp. 51–58. Arlington, VA: Strategic Analysis Inc. and the Augmented Cognition International (ACI) Society.

McCleary, R. A. (1950). The nature of galvanic skin response. *Psychological Bulletin*, 47(2), 97–117.

McCraty, R., Atkinson, M., Tomasino, D., and Bradley, R. T. (2009). The coherent heart: Heart-brain interactions, psychophysiological coherence, and the emergence of system-wide order. *Integral Review*, 5(2), 10–115.

McCraty, R., Atkinson, M., Tiller, W. A., Rein, G., and Watkins, A. D. (1995). The effects of emotions on short-term power spectrum analysis of heart rate variability. *American Journal of Cardiology*, 76(14), 1089–1093.

Mooshage, O., Distelmaier, H., and Grandt, M. (2002). Human centered decision support for anti-air warfare on naval platforms. In *Proceedings of RTO Human Factors and Medicine Panel (HFM) Symposium (RTO-MP-008)*, pp. 5-1–5-14. October 7–9, Neuilly-sur-Seine, France: NATO Research and Technology Organisation.

Morishima, Y., Okuda, J., and Sakai, K. (2010). Reactive mechanism of cognitive control system. *Cerebral Cortex*, 20(11), 2675–2683.

Murray, I. R. and Arnott, J. L. (1993). Toward the simulation of emotion in synthetic speech: A review of the literature on human vocal emotion. *Journal of the Acoustic Society of America*, 93(2), 1097–1108.

Nicholson, J., Takahashi, K., and Nakatsu, R. (2000). Emotion recognition in speech using neural networks. *Neural Computing and Applications*, 9(4), 290–296.

Niedermeyer, E. and Lopes da Silva, F. (1999). *Electroencephalography: Basic Principles, Clinical Applications and Related Fields*, 4th edn. Baltimore, MD: Williams and Wilkins.

Nunez, P. L. (2000). Toward a quantitative description of large-scale neocortical dynamic function and EEG. *Behavioral and Brain Sciences*, 23(3), 371–398.

Owens, J. M., Goodman, L. S., and Pianka, M. J. (1984). Interference effects of vocalization on dual task performance (Report No. NAMRL-1309). Pensacola, FL: Naval Aerospace Medical Research Laboratory.

Parasuraman, R. (1990). Event-related brain potentials and human factors research. In J. W. Rohrbaugh, R. Parasuraman, and R. Johnson (eds), *Event-Related Brain Potentials: Basic Issues and Applications*, pp. 279–300. New York, NY: Oxford University Press.

Parasuraman, R. (2003). Neuroergonomics: Research and practice. *Theoretical Issues in Ergonomics Science*, 4(1–2), 5–20.

Parasuraman, R. (2012). Mental workload, stress, and individual differences: Cognitive and neuroergonomic perspective. In C. D. Wickens and J. G. Hollands (eds), *Engineering Psychology and Human Performance*, 4th edn., pp. 346–376. Upper Saddle River, NJ: Pearson.

Parasuraman, R., Barnes, M., and Cosenzo, K. (2007). Adaptive automation for human-robot teaming in future command and control systems. *The International C2 Journal, 1*(2), 43–68.

Parasuraman, R. and Hancock, P. (2004). Neuroergonomics: Harnessing the power of brain science for HF/E. *Bulletin of the Human Factors and Ergonomics Society*, 47(1), 4–5.

Pfeffer, S., Decker, P., Maier, T., and Stricker, E. (2013). Estimation of operator input and output workload in complex human-machine-systems for usability issues with iflow. In D. Harris (ed.), *Engineering Psychology and Cognitive Ergonomics. Understanding Human Cognition*, pp. 167–176. Berlin, Germany: Springer.

Pilcher, J. J., Nadler, E., and Busch, C. (2002). Effects of hot and cold temperature exposure on performance: A meta-analytic review. *Ergonomics*, 45(10), 682–698.

Pleydell-Pearce, C. W., Whitecross, S. E., Butler, S., Warren, W. J., and Newton, P. D. (1999). Development of a cognition monitor (Report No. CHS8205). Farnborough, UK: Defence Evaluation and Research Agency.

Pope, A. T., Bogart, E. H., and Bartolome, D. S. (1995). Biocybernetic system evaluates indices of operator engagement in automated task. *Biological Psychology*, 40(1), 187–195.

Prendinger, H., Ma, C., and Ishizuka, M. (2007). Eye movements as indices for the utility of life-like interface agents: A pilot study. *Interacting with Computers*, 19(2), 281–292.

Prinzel, L. J., Freeman, F. G., Scerbo, M. W., Mikulka, P. J., and Pope, A. T. (1999). A closed loop system for examining psychophysiological measures for adaptive task allocation. *The International Journal of Aviation Psychology*, 10(4), 393–410.

Prinzel, L. J., Parasuraman, R., Freeman, F. G., Scerbo, M., Mikulka, P. J., and Pope, A. (2003). Three experiments examining the use of electroencephalogram, event-related potentials, and heart-rate variability for real-time human-centered adaptive automation design (Report No. NASA/TP-2003-212442). Langley, VA: NASA Langley Research Center.

Rosetta Stone. (2014). Rosetta stone—Language-learning software with speech recognition [Computer software]. Retrieved July 31, 2014 from http://www.rosettastone.com/speech-recognition.

Russell, C. A. (2005). Operator state estimation for adaptive aiding in uninhabited combat air vehicles (Report No. AFIT/DS/ENG/05-01). Wright Patterson Air Force Base, OH: Air Force Institute of Technology.

Sanders, M. S. and McCormick, E. J. (1993). *Human Factors in Engineering and Design*. New York, NY: McGraw-Hill.

Scerbo, M. W. (1996). Theoretical perspectives on adaptive automation. In R. Parasuraman and M. Mouloua (eds), *Automation and Human Performance: Theory and Applications*, pp. 37–63. Hillsdale, NJ: Lawrence Erlbaum Associates.

Scerbo, M. W. (2007). Adaptive automation. In R. Parasuraman and M. Rizzo (eds), *Neuroergonomics: The Brain at Work*, pp. 239–252. Oxford, UK: Oxford University Press.

Scerbo, M. W., Freeman, F. G., Mikulka, P. J., Parasuraman, R., Di Nocero, F., and Prinzel III, L. J. (2001). The efficacy of psychophysiological measures for implementing adaptive technology (Vol. 211018). Hampton, VA: National Aeronautics and Space Administration, Langley Research Center.

Schiflett, S. G. and Loikith, G. J. (1980). Voice stress analysis as a measure of operator workload (Report No. NATC-TM79-3SY). Patuxent River, MD: Naval Air Test Center.

Schmorrow, D. D. and Kruse, A. A. (2004). Augmented cognition. In W. S. Bainbridge (ed.), *Berkshire Encyclopedia of Human Computer Interaction*, pp. 54–59. Great Barrington, MA: Berkshire Publishing Group.

Schmorrow, D. D. and Stanney, K. M. (eds.). (2008). *Augmented Cognition: A Practitioner's Guide*. Santa Monica, CA: Human Factors and Ergonomics Society.

Schnell, T., Keller, M., and Macuda, T. (2007). Application of the cognitive avionics tool set (CATS) in airborne operator state classification. In D. D. Schmorrow, D. M. Nicholson, J. M. Drexler, and L. M. Reeves (eds.), *Foundations of Augmented Cognition*, 4th edn., pp. 3–12. Arlington, VA: Strategic Analysis Inc. and the Augmented Cognition International (ACI) Society.

Sciarini, L. W., Fidopiastis, C. M., and Nicholson, D. M. (2009). Towards a modular cognitive state gauge: Assessing spatial ability utilization with multiple physiological measures. *Proceedings of the Human Factors and Ergonomics Society Annual Meeting*, *53*(3), 146–150.

Sharpe, L., Tarrier, N., Schotte, D., and Spence, S. H. (1995). The role of autonomic arousal in problem gambling. *Addiction*, *90*(11), 1529–1540.

Singh, H., Bhatia, J. S., and Kaur, J. (2011). Eye tracking based driver fatigue monitoring and warning system. In *Proceedings of the 2010 India International Conference on Power Electronics*, pp. 1–6. January 28–30, IEEE.

Stanney, K. M., Hale, K. S., Fuchs, S., Baskin, A., and Berka, C. (2011). Training: Neural systems and intelligence applications. *Synesis: A Journal of Science, Technology, Ethics and Policy*, *2*(1), 38–44.

St. John, M., Kobus, D. A., and Morrison, J. G. (2003). DARPA augmented cognition technical integration experiment (TIE) (DARPA Technical Report No. 1905). San Diego, CA: SPAWAR Systems Center.

St. John, M., Kobus, D. A., Morrison, J. G., and Schmorrow, D. (2004). Overview of the DARPA augmented cognition technical integration experiment. *International Journal of Human-Computer Interaction*, *17*(2), 131–149.

Strauss, M., Reynolds, C., and Picard, R. W. (2005). The handwave bluetooth skin conductance sensor. In *Proceedings of the First International Conference on Affective Computing and Intelligent Interaction*, pp. 699–706. October 22–24, Berlin, Germany: Springer Berlin Heidelberg.

Thought Technology Ltd. (2010). GSR Temp 2X. Retrieved July 31, 2014 from http://thought-technology.com/index.php/hardware/gsr-temp-2x-b-f-system.html.

Trejo, L. J., McDonald, N. J., Matthew, R., and Allison, B. Z. (2007). Experimental design and testing of a multimodal cognitive overload classifier. In D. D. Schmorrow, D. M. Nicholson, J. M. Drexler, and L. M. Reeves (eds), *Foundations of Augmented Cognition*, 4th edn., pp. 13–22. Arlington, VA: Strategic Analysis Inc. and the Augmented Cognition International (ACI) Society.

Tremoulet, P. D., Craven, P., Wilcox, S., Regli, S. H., Barton, J., Stibler, K., Gifford, A. et al. (2007). Initial validation of tool for interface design evaluation with sensors (TIDES). In D. D. Schmorrow, D. M. Nicholson, J. M. Drexler, and L. M. Reeves (eds), *Foundations of Augmented Cognition*, 4th edn., pp. 41–50. Arlington, VA: Strategic Analysis Inc. and the Augmented Cognition International (ACI) Society.

Vartak, A. A., Fidopiastis, C. M., Nicholson, D. M., Mikhael, W. B., and Schmorrow, D. (2008). Cognitive state estimation for adaptive learning systems using wearable physiological sensors. In *Proceedings of the First International Conference on Biomedical Electronics and Devices, BIOSIGNALS*, pp. 147–152. January 28–31, IEEE.

Veltman, J. A. and Gaillard, A. W. K. (1996). Physiological workload reactions to increasing levels of task difficulty (TNO Report No. TM-96-A026). Soesterberg, The Netherlands: TNO Human Factors Research Institute.

Vidal, M., Turner, J., Bulling, A., and Gellersen, H. (2012). Wearable eye tracking for mental health monitoring. *Computer Communications*, *35*(11), 1306–1311.

Villringer, A. and Chance, B. (1997). Non-invasive optical spectroscopy and imaging of human brain function. *Trends in Neurosciences*, *20*(10), 435–442.

Wickens, C. D. (2002). Multiple resources and performance prediction. *Theoretical Issues in Ergonomics Science*, *3*(2), 159–177.

Wickens, C. D. (2008). Multiple resources and mental workload. *Human Factors: The Journal of the Human Factors and Ergonomics Society*, *50*(3), 449–455.

Wickens, C. D., Hollands, J. G., Banbury, S., and Parasuraman, R. (2012). *Engineering Psychology and Human Performance*, 4th edn. Boston, MA: Pearson.

Willis, A. R. (2005). Cognitive cockpit project presentation [PowerPoint slides]. In *Presented at the 2nd Situation Awareness and Team Cognition Workshop*. Laval, Canada: Laval University.

Wilson, G. F. and Russell, C. A. (2006). Psychophysiologically versus task determined adaptive accomplishment. In D. D. Schmorrow, D. M. Nicholson, J. M. Drexler, and L. M. Reeves (eds), *Foundations of Augmented Cognition*, 2nd edn., pp. 201–207. Arlington, VA: Strategic Analysis Inc.

Wilson, G. F. and Russell, C. A. (2007). Performance enhancement in an uninhabited air vehicle task using psychophysiologically determined adaptive aiding. *Human Factors: The Journal of the Human Factors and Ergonomics Society*, *49*(6), 1005–1018.

Wood, S. D. (2006). Automated behavior-based interaction customization for military command and control. Ann Arbor, MI: Soar Technology. Retrieved July 31, 2014 from Penn State website: http://citeseerx.ist.psu.edu/viewdoc/download?doi=10.1.1.116.4051&rep=rep1&type=pdf.

Young, P. M., Clegg, B. A., and Smith, C. A. P. (2004). Dynamic models of augmented cognition. *International Journal of Human-Computer Interaction*, *17*(2), 259–273.

7 Key Considerations for IAS Design*

7.1 OBJECTIVES

- Describe why it is necessary to consider a wide range of issues and constraints when designing an IAS for a specific application
- Describe how operational priorities can shape and influence IAS development
- Describe why ethical and legal considerations should be addressed early on in the design process
- Describe a range of possible adaptation types and options for adaptation triggers
- Describe an adaptation taxonomy that can be used as a road map to support IAS design

This book is intended as a resource for systems designers and developers to assist in IAS design from an interaction-centered point of view using coherent systems design guidance and methodologies. In Chapter 1, the world of current adaptive systems was discussed, including examples of good and poor systems design and associated human–machine interactivity and its implications. Explanation was also provided regarding the scope of human–machine and human–automation interaction, as well as some of the concepts behind developing good human interaction-centered designs. The information discussed in Chapters 2–6 regarding interface and automation technologies, conceptual architectures, analytical techniques, agent-based design methods, and the use of behavioral- and psychophysiological-based operator state monitoring approaches is intended to serve as guidance for designing and developing IASs. Part of the guidance provided in this chapter is based on previous chapters, and part is based on practical experience.

In previous chapters, the focus has been on identifying IAS capabilities from an ideal perspective—the technological capabilities of IASs have been discussed without considering the practical constraints associated with designing and developing systems. In this chapter, guidance is structured around an IAS implementation (or project-based) perspective, and practical constraints such as operational requirements, ethical and legal considerations, and project management issues are considered.

When designing an IAS, there will also likely be issues regarding access to suitable subject matter experts (SMEs) and experienced analysts to populate the task and operator models that need to be implemented within the IAS. Each project will also have its own unique constraints, such as the issues that arise when implementing IAS technologies within a particular target domain. For example, it is often the case that domains that are physically and cognitively demanding on operators—such as jet aircraft—are also equally demanding on IAS technologies due to factors such as space, weight, interference, and vibration. Additionally, a number of limitations inherent in the implementation of operator state monitoring within IASs, previously discussed in Chapter 6, must be considered when making decisions about which operator state measurements can be successfully monitored.

7.2   CONSIDERING DESIGN CONSTRAINTS

Guidance is important because it allows systems designers and developers to understand at the very beginning of an IAS development project what is possible, and not possible, in terms of adaptation options. When technological solutions that are impossible to implement due to issues of technological maturity, operating environment, and so on, have been eliminated, the remaining options are all possibilities.

Chapter 3 discussed the basic anatomy (i.e., main functional components) of an IAS. The framework is relatively generic (i.e., context independent) and scalable (i.e., adaptable for a particular application), and it allows an understanding of the broad scope of what is feasible in general conceptual terms. Chapter 4 discussed analytical techniques to help determine operational system and interaction-centered requirements based on working environment. After determining basic requirements, scope should be filtered in terms of ethical and legal issues, and implementation possibilities must be defined. For example, eye tracking is a generic capability that can be utilized in IASs, but the possibility of its actual implementation is determined by the conditions and environment within which it would be used—it is easier to implement eye tracking in a car context than in a jet cockpit context.

In any project, there will be constraints. Some constraints surface when considering the project from different perspectives and some are unique to the application domain for which the IAS is being developed. It is extremely important to identify these constraints as early in the design process as possible. Previous discussions have identified the following types of constraints:

- *Constraints from an operational perspective*: For example, what will the system be used for, and by whom?
- *Constraints from an implementation perspective*: For example, in what environmental conditions will the system be expected to operate?
- *Constraints from a technological perspective*: For example, what is the maturity of available technology and is it suitable for both the operational context and project constraints?

There are also less tangible but equally critical constraints that can ultimately determine the success or failure of an IAS, including operational priorities, ethical and legal considerations, and adaptation feasibilities. Figure 7.1 illustrates some of the questions that systems designers must answer in order to achieve good systems design.

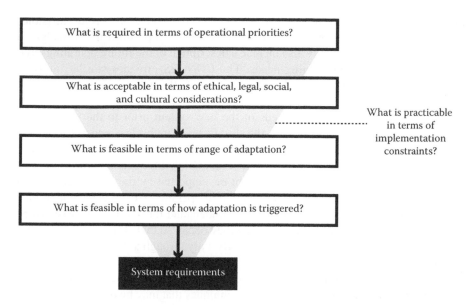

FIGURE 7.1   Sequence of determination for known project constraints that impact IAS design.

7.3   DETERMINING OPERATIONAL PRIORITIES

Chapter 2 discussed the history and evolution of intelligent adaptive interface (IAI) and intelligent adaptive automation (IAA) technologies as the two basic components of an IAS. Many different levels of adaptation and triggering opportunities were then described, all of which must be carefully considered to match the technology with the intentions of the workspace. The various roles and automation levels a particular IAS is expected to achieve provide the scope of its IAS design and development project.

The first consideration in designing an IAS is defining what is required of the system based on operational priorities. What are the operational or user requirements that need to be met by the IAS? In the case of the Pilot's Associate, Rotocraft Pilot's Associate, and Cognitive Cockpit, discussed in Chapter 3, pilots were overwhelmed at various points in the mission and needed assistance to increase operational effectiveness and pilot safety. In the case of driverless cars, user requirements might include improved safety; a reduction in running cost, which is derived from increased fuel efficiency; and a better work–life balance, which is derived from more productive commuting time. Generally, operational and user priorities are related to one or more of the following top-level goals:

- Error reduction
- Performance improvement
- Efficiency increases
- Operator need or preference
- Health and safety

The high-level goals of a new system are often defined in a statement of operating intent (SOI) or concept of operations (CONOPS) document. The SOI is developed using input from a wide range of SMEs and identifies the intended roles, missions, tasks, and usage of the new system in sufficient detail to permit further analytical studies. This information may also include a description of the future operating environment and anticipated training needs. The CONOPS provides similar information, but is far more detailed, and might also include the definition of the roles and responsibilities of operators. Typically, these documents are created by the operational community to define the needs of the new system prior to the involvement of systems designers. They provide the design team with clear specifications for the top-level goals of the project. In order to outline the design of the IAS to meet the top-level goals described within the SOI and CONOPS documentation, it is first necessary to clarify how these goals might be supported by the implementation of IAS technologies. For example, if a reduction in operator errors has been identified as an operational priority, it is important to understand the sources of these errors, how they occur, why they occur, and when they occur, so that strategies for operator support by the IAS can be identified. Similarly, only through a thorough understanding of the mission or work activities of the operator can strategies for improving performance or increasing efficiency by implementing IAS technologies be identified.

Chapter 4 discussed several analytical techniques that may be used to decompose high-level goals into more detailed IAS implementation strategies and extract the information necessary to populate the various IAS models to a sufficient level of detail. These techniques are drawn from the HF field and focus on identifying the components of required cognitive work and decision making. They seek to better understand the work of the operator, the information required, and the decisions and actions that must take place. This information is useful in multiple ways: (a) it can be used to select the functions that are the best candidates for automation; (b) it can inform the OMI design; and (c) it can be used to develop simulations and test environments for IAS evaluation.

Although they may differ in their details and origins, these techniques share many of the same goals; however, the selection of a particular technique, or techniques, should fit best with the contextual design problem and the type of solution being designed. Figure 7.2 illustrates a decision tree for selecting analytical techniques based on project constraints that influence the development process. All constraints should be considered as a whole, and in practice it is likely that trade-offs among the constraints will be required. Project constraints are described in terms of:

- *Time for analysis*: The time available to conduct the analysis as a function of the project budget or project schedule.
- *Tasks for analysis*: Types of tasks that need to be analyzed. Analytical techniques are subdivided into those that are best to be suited to analyze simple, complex, and cognitive tasks.
- *Analysts*: Project personnel required to conduct the analysis. Analytical techniques are subdivided into those that are simple to learn and apply, and those that are difficult to learn and apply.

Figure 7.2 also illustrates that the analytical techniques tend to fall into two distinct categories: those that can be used for complex tasks, but are time consuming

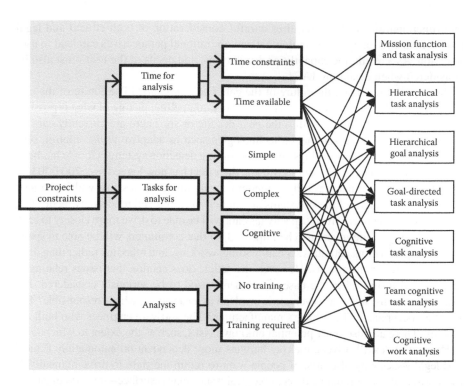

FIGURE 7.2 Decision tree to aid in the selection of appropriate analytical techniques based on known project constraints.

and require well-trained analysts; and those that can be used for simple tasks, in less time and with less well-trained analysts.

Regardless of the analytical technique selected, an important first step is to expand on the SOI and CONOPS documentation (or similar artifacts) by specifying the operators and their likely characteristics; the anticipated system functions and features; and the likely environments and scenarios in which the system will operate. The scenarios describe the sequence of actions and events associated with the execution of a particular work activity and are often written as a narrative, such as a detailed description of a typical day in the life of an operator using the future system within the operating environment specified by the CONOPS. This information provides a foundation for conducting detailed analysis activities. The scenarios will also underpin subsequent high-level IAS design and validation activities.

7.4 CONSIDERING ETHICAL, LEGAL, SOCIAL, AND CULTURAL ISSUES

Technology has matured to the point that it is no longer dependent on human involvement to perform activities such as driving cars or tracking and firing at enemy targets. Systems designers must now consider "should we?" rather than "can we?" Tension between what is possible and what is acceptable in terms of the range of capabilities and functionalities in IASs requires careful consideration of both ethical and legal issues; focusing solely on technological and operational perspectives can lead to user rejection. As a result, the capabilities, limitations, and needs of the user must also be considered when designing IASs.

In a special report on the future of the car, *The Economist* (The future of the car: Clean, safe and it drives itself, 2013) paints a very compelling picture of what is possible in terms of car automation within the next decade or so. There are currently cars on the market that feature automation technologies such as adaptive cruise control, self-parking, lane departure warnings, and auto-braking detection systems. Google has been working on self-driving cars since 2010 and their fleet has now logged over 700,000 km. Industry experts predict that fully autonomous cars might be possible within the next 20 years. Advantages of these systems are numerous: improved road safety, better fuel efficiency, and increased road capacity (as cars will be able to drive more closely to each other). Quality of life might also be improved, in that commuters will be able to spend their time performing activities such as reading, working, and relaxing, rather than driving in stressful environments. However, the article does caution that issues relating to legality and insurance in the case of accidents need to be seriously considered. For example, if a fully automated vehicle is involved in an accident, who is responsible? The vehicle's occupants? The vehicle's manufacturer? The software engineer who built the system? There are also other potential unknown risks, such as the extent to which technologies will deskill drivers, making humans more dependent on automation. Ethical and legal issues may also arise if people want to retain the right to drive manually. If these wants are accommodated, technological issues may resurface—how do manually and automatically driven cars operate on the same highway? For a review of HF concerns related to driverless cars, see Cummings and Ryan (2013).

While the ethical and legal ramifications of driverless cars seem extensive, other IAS designs require even more consideration. The British Broadcasting Corporation (Hughes, 2013) cites a campaign calling for an international ban on "killer robots" due to concerns about the ethical and legal implications of fully autonomous aircraft drones. The drones had been operated by NATO forces in Afghanistan as human-in-the-loop systems, their targets selected and fired on by human command. Campaigners argued that it was only a matter of time until fully autonomous human-out-of-the-loop systems capable of selecting targets and firing on them without human involvement would be developed. They also argued that giving machines increased power over human life and death would be an unacceptable application of technology, and would pose fundamental challenges to human rights and humanitarian laws.

More than 60 years after Isaac Asimov penned three "laws" in his collection of short stories *I, Robot*, they still reflect human understanding and expectation regarding the relationships of humans and autonomous machines (Murphy and Woods, 2009). Two of those laws, and Murphy and Woods' proposed amendments, are extremely relevant to the ethics discussions systems designers face when they develop new IASs.

The first of Asimov's laws, "A robot may not injure a human being, or through inaction, allow a human being to come to harm," seems like simple and ethical common sense, but it is not so easily maintained in real-world situations, as is demonstrated by the weaponized border patrol robots operating in South Korea (Murphy and Woods, 2009). The main challenge of implementing Asimov's first law is that it places the responsibility for the safety of the operator with the machine. From a legal point of view, the machine can be viewed as a product and the operator or manufacturer can be held responsible for its actions. Therefore, the notion of legal and professional responsibility of those who design and deploy autonomous machines must also be considered. As a result, Murphy and Woods (2009) proposed amending the original law to "A human may not deploy a robot without the human–robot work system meeting the highest legal and professional standards of safety and ethics."

Asimov's second law, "A robot must obey orders given to it by human beings, except where such orders would conflict with the first law," speaks to morality. Murphy and Woods (2009) argue that this law ignores the complexity and dynamics of relationships and responsibilities between human and machine. The alternative law they propose is "A robot must respond to humans as appropriate for their roles."

Although the majority of ethical issues are human centered, Whitby (2008) has proposed that machines have basic rights and their designers must ensure that these technologies are not mistreated by their operators by designing out unethical human behavior using human–human social interaction knowledge as a template.

Generally speaking, IASs are not benign tools; they can engender social relationships between themselves and their users. In fact, many researchers have deliberately engineered knowledge about social relationships into the design of intelligent systems so that they are sensitive to individual cultural, social, and contextual differences, and follow good etiquette based on social norms that apply to the user (Sheridan and Parasuraman, 2006). Miller et al. (2007) developed simulated game agents to interpret etiquette directed at them from the user and generated polite behaviors in response using Brown and Levinson's (1987) seminal work on politeness across a range of cultures: for example, using the word "please" to mitigate or soften direct expressions, such as "Please pass the salt."

IASs used in educational settings, further discussed in Chapter 8, exploit knowledge of human–human social relationships to enhance student learning experiences (Mishra and Hershey, 2004). For example, they use personalized messages, affective feedback (e.g., praise rather than criticism), and even humor. Although there are challenges associated with representing social knowledge within IASs, Miller et al. (2007) suggest that a wide range of social interactions could be modeled using quantitative computational tools.

7.5 DETERMINING AUTHORITY ROLES

Interaction-centered IASs share responsibility, authority, and autonomy with operators over many system behaviors. As discussed in Chapter 1, the main motivation for creating IASs is to reduce operator workload and information overload. However, while operators wish to remain in control, and it is desirable for them to do so, today's complex systems do not allow operators to be fully in charge of all system functions—especially not in the same way as they used earlier workstations.

The question of who is in charge was addressed by Moss et al. (1984), who suggested that task and decision allocation should be completed according to the mission goals and the capabilities of the human and the machine. When tasks can only be performed by either the human or the machine, task allocation is simple. When both can perform a task, the task should be allocated by considering which one would perform the task better and what the mission impact of the allocation would be. Moss et al. (1984) suggest that the large data-handling capabilities of machines suit them for processing large amounts of sensor data. Knowledge of mission goals and operator information requirements allows the data to be collapsed intelligently into a form readily interpreted by the operator, through the IAI. The operator uses this fused data to make high-level decisions and decisions with uncertain results—tasks at which humans are superior to machines—to accomplish system mission or work goals as effectively as possible. Under these circumstances, authority allocation is a dynamic, goal-oriented process, dependent on the state of the human team members and the status of the machine team members relative to their environment.

Experience with IASs has consistently shown that the concept of human–machine interaction is highly influenced by a basic sociological phenomenon—operators of complex systems almost always want to remain in charge of the machine. For example, the Pilots Associate research program developed a list of prioritized goals for a good cockpit configuration manager. Two of the top three items on the list were "Pilots remain in charge of task allocation" and "Pilots remain in charge of information presented" (Miller et al., 1999).

A survey conducted by NASA (Tenney et al., 1995) on civil pilot opinions on high-level flight deck automation issues found that participants were nearly unanimous in proclaiming that the pilot should still be responsible for flying the aircraft in the future. They also showed a preference for automation that assists the pilot in problem solving, instead of automation that automatically solves problems. The majority of participants felt that the main need for further automation was to alleviate additional cognitive demands imposed on them in time-constrained decision-making situations and preferred that automated systems advise, rather than take charge. Interestingly, a more recent study on pilot attitudes toward automation found that increased automation over the past two decades resulted in pilots sometimes relying too much on automated systems and reluctance to take control from the machine (Flight Deck Automation Working Group, 2013).

A study conducted to ascertain the opinions of UK Royal Air Force crew members on automation and decision support to reduce workload, improve mission effectiveness, and improve SA showed significantly different opinions as to which systems should have automation and to what level (Banbury, 1997). The general consensus was in favor of high levels of automation; however, aircrew also wanted to remain in the decision-making loop and to have an interactive role in the HMS. Single-seat aircrew preferred increased automation of the defensive subsystems to the point where little human involvement was required. This suggests that the single-seat crew had fewer cognitive resources available to make appropriate decisions and effectively operate defensive systems. An earlier study on aircrew attitudes toward cockpit automation found that aircrew wanted high levels of automation to the point where they could retain ultimate executive control (Enterkin, 1994). The aircrew reported an underlying reluctance to place complete trust in automated systems and a desire to resume control if the system was in error or nonfunctional.

The Defense Science Board task force on the role of autonomy in Department of Defense systems (2012) concluded that, while the potential of autonomy is great, there have been many obstacles to broader acceptance of unmanned systems, and, specifically, the autonomous capabilities needed to realize the benefits of autonomy in military applications. Not only do operators not fully trust that these systems will operate as intended; they also mistake machine autonomy as the capability to think and act independently of the human operator. The task force recommended that the design and operation of autonomous systems be framed within the context of human–machine interaction. Thus, even 20 years later, the issues tackled during the early electronic crew member conferences discussed in Chapter 3 still resonate and shape policy, despite great strides in the technological capability of these systems.

7.6 DETERMINING RANGE OF ADAPTATION

Chapters 1 and 2 discussed the evolution of IASs over the past few decades and demonstrated the possibilities for human–machine interaction, especially in terms of how the machine can adapt to operator needs. Recent advances in real-time operator state assessment, discussed in Chapter 6, have facilitated the development of very sophisticated adaptive systems, such as those discussed in Chapter 3. The range of functions that can now be performed by machines provides systems designers with many possibilities when implementing IAS technologies for a specific application. For example, Feigh et al. (2012) describe four types of adaptation that have been used in a wide range of applications: (a) function allocation, (b) task scheduling, (c) interaction, and (d) content.

7.6.1 MODIFICATION OF FUNCTION ALLOCATION

Chapter 2 discussed function allocation, which is the process of assigning functions and tasks between the operator and the machine. IASs have dynamic function allocation—functions are continually reassigned to either operator or machine while the system is in use. As discussed in Chapter 3, this dynamic function allocation can be either "explicit" or "implicit"; explicit allocation allows the operator to maintain control over task allocation, whereas implicit allocation gives the machine control of task allocation.

The distribution of responsibility and authority is inherent within function allocation. Feigh et al. (2012) make the distinction between task sharing, in which tasks are performed by both human and machine, and task off-loading, in which tasks are moved from the human to the machine. Dynamic function allocation usually involves a combination of the two. For example, in situations of relatively low workload the operator might prefer to share functions with the machine, and if operator workload increases due to situation complexity the operator might choose to off-load a subset of functions to the machine in order to concentrate fully on mission-critical tasks.

Decisions regarding function allocation are complex and depend on many variables. As discussed in Chapter 1, several researchers have attempted to label the range of possible function allocation using a continuum; at the lowest level of automation the human performs all tasks, and at the highest level the machine performs all tasks (Sheridan and Verplank, 1978). Building on this, continuums corresponding to cognitive tasks, such as monitoring, generating, selecting, and implementing (Endsley and Kaber, 1999), or specific information processing stages, such as sensory processing, perception and working memory, and decision-making and response selection (Parasuraman et al., 2000), have been created. Notably, the Defense Science Board task force on the role of autonomy in Department of Defense systems (2012) cautioned against these types of frameworks, which focus on the machine's capabilities at the expense of defining the scope of human–machine collaboration.

Miller and Parasuraman (2007) are also cautious about this type of framework, and have argued the need for a fifth stage of function decomposition—delegation. They believe there are multiple options for allocating the hierarchical sequences of activities that comprise functions to either human or machine depending on context. In their understanding, a single IAS could occupy multiple points on a levels of automation continuum.

7.6.2   MODIFICATION OF TASK SCHEDULING

Adaptation occurs when the system dynamically changes task timing, task priority, and task duration. This type of adaptation is especially useful for operators that work in multitasking environments and endure frequent interruptions. Adaptation of task scheduling can be used to reduce downtime and optimize performance. For example, tasks that require specific information or resources to be available, or require other tasks to be completed first, are ideal candidates for IAS support.

Task timing is a challenge for humans working in complex, event-driven environments (Ho et al., 2004), as they often do not plan ahead when experiencing high levels of workload (Tulga and Sheridan, 1980). The goal of supporting the operator in these cases is to schedule tasks to optimize resources and reduce downtime.

Task priority can also be used to schedule operator tasks. For example, the impact of a threat on safety or mission objectives can be used to dynamically modify the order in which tasks are carried out, or whether a particular task is even carried out at all. Completion times or time allocated for task accomplishment can also be modified. For example, deadlines can be shortened in low-workload situations, or for safety-critical tasks, and extended for low-priority tasks. Systems designers should note that the dynamic modification of task scheduling may require interrupting the operator, such as in the case of initiating a new high-priority task before the operator completes an active low-priority task. There is a large body of research demonstrating a "switch cost" associated with interrupting the task-at-hand with a new task. Wickens et al.'s (2012) *Engineering Psychology and Human Performance* contains a recent review. To minimize these issues, it is extremely important that systems designers identify good and bad times to interrupt and reschedule tasks involving the operator and incorporate this knowledge into their IAS design. This knowledge is obtained from a thorough analysis of the cognitive requirements of the range of functions and tasks that the operator is expected to undertake within a system using the analytical techniques discussed in Chapter 4.

7.6.3 MODIFICATION OF INTERACTION

An IAS adapts when it dynamically changes how it interacts with its operator, such as changing how information is exchanged between the operator and the machine, and when and where the exchange takes place. These modifications might include how the information is presented to the human on the OMI: for example, using high-lighting or context-dependent menus to direct attention to critical information or streamline interaction (Feigh et al., 2012), or changing the modality of the information presentation, such as from visual to verbal, to maximize the amount of information that can be presented to the operator (Wickens et al., 2012).

Human–machine interaction can also be modified directly to govern whether the information exchanged between human and machine is given or requested (Entin and Entin, 2001), and who is in control of the interaction. Human operators might delegate the performance of a particular function to a machine, while retaining final authority over the task; this type of interaction is common for safety-critical functions. As discussed in Chapter 2, if there is a risk of an operator becoming incapacitated, the machine partner would be able to take authority to ensure human survival.

7.6.4 MODIFICATION OF CONTENT

The final type of adaptation occurs when the machine dynamically changes what information it presents to the operator, including what categories of information are presented and at what level of detail or abstraction. For example, a car GPS could automatically adjust the scale of the map according to the speed of the vehicle; speeds common for highway use could prompt the GPS to zoom out, as the driver would not need detailed local information, and slower speeds could prompt a more detailed map, as the driver would be more likely to want information that included street names and nearby points of interest. Information quality and quantity can be modified dynamically according to context.

7.7 DETERMINING ADAPTATION TRIGGERS

Chapter 3 discussed a design process for IASs. In many ways, this design process is very similar to other software design processes. The key difference in IAS design lies in the fact that IASs have a closed-loop feed-forward feature, as well as feedback capabilities, which enables well-designed systems to truly act as active partners and assist their human counterparts.

To achieve the goal of creating an IAS as an active partner and to provide the machine with an understanding of its human partner's capabilities and limitations in a specific situation at a particular moment, a generic IAS conceptual architecture was developed. The generic conceptual architecture discussed in Chapter 3 specified the requirements of the situation assessment module, the operator state assessment module, and the approach for adapting the information sent to the OMI through the adaptation engine module. In particular, it was noted that IASs are capable of monitoring several parameters of external situation status and internal operator state to determine the type, degree, and duration of support that can be deployed.

To facilitate deployment, Feigh et al. (2012) describe five categories of adaptation triggers: (a) operator based, (b) system based, (c) environment based, (d) task or mission based, and (e) location or time based.

7.7.1 OPERATOR-BASED TRIGGERS

In their simplest form, operator-based triggers are reliant on human request; the operator simply engages or disengages the automation as needed. For example, a driver would engage cruise control on an empty highway but disengage cruise control when maneuvering through busy traffic. More complex automation is based on delegation methods. Here, the human operator delegates tasks to the machine as they would to a human subordinate (Miller and Parasuraman, 2007). The advantages of this type of operator-based trigger are that the operator decides which tasks are delegated to the machine, how the tasks should be accomplished, and how much oversight is required.

For more adaptive systems, in which the engagement and disengagement of the automation can be initiated by the machine, another form of operator-based trigger is needed. This trigger is based on the real-time assessment of operator performance against known mission goals and cognitive state. Cognitive state, including fatigue, stress, and visual or verbal workload, can be determined using the wide range of psychophysiological measurements discussed in Chapter 6. Automation can be initiated when the machine detects high levels of operator workload, and turned off when it detects low levels of operator workload. To be useful, operator-based triggers require an accurate and comprehensive understanding of human cognition and behavior, as determined by the operator state assessment module discussed in Chapter 3.

7.7.2 MACHINE-BASED TRIGGERS

Triggers can be also based on current or anticipated machine status. Machine-based triggers might include physical parameters such as speed, heading, height, acceleration, or deceleration. For example, an adaptive vehicle navigation display could use machine status information, such as speed and heading, to adapt the amount of information displayed to the driver. Triggers can also be based on machine modes, which typically correspond to a set of situation-specific machine functions. For example, the cruise control mode on a vehicle incorporates the automation of several vehicle functions. If those functions were individually automated it would be difficult for the operator to monitor them all at once. Adapting automation based on machine mode triggers makes it much easier for the operator to effectively monitor the machine. Both types of machine triggers require an accurate and comprehensive machine model and are represented in the situation assessment module discussed in Chapter 3.

7.7.3 ENVIRONMENT-BASED TRIGGERS

Triggers can be based on environmental (i.e., external situation) status, such as ambient temperature, barometric pressure, light, humidity, or wind speed. Adaptations might also be triggered by discrete environmental events, such as an automatic car braking system triggered by the sharp speed reduction of the car in front. Both types of environmental trigger require an accurate and comprehensive world model, and are represented in the situation assessment module discussed in Chapter 3.

7.7.4   Mission- and Task-Based Triggers

Events that occur during the execution of an operator mission or task can also be used as adaptation triggers. In the case of mission-based triggers, operator actions during mission execution can be compared to the expected mission actions based on prior knowledge of mission goals, plans, and intent. The completion or failure of mission objectives can also be used as a trigger for adaptation. IASs that require a high fidelity of adaptation require a more detailed understanding of the tasks that the operator needs to accomplish in order to satisfy high-level mission goals. For example, the cockpit information manager of the Rotorcraft Pilot's Associate program discussed in Chapter 3 differentiates between planned, active, and completed tasks to drive adaptation support. Chapter 4 discussed a range of analytical techniques that can be used to decompose missions and tasks to the level required to populate the task model discussed in Chapter 2.

7.7.5   Location- and Time-Based Triggers

Location can be a simple and effective trigger for adaptation, as an absolute location or relative location from a particular mission event or environmental feature can be used. For example, a machine can automatically take control of an aircraft if the relative location of the ground to the aircraft is below a specific threshold.

Time can also be used to trigger automation. Adaptation cycles are defined as the frequency with which automation is turned on or off over a period of time. There is a continuum of short to long cycles of adaptive automation, and what constitutes short or long cycles is dependent on the particular task being performed. Note that the allocation of tasks from human to machine for brief time periods can produce a potentially uncontrollable oscillation between manual and automated control. This condition is referred to as automation cycling, which is more formally defined as the frequency of automation change over a specific time period (Hilburn et al., 1993; Parasuraman et al., 1996).

If uncontrolled, automation cycling can prove detrimental to overall performance. Short episodes of automation can prove so detrimental that the operator might shut the system off. The failure to understand potential operator responses to short episodes of automation can obviate the fundamental purpose of IASs. Systems designers should keep the following in mind when thinking about time-based adaptation triggers:

- Excessively long or excessively short adaptation cycles can limit the effectiveness of IASs in enhancing operator performance (Hilburn et al., 1993).
- Occasional brief reversions to manual control can counter some monitoring inefficiencies typically associated with long cycle automation (Hilburn et al., 1993).

- Supervision of dynamic tasks is significantly worse than supervision of more stable tasks (Parasuraman et al., 1996).
- Detection of automation failures is substantially degraded in systems containing static automation when the allocation of tasks between operator and system remains fixed over times of approximately 20 min (Parasuraman et al., 1996).

7.7.6 SELECTING ADAPTATION TRIGGERS

Chapter 6 discussed monitoring operator state, which is a key method for triggering IAS adaptation. Recent developments in biofeedback sensor technology make advanced operator state monitoring more feasible and affordable than it has been in the past. However, to effectively incorporate these technologies, systems designers first need to understand what they measure and what that means for operator state inferences, as well as the various challenges that may be faced when implementing the finished design. Figure 7.3 illustrates a decision tree for identifying appropriate operator state monitoring approaches based on the required capabilities of a

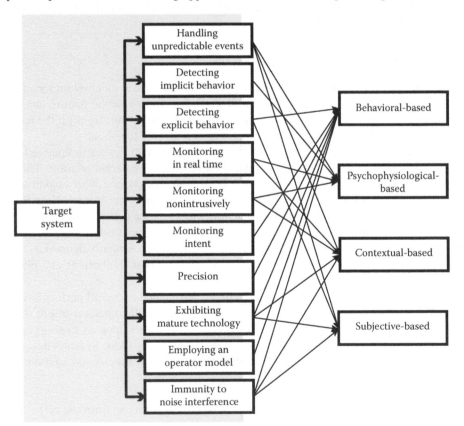

FIGURE 7.3 Decision tree to aid in defining appropriate operator state monitoring approach requirements. Operator state monitoring approaches are broadly classified as behavioral-based, psychophysiological-based, contextual-based, and subjective-based.

to-be-developed IAS. Operator state monitoring approaches are broadly classified as behavioral-based, psychophysiological-based, contextual-based, and subjective-based. All requirements should be considered together, and certain requirements make use of multiple monitoring approaches. In general, operator state monitoring approaches should be capable of:

- *Handling unpredictable events*: The system must assess operator state for events that occur outside normal operations.
- *Detecting implicit behavior*: The system must monitor operator state, including cognitive activities, with few or no observable operator actions.
- *Detecting explicit behavior*: The system must monitor operator state through observable operator actions, such as using vehicle controls.
- *Monitoring in real time*: The system must be able to determine operator state in real time.
- *Monitoring nonintrusively*: The system must ensure that the operator is not aware of, or impeded by, the monitoring system.
- *Monitoring intent*: The system must be able to determine operator intent, such as operator goals and objectives.
- *Precision*: The system must be able to determine operator state with precision, and not limit itself to whether the operator is overloaded or underloaded.
- *Exhibiting mature technology*: The system must contain technology that is ready to be fielded, or is in development.
- *Employing an operator model*: The system must use a highly accurate or customized operator model.
- *Immunity to noise interference*: The system must be immune to electrical–magnetic noise interference. Note that EEG monitoring is very susceptible to noise interference and as a result is very difficult to deploy in aircraft.

7.8 DEVELOPING AN ADAPTATION TAXONOMY

The information discussed within this book about conceptual architectures, analytical techniques, agent-based design methods, and the use of behavioral- and psychophysiological-based operator state monitoring approaches provides guidance for the design and development of IASs. Over the course of the last few chapters, the strengths and limitations of several analytical and operator state monitoring approaches have also been highlighted. This section outlines and describes a road map for the successful development of IASs, and an integral part of this road map is the creation of an adaptation taxonomy.

The development and implementation of IASs can be guided by a taxonomic approach that establishes a range of available options for system capabilities and functionalities. In addition, a taxonomic approach can assist in the creation of an audit trail for the system design by allowing all design decisions to be traced back to a set group of basic assumptions. A taxonomic approach also provides a road map for development, as it lets the development team focus on specific implementations after identifying the range of possibilities.

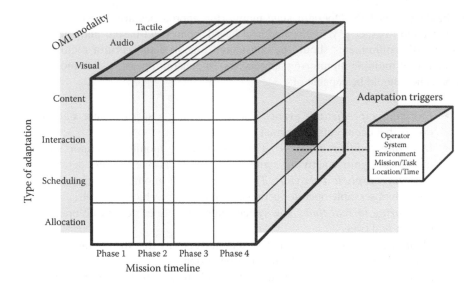

FIGURE 7.4 Illustration of an IAS adaptation taxonomy. The taxonomy has three axes: mission timeline, type of adaptation, and OMI modality.

The basic role of a taxonomy is to "describe the structure and relationships of the constituent objects in regard to each other and to similar objects, and to simplify these relationships in such a way that general statements can be made about classes of objects" (Fleishman and Quaintance, 1984). For example, a taxonomy developed for an IAS to support pilots might detail the following areas: (a) the role of the human; (b) the role of the system; (c) the level of automation or support possible; and (d) the operational requirements of the scenario in which both the human and the decision aid are expected to operate. In creating this taxonomy, responsibilities can be allocated between the human and machine for any given mission segment. Figure 7.4 illustrates the adaptation taxonomy approach as having three axes:

- *Mission timeline*: This axis captures the understanding that the types of decisions the operator will face over a mission or work activity will change, and, as a result, so will the type of support required from the IAS. For further information, see "Mission Timeline View" in *Task Force Report: The role of autonomy in DoD systems* (2012). For example, aircraft takeoff and landing require different types and levels of support. As a result, it is possible, and often desirable, that the operator and the machine interchange roles across the mission or work activity to meet IAS operational goals. Thus, the operational requirements of the mission or work activity within which the human and machine are expected to operate are represented on this axis by the temporal sequence in which they occur. For example, in the case of a commercial pilot, the mission phases could be flight planning, push back/

taxi/takeoff/climb, cruise, and descent/final approach/landing. Each phase can be decomposed into more detailed subtasks to the point where adaptation activities can be identified to support specific operator tasks or goals.

- *Type of adaptation*: This axis captures the types of decisions that need to be made by the operator and what support is required by the machine. For further information, see "Cognitive Echelon View" in *Task Force Report: The role of autonomy in DoD systems* (2012). The type of support provided by the machine can be categorized as one of the four adaptation methods described earlier in this chapter: (a) modification of allocation of function, (b) modification of task scheduling, (c) modification of interaction, and (d) modification of content. Additionally, the allocation of roles between the operator and the machine is defined.
- *OMI modality*: This axis represents the modality of the OMI through which interaction can occur. It includes the visual channel (e.g., dials, gauges, visual displays), the auditory channel (e.g., tones, frequency, alerts, voice warnings), and the haptic channel (e.g., spatial information presented using tactile actuators such as those found in pagers or cell phones). One or more of these modalities can be used for a particular instance of machine support during the mission or work activity.

By defining all the relevant factors, the taxonomic approach allows the design team to identify specific adaptation methods for any given mission task through specific OMI channels. As well, the appropriate adaptation triggers can be identified for each mission timeline, type of adaptation, and OMI instance.

A taxonomic approach also allows the development team to construct an appropriate mission scenario that encompasses the range of system capabilities and functionalities identified by the taxonomy. The mission scenario is used as a precursor to the functional decomposition of tasks, goals, and functions in analysis activities, and to determine measurements for effectiveness and performance in verification activities. The taxonomic approach also allows the development team to quickly determine IAS capabilities and functions in terms of priority and feasibility. This lets the development team maximize the impact of the IAS on operational performance and reduce development risks, such as depending on immature technology, within the time and budgetary constraints of the project.

The development of an adaptation taxonomy at the start of a project is a critical step in supporting effective and efficient IAS design and development. Chapter 1 discussed a number of automation taxonomies to illustrate the range of possible human–machine interaction and noted, in particular, how functions such as monitoring, acquiring, and analyzing information, and making decisions and acting upon them, can be allocated to either the operator or the machine. Figure 7.5 illustrates how the development of an adaptation taxonomy can be integrated into an IAS development road map, and links Chapters 3–6 with specific steps.

The first step of the road map requires determining the operational priorities of the IAS. As discussed earlier in this chapter, these are the high-level objectives of the project—the problems that the system is expected to solve—such as improving HMS performance by reducing operator workload during time- or safety-critical

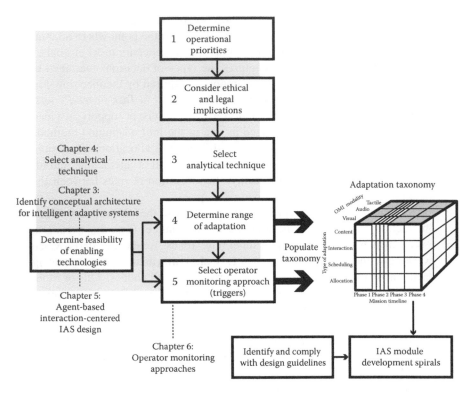

FIGURE 7.5 Development road map for the conceptual design of an IAS. Chapters containing relevant information and guidance are noted where applicable.

phases of the mission or work activity. The identification of these priorities can be assisted, in part, by using specific frameworks. For example, a framework developed by the Defense Science Board Task Force (2012) takes into account factors that influence the high-level objectives of the project, such as how well the IAS balances the need for optimal performance in expected missions or work activities with the need for resilience and adaptability in new missions or unexpected conditions. When the operational priorities of the to-be-developed IAS are understood and the ethical and legal implications of the proposed system have been considered, the appropriate analytical techniques can be identified using the decision tree illustrated in Figure 7.2. At this point, the high-level structure of the adaptation taxonomy axes can begin to be defined.

Once the analysis is under way, the range of adaptation across the mission timeline can be determined in sufficient detail to allow the design team to identify the conceptual architecture required to support the development of the IAS modules. Ideally, the high-level structure is defined in terms of both the models and modules required to support the range of adaptation being considered.

Next, systems designers must move from understanding what range of adaptation is required to understanding what range of adaptation can actually be implemented. This critical transition is made by considering the technological capabilities required to support the adaptation, discussed in Chapter 5; the ethical and legal implications of specific adaptations, identified earlier; and the impact of those implications on operator acceptance.

Further analysis should then be conducted to populate the adaptation taxonomy across all three axes. Additionally, the adaptation triggers for each tiny taxonomy cube need to be identified by selecting the most appropriate operator monitoring approach.

During the implementation process, including the software build stages, or "spirals," systems developers must consider and address design guidelines. For example, there are many resources for general guidance—both standards and style guides— for OMI design: defense-related standards include: (a) *NATO STANAG 3994— Application of Human Engineering to Advanced Aircrew Systems* (North Atlantic Treaty Organization, 2007), and (b) *MIL-STD 1472G Department of Defense Design Criteria Standard—Human Engineering* (Department of Defense, 2012). OMI design style guides are also produced by many large software vendors, such as Microsoft and Apple, and are freely available. Guidance relevant to IASs can be found in the following documents:

- *Human Factors Design Standard (HFDS)* (Ahlstrom and Longo, 2003): The HFDS provides reference information to assist in the selection, analysis, design, development, and evaluation of new and modified Federal Aviation Administration (FAA) systems and equipment, and contains chapters on automation and the human–computer interface. Additional information has been included to help users better understand trade-offs involved with specific design criteria. HDFS also covers a broad range of HF topics pertaining to automation, maintenance, displays, printers, controls, visual indicators, alarms, alerts, voice output, input devices, workplace design, system security, safety, the environment, and anthropometry (i.e., body measurement).
- *Aviation Human–Computer Interface (AHCI) Style Guide* (Veridian, Veda Operations, 1998): The AHCI Style Guide presents guidelines to assist in the selection, analysis, design, development, and evaluation of tri-service military aircraft cockpits, and emphasizes army aviation. The guidelines are intended to complement and extend those published in other Department of Defense (DoD) HCI Style Guides.
- *DoD Human Computer Interface Style Guide* (Defense Information Systems Agency, Center for Standards, 1996): The DoD Human Computer Interface Style Guide provides reference information to assist in the selection, analysis, design, development, and evaluation of new and modified decision aid systems and equipment. It covers a broad range of HF topics that pertain to decision aid design and implementation, and is largely based on user-centered design principles.

Systems designers are encouraged to consult other publications for general advice pertaining to user-centered design principles, such as Salvendy's (2012) *Handbook of Human Factors and Ergonomics*, and the cognitive and experimental psychology and human performance theories underpinning them, such as Wickens et al.'s (2012) *Engineering Psychology and Human Performance*. As discussed in Chapter 1, it is not the intent of this book to give advice pertaining to design and implementation, but rather to guide the early design and development process from an interaction-centered perspective. This road map, therefore, ends at the point where the system is designed at a conceptual level. Chapter 8 provides examples of the road map in action, and presents two case studies—one on the development of an IAI for the control of multiple UAVs, and one on the development of an intelligent tutoring system.

7.9 SUMMARY

This chapter introduced some common practical constraints associated with the design and development of IASs, including operational requirements and project management considerations. Common top-level goals for IAS design were noted, and common types of definition-focused documentation were outlined, specifically SOI and CONOPS. The ethical, legal, social, and cultural issues associated with IAS design were discussed, and intelligent automation technologies were offered to highlight specific issues that may arise in the future.

Guidance for IAS design was provided based on practical experience, and decision trees for selecting appropriate analytical techniques and operator state monitoring approaches were illustrated. Systems designers and developers were advised to learn from the experience of others, as knowing what has happened in the past will allow them to understand from the beginning of an IAS project what is possible and what is not possible.

The range of adaptation available to IASs was discussed, and common modification types, such as function allocation and task scheduling, were noted. Various adaptation triggers that can be used, such as operator-based and task-based triggers, were also addressed.

A road map for successful IAS development was outlined, and an adaptation taxonomy was created. This taxonomy has three axes: mission (or work activity) timeline, type of adaptation, and OMI modality; a well-developed taxonomy allows developers to maximize IAS impact on operational performance and minimize development risks within the time and budgetary constraints of the project.

Additionally, suggestions regarding well-known design guidelines, in the form of standards and style guides, were offered. The provided road map was used to guide the development of two IASs discussed in Chapter 8.

REFERENCES

Ahlstrom, V. and Longo, K. (2003). Chapter 3—Automation. In *Human Factors Design Standard* (Report No. DOT/FAA/CT-03/05 HF-STD-001). Atlantic City International Airport, NJ: DOT/FAA Technical Center.

Banbury, S. (1997). The scope for decision support systems to improve FOAS mission effectiveness (Report No. DERA/AS/SID/566/CR97377). Farnborough, UK: Defence Evaluation and Research Agency.

Brown, P. and Levinson, S. (1987). *Politeness: Some Universals in Language Usage*. Cambridge, UK: Cambridge University Press.

Cummings, C. L. and Ryan, J. (2013). Shared authority concerns in automated driving applications. *Journal of Ergonomics*, *S3*(001). Retrieved July 31, 2014 from http://www.omicsgroup.org/journals/shared-authority-concerns-in-automated-driving-applications-2165-7556.S3-001.pdf.

Defense Information Systems Agency, Center for Standards. (1996). Department of Defense technical architecture framework for information management, vol. 8. DoD human computer interface style guide [Version 3.0]. Fort Meade, MD: Defense Information Systems Agency.

Defense Science Board. (2012). *Task Force Report: The Role of Autonomy in DoD Systems*. Washington, DC: Office of the Secretary of Defense. Retrieved July 31, 2014 from Office of the Under Secretary of Defense for Acquisition, Technology and Logistics: http://www.acq.osd.mil/.

Department of Defense. (2012). *MIL-STD 1472G Department of Defense Design Criteria Standard—Human Engineering*. Redstone Arsenal, AL: U.S. Army Aviation and Missile Command.

Endsley, M. R. and Kaber, D. B. (1999). Level of automation effects on performance, situation awareness and workload in a dynamic control task. *Ergonomics*, *42*(3), 462–492.

Enterkin, P. (1994). Automation issues for a future offensive aircraft (Report No. DRA/AS/MMI/TR94059/1). Farnborough, UK: Defence Evaluation and Research Agency.

Entin, E. E. and Entin, E. B. (2001). Measures for evaluation of team processes and performance in experiments and exercises. In *Proceedings of the 6th International Command and Control Research and Technology Symposium*. June 19–21, Washington, DC: Command and Control Research Program.

Feigh, K. M., Dorneich, M. C., and Hayes, C. C. (2012). Toward a characterization of adaptive systems: A framework for researchers and system designers. *Human Factors: The Journal of Human Factors and Ergonomics Society*, *54*(6), 1008–1024.

Fleishman, E. A. and Quaintance, M. K. (1984). *Taxonomies of Human Performance: The Description of Human Tasks*. Orlando, FL: Academic Press.

Flight Deck Automation Working Group. (2013). Operational use of flight path management systems. Final Report of the Performance-based operations Aviation Rulemaking Committee/Commercial Aviation Safety Team Flight Deck Automation Working Group: Federal Aviation Authority.

Hilburn, B., Molloy, R., Wong, D., and Parasuraman, R. (1993). Operator versus computer control of adaptive automation (Report No. NAWACADWAR-93031-60). Warminster, PA: Naval Air Warfare Center.

Ho, C. Y., Nikolic, M. I., Waters, M. J., and Sarter, N. B. (2004). Not now! Supporting interruption management by indicating the modality and urgency of pending tasks. *Human Factors: The Journal of Human Factors and Ergonomics Society*, *46*(3), 399–409.

Hughes, S. (2013). Campaigners call for international ban on "killer robots". *BBC News*. Retrieved July 31, 2014 from http://www.bbc.co.uk.

Miller, C. A. and Parasuraman, R. (2007). Designing for flexible interaction between humans and automation: Delegation interfaces for supervisory control. *Human Factors: The Journal of Human Factors and Ergonomics Society*, *49*(1), 57–75.

Miller, C. A., Pelican, M., and Goldman, R. (1999). Tasking interfaces for flexible interaction with automation: Keeping the operator in control. In *Proceedings of the 4th International Conference on Intelligent User Interfaces*. New York, NY: ACM.

Miller, C. A., Wu, P., and Funk, H. (2007). A computational approach to etiquette and politeness: Validation experiments. In D. Nau and J. Wilkenfeld (eds), *Proceedings of the First International Conference on Computational Cultural Dynamics*, pp. 57–65. August 27–28, Menlo Park, CA: AAAI Press.

Mishra, P. and Hershey, K. A. (2004). Etiquette and the design of educational technology. *Communications of the ACM*, *47*(4), 45–49.

Moss, J., Reising, J. M., and Hudson, N. R. (1984). Automation in the Cockpit: Who's in Charge? In *Proceedings of the 3rd SAE Aerospace Behavioral Engineering Technology Conference*, pp. 1–5. Warrendale, PA: SAE International.

Murphy, R. R. and Woods, D. D. (2009). Beyond Asimov: The three laws of responsible robotics. *IEEE Intelligent Systems*, 24(4), 14–20.

North Atlantic Treaty Organization. (2007). *STANAG 3994 AI Application of Human Engineering to Advanced Aircrew Systems*, 3rd edn. Brussels, Belgium: North Atlantic Treaty Organization.

Parasuraman, R., Mouloua, M., and Molloy, R. (1996). Effects of adaptive task allocation on monitoring of automated systems. *Human Factors: The Journal of Human Factors and Ergonomics Society*, 38(4), 665–679.

Parasuraman, R., Sheridan, T. B., and Wickens, D. C. (2000). A model for types and levels of human interaction with automation. *IEEE Transactions on Systems, Man, and Cybernetics, Part A*, Systems and Humans, 30(3), 286–297.

Salvendy, G. (2012). *Handbook of Human Factors and Ergonomics*. Chicago, IL: Wiley.

Sheridan, T. B. and Parasuraman, R. (2006). Human-automation interaction. In R. S. Nickerson (ed.), *Reviews of Human Factors and Ergonomics*, vol. 1. Santa Monica, CA: HFES.

Sheridan, T. B. and Verplank, W. L. (1978). Human and computer control of undersea teleoperators (Report No. N00014-77-C-0256). Cambridge, MA: MIT Cambridge Man-Machine Systems Lab.

Tenney, Y. J., Rogers, W. H., and Pew, R. W. (1995). Pilot opinions on high level flight deck automation issues: Toward the development of a design philosophy (Contractor Report No. 4669). Hampton, VA: NASA.

The future of the car: Clean, safe and it drives itself. (2013). *The Economist*. Retrieved July 31, 2014 from http://www.economist.com/.

Tulga, M. K. and Sheridan, T. B. (1980). Dynamic decisions and work load in multitask supervisory control. *IEEE Transactions on Systems, Man, and Cybernetics*, 10(5), 217–232.

Veridian, Veda Operations. (1998). Aviation Human-Computer Interface (AHCI) Style Guide. (Report No. 64201-97U/61223). Arlington, VA: Veridian Engineering. Retrieved July 31, 2014 from http://www.deepsloweasy.com.

Whitby, B. (2008). Sometimes it's hard to be a robot: A call for action on the ethics of abusing artificial agents. *Interacting with Computers*, 20(3), 326–333.

Wickens, C., Hollands, J., Banbury, S., and Parasuraman, R. (2012). *Engineering Psychology and Human Performance*, 4th edn. Upper Saddle River, NJ: Pearson.

Section III

Practical Applications

Practical Applications

8 Case Studies*

8.1 OBJECTIVES

- Present two case studies as worked examples of the IAS development road map
- Describe the process of designing a UAV operator interface
- Describe the process of developing an intelligent tutoring system
- Describe the results of each case study to evaluate the impact of IASs on their target audience

This chapter presents two IAS development case studies based on practical experience. The intent is not to give advice on the detailed design and implementation of these IASs, but to walk through the steps that were taken while illustrating the interaction-centered design perspective that was applied in both cases. The first case study describes the development of an intelligent adaptive interface (IAI) to control multiple UAVs. The design was implemented in support of the Canadian Armed Forces (CAF) UAV operator machine interface (OMI) requirements. The second case study describes the development of an intelligent tutoring system called QuestionIT. QuestionIT was developed to help CAF personnel acquire the skills necessary to effectively question witnesses following improvised explosive device (IED) attacks. These case studies are worked examples of the IAS development road map discussed in Chapter 7. They also describe a number of practical constraints that were encountered, which shaped and influenced how the two IASs were eventually developed. These constraints were by no means unusual; similar issues and constraints are likely to be encountered by any project team tasked to develop an IAS for a real-world application.

8.2 DESIGNING AN UNINHABITED AERIAL VEHICLE OPERATOR–MACHINE INTERFACE

UAVs are becoming more prevalent in military and police operations, and are also being developed for commercial applications (e.g., drones delivering products to customers). UAV operators, however, face high-workload conditions created by high data levels, multiple dynamic sensor views, and the remote operation characteristics of UAV control. In applications where one operator controls multiple UAVs, these challenges are magnified.

Case Study 1 takes a close look at the IAS design methodology used in a three-year project by Defence Research and Development Canada (DRDC). The aim of this project was to develop, demonstrate, and prioritize enabling technologies that can be applied to OMIs to support reduced personnel requirements and enhanced performance in complex military systems used by the CAF, particularly multiple UAV control from an airborne platform. This project laid the foundation for the production of preliminary design guidelines for IAIs. IAI has been defined and discussed in Chapter 2.

This case study provides IAS designers with the information necessary to understand the development, implementation, and outcomes of the project; systems designers can find more detailed descriptions of this particular IAI for UAV control in other work (Hou and Kobierski, 2005, 2006; Hou et al., 2007).

8.2.1 The Importance of IAI for UAV Control

UAV control is operator intensive and can involve high-workload levels. As the quantity and variety of collected data increase, the UAV operator's workload increases proportionally. Moreover, the allocated data must be integrated or converted into information and then disseminated to decision-making operators. Accordingly, data collection, data fusion, information management, information distribution, intelligence collecting, and data-related decision making have threatened to become a bottleneck. This situation is made even more complex by increases in joint operations and the advancement of rapid and flexible warfare. Feedback from UAV operators indicates that improvements in the OMI aspect of these systems would provide significant increases in system performance and effectiveness. These gains include more effective UAV control, and more efficient data management and dissemination of associated information. The effective application of automation technology in decision-making processes is a key concern of both tactical commanders and UAV system managers. As a result, it is important to continue to investigate the supporting technologies that optimize operator interaction with software agents in order to satisfy mission requirements.

As discussed in Chapter 2, IAIs are intended to reduce the impact of OMI complexity and operator workload—IAIs are OMIs that improve the efficiency, effectiveness, and naturalness of human–machine interaction by acting adaptively and reacting proactively to external events based on internal task requirements. In the specific context of UAV control, an IAI is a subsystem of the UAV ground control station (GCS), and is represented by a GUI driven by software agents. The UAV IAI supports the decision-making and action requirements of operators under different levels of workload and task complexity. It manifests itself by presenting the right information, including action sequence proposals, or by performing the right actions, at the right time. In addition to reducing workload for operators involved in UAV missions, IAIs have the potential to reduce personnel requirements (e.g., moving from a ratio of four operators controlling one UAV to one operator controlling four UAVs).

8.2.2 IAI Project Scope

DRDC acknowledged that designing effective OMIs was a key factor in mission success and initiated a project to develop and evaluate IAIs for the control of multiple

UAVs. The selected environment involved UAV operations in support of counterterrorist activities. The IAI was modeled as part of UAV tactical workstations for a modernized CP140 Canadian maritime patrol aircraft. This work was divided into three phases:

- *Phase I*: Phase I focused on concept development and performance modeling, including the development of a methodology to analyze UAV operations in a mission scenario. The scenario reflected a portion of a CAF UAV experimental program. The analytical results were used to develop a human–machine task network model that was then implemented in an integrated performance modeling environment (IPME) for simulation purposes. The model had two modes for operators working with OMIs to control multiple UAVs. One mode assumed that operators used conventional OMIs. The other mode assumed that operators used IAIs with automated decision aiding (i.e., software agents). The difference between mission activities with and without these IAI agents was reflected in the simulation results, such as the time to complete critical task sequences and task conflict frequency.
- *Phase II*: Phase II focused on the design and implementation of IAI prototypes, which incorporated six agents representing the following system function groups: (a) intercrew communications; (b) route planning; (c) route following; (d) screen management; (e) data-link monitoring; and (f) UAV sensor selection. A synthetic environment was developed based on the North Atlantic Treaty Organization (NATO) standardization agreement (STANAG) 4586 interface software protocol. The experimental environment had three UAV GCSs replicating CP140 tactical compartment workstations, with a set of displays and controls appropriate for each of the UAV crew members: (a) UAV pilot (UP); (b) UAV sensor operator (UO); and (c) tactical navigator (TN). The experimental environment also had an integrated video and audio data collection suite to facilitate the empirical assessment of IAIs.
- *Phase III*: Phase III focused on experimentation, in which empirical evaluations were conducted to examine operator workload and interface adaptability with mock-up UAV GCSs. Eight crews totaling 24 operational CP140 members participated in the experiment. Each crew completed a two-day experiment that assessed operator interfaces with and without IAI agents.

8.2.3 IAI Design and Development Road Map

The development of an adaptation strategy at the commencement of the project is an effective tool to support the design and development of an IAS, and a useful road map was discussed in Chapter 7. It included several steps that needed to be performed in a sequential and logical manner. This section provides an example of the same development road map and taxonomy for developing UAV IAIs.

8.2.3.1 Determining Operational Priorities

As discussed in Chapter 7, the first step of the road map determines the operational priorities of the IAS; in this context, operational priorities are the high-level objectives of UAV missions. In order to identify these priorities, a stakeholder analysis was conducted to investigate the requirements of stakeholders in the CAF UAV operational community (e.g., program management office, systems and material requirements office, operators, etc.) for the development and evaluation of IAI technologies into the CAF UAV experimental program. To understand mission goals, focus group interviews were conducted with three SMEs involved in the CAF UAV experimental program who had extensive CP140 and UAV operational experience. These activities resulted in a composite mission scenario that described a fictitious counterterrorist security arrangement, which was used for further analysis in the IAI design process and prototype evaluations.

The composite mission scenario was a minute-by-minute description of how events unfolded. Figure 8.1 illustrates the chosen scenario, which involved an ongoing fisheries patrol and approximately 200 vessels in the vicinity of the nose and tail of the Canadian Grand Banks. A CAF patrol frigate was on scene with two vertical takeoff UAVs (VTUAVs) and a maritime helicopter. Overhead, there was a CP140 patrol aircraft equipped with 16 mini-UAVs and its own sensor suite. A medium-altitude long-endurance (MALE) UAV was also on scene under the control of the

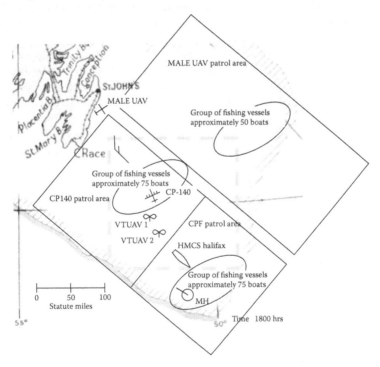

FIGURE 8.1 UAV mission scenario overview. (From Hou, M. and Kobierski, R. D., Intelligent adaptive interfaces: Summary report on design, development, and evaluation of intelligent adaptive interfaces for the control of multiple UAVs from an airborne platform (Report No. TR 2006-292), Defence Research and Development Canada, Toronto, Canada, p. 11, 2006.)

regional operations center located in Halifax, Canada. In total, the scenario contained 16 mini-UAVs, two VTUAVs, and one MALE UAV.

The CP140 UAV tactical crew (i.e., UAV pilot, UAV sensor operator, and tactical navigator) were responsible for controlling UAV assets. Figure 8.2 illustrates how the tactical navigator occupied his normal position, with the sensor operator and pilot on his left. As UAV team leader, the tactical navigator was given the most complex role. In addition to coordinating the crew on board the CP140, the tactical navigator played the primary role in understanding the overall tactical situation, planned appropriate responses to that situation, and delegated work to the other two UAV crew members to execute those responses. The UAV pilot was given the least complex role and was responsible for the safe and appropriate conduct of all UAVs under the crew's control. The pilot could plan the flight path of individual UAVs, but the tactical navigator was ultimately responsible for the task. The UAV sensor operator was responsible for managing the information being returned by sensors on UAVs under the crew's control and for relating findings based on sensor data back to the rest of the crew as appropriate. The resulting environment was suitable for assessing the efficacy of the IAI across three levels of workload and job complexity.

From the stakeholder analysis and SME focus group studies, it was understood that the control of multiple UAVs is difficult and that the workloads associated with the roles of UAV pilot, UAV sensor operator, and tactical navigator are different due to individual job complexities. The workload could also be influenced by the number of UAVs the crew was controlling, including onboard assets. In the scenario, workload was assigned at three levels: (a) low (i.e., one UAV); (b) medium (i.e., two UAVs); and (c) high (i.e., five UAVs). However, no advanced technology existed to help reduce workload and improve SA when controlling multiple assets. The goal of the project was to develop an IAI prototype as a part of the UAV GCS within the context of the counterterrorist scenario, and then assess the effectiveness of the IAI prototype in reducing workload and improving SA.

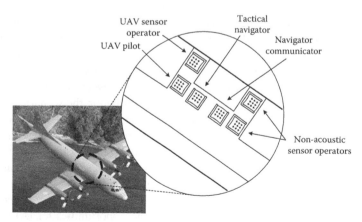

FIGURE 8.2 UAV crew positions in CP140 tactical compartment. (From Hou, M. and Kobierski, R. D., Intelligent adaptive interfaces: Summary report on design, development, and evaluation of intelligent adaptive interfaces for the control of multiple UAVs from an airborne platform (Report No. TR 2006-292), Defence Research and Development Canada, Toronto, Canada, p. 5, 2006.)

8.2.3.2   Ethical and Legal Implications

As automation technologies become critical elements of military activities, a revolution of automated flying machines is upon us. This is a revolution in warfare, much as the invention of the atomic bomb was a revolution in the first half of the twentieth century. Automated systems, including UAVs, are becoming increasingly common in all theater of operations (i.e., land, sea, air, and urban). These robotic systems do not change the "how" but the "who" of war fighting; UAVs redefine both the experience and identity of "warrior." Missions can now be enabled through broad networks with focus on operational effectiveness and ever-increasing emphasis on the design of agile, intelligent, sociotechnical systems. These systems allow more effective and efficient system acquisitions and operation concept development. Human roles, however, remain paramount, and the interaction of human and machine is critical to achieving mission success; effective human interaction with automated systems is vital.

Keeping social, ethical, moral, and legal issues associated with human–machine interaction in mind, the IAI project team consulted with CAF UAV SMEs to determine what missions the UAV experimental program were planning to conduct and what types of scenarios should be used for the project. The feedback from the SMEs was that the project team should avoid classified material, in terms of both the type of scenario and the inclusion of any material relating to the technical details of UAV systems. These constraints required that the scope of the IAI project be restricted to the design and evaluation of simulation-based IAI technologies only. UAV missions were limited to search, surveillance, reconnaissance, target acquisition, target engagement, and combat assessment, and a domestic counterterrorist security scenario was chosen to avoid any issues relating to the use of classified material.

8.2.3.3   Selecting an Analytical Technique

As discussed in Chapter 4, selecting an appropriate analytical technique requires knowledge of a number of factors, including project scope, timeline, budget, the nature of tasks, the subject matter to be analyzed, the experience of the project team, and the level of effort deployed toward the analysis work. In this case, the project team had sufficient time and budget to conduct a thorough analysis. A number of experienced SMEs and analysts were also available to help understand the complex and highly cognitive nature of UAV operations, especially the interaction between a team of UAV crew members and the multiple UAVs and other assets described in the scenario.

The goal of this IAI project was to investigate the efficacy of IAIs in controlling multiple UAVs; of particular interest was the identification of goals and associated tasks that could be candidates for IAI agents. Consequently, the analytical technique chosen to model this UAV control problem was Hierarchical Goal Analysis (HGA) based on Perception Control Theory (PCT) discussed in Chapter 4. Using standard mission, operation, and goal analysis procedures, a PCT-based HGA provided a more detailed understanding of implementation issues and opportunities for IAI tasks that could be automated. PCT-based HGA lends itself well to describing situations in which humans must interact with various IAI agents to effectively control multiple UAV systems while facing many external influences, such as changes in workload, task complexity, and interface conditions. It was seen as a good match with a goal-oriented, interaction-centered IAS design approach for the UAV control correspondence domain. Additionally, by building task network models, simulations of the new designs could be run to anticipate performance effects. Function flow diagrams were developed to understand the primary tasks that UAV operators would perform and their logical interconnections; these diagrams also proved to be the most efficient tool for beginning the conversion of the mission analysis to a modeling structure to assess IAI concepts.

Once the function flow diagrams were completed, they were restructured into a top-down HGA for the identification of potential IAI agents. The function flow diagrams were then used with the composite scenario to create an extensive task list as part of a complete mission, function, and task analysis (MFTA). The task list was used to compile the lowest level of the HGA. An exercise was then completed to link the bottom-up definition of the HGA to the top-down HGA. This approach was very effective and allowed the project team to understand how the various UAVs would be used in the scenario. The result was much more complete than decomposing the HGA from the top level, and then hoping it would align with the scenario when completed. The lowest-level HGA tasks were considered actions of the lowest goals of the hierarchy, and the sensations and perceptions of the PCT feedback loop were the appropriate feedback of the task analysis. The feedback from multiple PCT loops was used to define the attributes of the functions built into the IAI agents. Figure 8.3 illustrates a decision tree for the analysis of the requirements of UAV control systems.

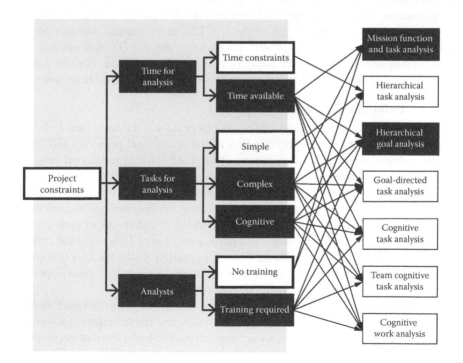

FIGURE 8.3 Decision tree highlighting the analytical techniques applicable to UAV IAI control station requirements. An unmarked decision tree can be found in Chapter 7.

Operational sequence diagrams (OSDs) were developed using the composite scenario, task lists, and task data (e.g., task completion times, visual, cognitive, auditory, and psychomotor ratings, initiating conditions, etc.). These diagrams were created to support the production of workload plots, which were used to determine whether there were any predicted advantages to the use of an IAI as part of the UAV control system. Additionally, the information flow and cognitive processing analysis conducted using the OSDs provided a clear indication of which operator tasks were the best candidates for conversion to IAI agents (i.e., automated decision aiding).

OSDs were also used to show the flow of information and logical interconnection of operator functions through the system in relation to the mission timeline. The visual representation of OSDs indicated actions; inspections; data transmission, reception, and storage; time delays; and decisions of the mission scenario. OSDs proved particularly useful for analyzing highly complex UAV GCSs that required many time-critical information-decision-action functions by multiple operators.

With the scenario written, a series of OSDs were prepared to facilitate a function and goal/task analysis of the envisaged UAV GCSs. The OSDs also guided the development of a network model that could be used to make performance predictions and identify potential IAI agents. These OSDs also created an inventory of all bottom-up lowest-level tasks in a temporal sequence.

Figure 8.4 illustrates a small portion of an OSD demonstrating the operational network among three UAV operators. It shows the advantages of combining more than one analytical technique. PCT and HGA provide a good control theoretic description of the domain, while MFTA and OSDs can provide clear task flow models. This process was similar to a cognitive task analysis (CTA), as a complete information flow and cognitive processing analysis were conducted using the OSDs combined with the workload analysis. Due to the study of multiple ongoing tasks, cognitive attentional demand ratings were included and task conflict parameters were reviewed.

8.2.3.4 Determining Range of Adaptation

After the scenario and associated OSDs were in place and the analytical technique was determined through an understanding of the operational priorities and social, ethical, cultural, and legal considerations of the CAF UAV experimental program, the next step, according to the IAS design road map, was to start defining the IAI's range of adaptation. Types of adaptation relevant to IAI design and implementation include: (a) function allocation; (b) task scheduling; (c) operator–agent interaction; and (d) interaction content. Once the types of adaptation and operator–agent interaction were clearly defined, the design and implementation of IAIs using the conceptual architecture discussed in Chapter 3 could be started. Using the IAS framework discussed in Chapters 3 and 5, environment, expert, knowledge, agent, operator, machine, task, and communication models were created.

Next, a hierarchical decomposition of the mission, operations, and goal analysis procedures was completed in order to better understand implementation issues and opportunities for agent tasks. This was effectively an HGA process that decomposed goals for all three UAV operators according to a means-end hierarchy. The needs of HGA were typically satisfied at the fourth or fifth level in this exercise. First, goal decomposition was performed in a top-down fashion from the highest level

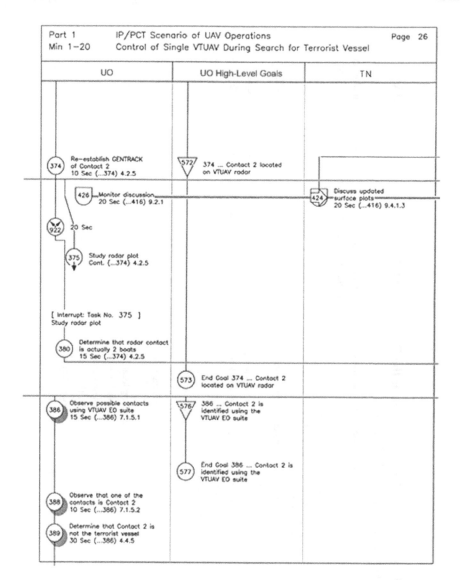

FIGURE 8.4 Portion of an UAV control operational sequence diagram. It shows the advantages of combining multiple analytical techniques.

(i.e., goal = counterterrorist mission is completed) to low levels (e.g., goal = VTUAV sector search is planned). Then, a more stringent bottom-up approach was completed by studying detailed mission activities in OSDs. As a result, more goals were added in the top-down analysis list to complete the generation of the HGA inventory. Table 8.1 shows a small portion of the HGA results.

To design the IAI, consultation with experienced UAV SMEs based on an expert model developed for the project was used with extensive PCT analysis and HGA to identify potential candidate goals that could be suitable for IAI agents. This served the

TABLE 8.1

Subset of the Hierarchical Goal Analysis (HGA) Outputs for the Design of an Intelligent Adaptive Interface (IAI) for Multiple Uninhabited Aerial Vehicle (UAV) Control

Goal Number	Level 1	2	3	4	5	Goal/objective and Subgoals/Subobjectives	IAI Agent Candidate	Influenced Variable	Role Assignment	Completion Time (sec)
Top	I want to perceive the (...) conduct of the terrorist patrol mission									
9		... communications are conducted and maintained								
9.1			... directions (instructions) are received							
9.1.1				... directions are received from other crew members			Yes	Directions	UP	4–25
9.1.2				... directions are received from other units			Yes	Directions	UP	5–14
9.1.3				... directions are received from tasking agency				Directions	TN	20
9.2			... information is received							
9.2.1				... information is received from other crew members						
9.2.1.1					... visual contact is established by flight crew					
9.2.1.2					... VTUAV refueling location		Yes	Location	TN	5
9.2.1.3					... VTUAV calculated time on task			Time	TN	5
9.2.1.4					... the flight crew's message that contact is identified			Message contents	UO	5
9.2.1.5					... the pilot's message that ac is turning to waypoint					

process of deciding function allocation between IAI agents and operators. A subset of HGA outputs is shown in columns 1 and 2 of Table 8.1. Goal suitability for IAI agents is shown in column 3 and the key influenced control variable is identified in column 4. The associated roles assigned to one of the three UAV crew members and the times needed to achieve the goals are shown in columns 4 and 5. This information also facilitated decision making regarding function allocation and task scheduling for IAS adaptation.

At the end of the analysis process, IAI agent candidates were selected based on whether the identified goals and associated tasks could be automatically implemented for the purpose of workload reduction and SA improvement. These IAI agents were functional components of the UAV GCSs. They were designed to provide decision support to the UAV crew and take over certain crew tasks associated with high workload. For example, the system could track keystrokes and cursor movement and would be able to deduce that an operator was attempting to complete multiple concurrent tasks, such as UAV flight control and surface plot manipulation. At that point, the IAI could adapt to the situation and provide partial or complete assistance to the busy operator. For example, if the operator was unable to assign all airborne UAVs a task due to high-workload demands, the IAI could automatically evoke potential UAV search or loiter patterns. This type of agent was chosen as the UAV route planner agent. The feasibility of implementation of selected agents was confirmed by military SMEs following predefined operational rules. Similarly, many other agents were chosen for the IAI to optimize operator–agent interaction at the managing, working, and junior levels as discussed in Chapter 5.

Following the IAS design guidelines discussed in Chapter 7 and considering how IAI agents could adapt to the needs of operators through modifying function allocation, task scheduling, interaction optimization, and information content manipulation, six working-level IAI agents were designed and implemented in the IAI prototype:

- *Screen manager*: A working agent that served the managing agent. According to predefined rules, this agent managed a shared tactical plot (TACPLOT) among the three UAV operators whenever new high-priority events occurred. This included automatically panning the TACPLOT to locations of interest and zooming in or out. Figure 8.5 illustrates how a text label was added to the TACPLOT on UAV icon 5 to identify the contact designed by a junior agent and to indicate when this agent is exercising UAV flight control (P) or sensor geotracking (S).
- *Interaction communicator*: A working agent that served the senior interaction agent. To optimize operator–agent interaction, this agent relayed all information and knowledge generated by the sensing, tasking, modeling, and managing agents. It communicated events and actions and provided feedback to operators through a message window. The window was designed to allow all messages to be displayed in the primary display, which meant that it attained the crew's attention faster and more reliably. Figure 8.6a illustrates the tactical navigator's operator interface, featuring an interaction communicator message box in the bottom left corner of the primary display. This message box is used by all three operators. The

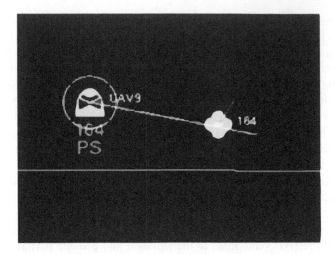

FIGURE 8.5   IAI readouts in TACPLOT. (Redrawn from Hou, M. and Kobierski, R. D., Intelligent adaptive interfaces: Summary report on design, development, and evaluation of intelligent adaptive interfaces for the control of multiple UAVs from an airborne platform (Report No. TR 2006-292), Defence Research and Development Canada, Toronto, Canada, p. 9, 2006.)

display method is similar to the pop-up chat windows used in web-based e-mail services. Figure 8.6b illustrates how the message window continuously shows active UAVs, allocated tracks, and information about how the agents supported the UAVs.

- *Route planner*: A working agent that served the senior modeling agent. If a UAV was used to investigate an unknown or hostile contact, this agent worked with the inference agent to compute the most direct route and activate that route for the UAV. The allocation of tracks to UAVs was based on a search for the closest unknown or hostile contact. Based on an expert model, additional logic ensured that no more than one UAV could be engaged with a single unknown contact. More than one UAV could, however, be engaged with hostile contacts. Once a route was planned, the route was passed through the modeling agent to the tasking agent.
- *Route follower*: A working agent that served the senior tasking agent. This agent piloted the UAV on the active route provided by the route planner. The route included flight altitude, speed management, and self-preservation in close proximity to the track. The agent entered an orbital flight pattern around the track once the UAV reached sensor identification range. As illustrated in Figure 8.6b, the agent updated the IAI on UAV status and its tracks in the message window, through the interaction communicator.
- *Sensor manager*: A working agent that served the senior sensing agent. Once a UAV was close enough to a track to engage an electronic optic sensor, the agent took over sensor management. This included pointing the sensor and establishing a stable lock on the moving target when the track was

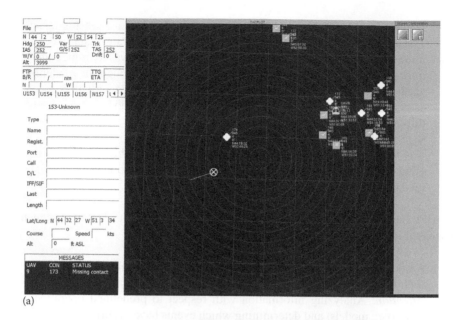

(a)

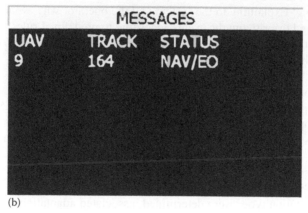

(b)

FIGURE 8.6 Operator interface featuring an interaction communicator message window. Events, actions, and feedback for operators are provided through this IAI window. (a) Tactical navigator's interface featuring an IAI message window. (b) IAI message window displays all active UAV activities. (From Hou, M. and Kobierski, R. D., Intelligent adaptive interfaces: Summary report on design, development, and evaluation of intelligent adaptive interfaces for the control of multiple UAVs from an airborne platform (Report No. TR 2006-292), Defence Research and Development Canada, Toronto, Canada, p. 9, 2006.)

within visual range. The agent passed the information to the screen manager to manage the shared tactical plot. The information was also passed through the modeling agent to the route planner for route planning and through the interaction agent to the interaction communicator for display in the message window.

- *Data-link monitor*: A working agent that served the senior sensing agent.
  The agent monitored the flight pattern and other UAV status data to determine whether data links were working. If not, the agent was to immediately
  inform operators. It worked with the inference agent through the modeling
  agent to update the system about current communications with all mission
  assets. The agent also communicated with the interaction communicator
  through the interaction agent and displayed red emergency text in the message window if anything went wrong. If an emergency situation arose, it
  was to turn on an alarm to draw operator attention. These actions are a
  good example of how to optimize operator–agent interaction using multiple
  modalities.

All IAI agents developed for the UAV GCSs follow the agent adaptation sequence
discussed in Chapter 5:

1. *Knowledge acquisition*: Gathering status information about all active UAVs,
   their tracks, the states of operators, and the current display configuration
2. *Attention*: Analyzing information with respect to predefined operational
   rules (i.e., models) and determining which events have occurred
3. *Reasoning*: Explaining to the system and the operator why events are happening and prioritizing the events according to predefined rules
4. *Decision making and action*: Executing predefined tasks for each event following the prioritization order

The operator–agent interaction model and IAI hierarchical design framework discussed in Chapter 5 were conceived at the beginning of the UAV GCS IAI design
process. The operational scenario and HGA helped identify the types and number
of different IAI adaptation and IAI agents to be designed and implemented in the
UAV GCSs.

8.2.3.5 Selecting Adaptation Triggers

After IAS adaptation types were determined, associated adaptation-triggering conditions were identified. Operator state monitoring approaches and categories of
adaptation triggers were based on the IAS design guidelines offered in Chapters 6
and 7. Chapter 6 discussed four categories of operator state monitoring approach: (a)
behavioral; (b) psychophysiological; (c) subjective; and (d) contextual. Chapter 7 discussed five adaptation-triggering categories: (a) operator-based; (b) machine-based;
(c) environment-based; (d) mission- and task-based; and (f) location- and time-based.
Keeping these guidelines in mind, and considering the nature of the UAV control
scenario, the project team determined the most appropriate operator state monitoring approaches for triggering the type and timing of the IAI contents presented to
the operators. Operator state monitoring was a critical component of the adaptation
mechanism through which the IAI could adapt the content and presentation of the
mission situation to optimize operator–agent interaction and overall system performance. To identify operator state monitoring requirements, this process was adopted
from the decision tree illustrated in Chapter 7. Figure 8.7 illustrates how the four

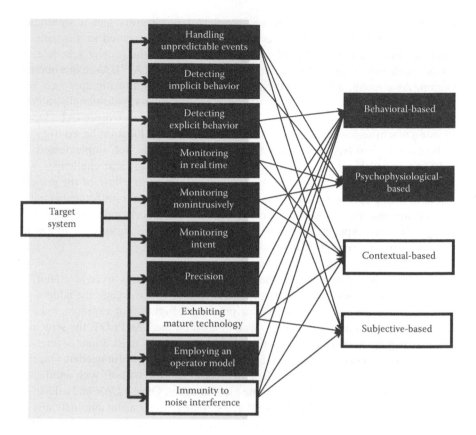

FIGURE 8.7 Decision tree highlighting the operator state monitoring approaches applicable to UAV IAI design. An unmarked decision tree can be found in Chapter 7.

categories of operator state monitoring approaches noted above were also needed to meet these requirements.

From the decision tree, a combination-based approach for operator state monitoring was identified. A more in-depth review of state-of-the-art behavioral-based and psychophysiological-based approaches to the monitoring of operator behavior and state for the UAV GCS was then conducted. Techniques and technologies relating to eye tracking, psychophysiological indices of workload and stress, and performance tracking that were successfully used or could theoretically be used to augment IAIs were reviewed. The review provided straightforward and applicable recommendations leading to the acquisition or development of technologies that could be implemented into the IAI. The need and feasibility of implementing subjective and contextual operator state monitoring technologies into the IAI was also reviewed. This maximized the impact of the IAI on UAV control within the time and budgetary constraints of the project, while reducing development risk due to issues such as dependence on immature technology.

As discussed in Chapter 7, for projects such as the UAV IAI, it is important to consider what is possible from an implementation perspective. Schedule constraints, budgetary constraints, and system environmental constraints had a significant impact on which IAS technologies could actually be implemented in this case. In particular, there were a number of constraints relating to the project schedule, as well as constraints relating to the facilities used for the CAF UAV experimental program. As a result, a behavioral-based operator state monitoring approach was used to trigger IAI adaptation using operator-based triggers. Additionally, a contextual-based operator state monitoring approach was used to trigger IAI adaptation using machine-based triggers, task-based triggers, and location-based triggers. Psychophysiological-based monitoring technologies were not implemented in the IAI, as a combination of technical difficulties and logistical challenges was associated with implementing these technologies during this phase of the project. Subjective-based operator state monitoring approaches were not implemented in the IAI, as it was not feasible to ask operators for their own workload readings before or during the UAV experimental trials. Based on these considerations, the following adaptation triggers were identified:

- *Operator-based triggers*: In the IAI ON (i.e., IAI agents were activated) mode of the UAV GCS, operators could engage or disengage the adaptation by simply clicking a cursor or pressing a button. For example, once an operator identified a contact as an interest on the TACPLOT, the screen manager agent could automatically pan the TACPLOT to locations of interest and zoom in or out to provide the information the operator needed. Once the task was completed, the adaptation could be switched off with another click. A second example is the route planner agent. Once a UAV and a destination were identified by an operator, the route planner agent automatically computed the most direct route and activated that route for the UAV based on logic specified by an expert model.
- *Machine-based triggers*: Triggers can also be based on the anticipated status of the machine. For example, after the route planner agent provided an active UAV route to the route follower agent, including flight altitude, speed management, and self-preservation in close proximity to the track, and the UAV reached sensor identification range, the route follower agent entered an orbital flight pattern around the track. Through the interaction communicator, the route follower agent also updated the IAI about the UAV's status and its tracks in the message window. Another example of a machine-based trigger is seen in the data-link monitor agent. If a UAV was shut down, this agent immediately informed operators by displaying red emergency text in the message window and turning on an alarm to draw attention.
- *Environment-based triggers*: The IAI experimental setup was located in an office environment. As a result, environment-based triggers were not developed.
- *Mission and task-based triggers*: The tasks that operators needed to accomplish to satisfy high-level mission goals were well understood by the project team through scenario generation, OSD development, HGA, and IAI agent identification. This understanding was built into the adaption

triggering conditions based on the differentiation among planned, active, and completed tasks. There were many mission and task-based triggers and associated IAI agents built into different parts of the scenario to support adaptation. For example, if the IAI detected an operator working on multiple tasks, and the operator's workload was high enough to miss a high-level mission goal, the IAI would automatically shed some planned tasks and give the operator more time to complete active tasks working toward achieving the high-level goal.

- *Location and time-based triggers*: Location-based triggers were also built into the UAV GCS. For example, based on airspace engagement rules and for safety reasons, UAVs were not allowed to fly within a 0.5 nautical mile radius of a contact. Once a UAV and target were identified, the route planner agent automatically computed the most direct route without violating the airspace engagement rule so that the UAV would automatically be guided to fly by the contact without entering the radius.

8.2.3.6 Testing Design Concepts

Following the identification of (a) IAI agents; (b) the range of IAI adaptation; and (c) IAI adaptation strategy, a task network model was developed in IPME to test IAI design concepts. IPME is a discrete event simulation framework used to model and assess human and system performance (Dahn and Laughery, 1997). IPME has various models, including an environment model, a crew model, a task network model, a performance-shaping model, and optional external models. Combined with IPME's scheduling algorithm, these models can help an analyst predict workload and operator performance. The information processing PCT (IP/PCT) mode uses an algorithm and an information processing scheduler to adjust operator performance based on various factors, such as task conflicts, task criticality, and time pressure.

The UAV operational task network model was validated by confirming that it was consistent with the OSDs. An associated IPME database was also generated while producing the HGA and OSDs. In the database, each goal was allocated to an operator or a machine component with a descriptive label, and a sequential operational network model was developed. Although the network model was a UAV operator model, external events, such as other aircrew activities, UAV activities, and other unit activities, had to be established to allow the network to function as a closed-loop feedback system. These were prepared and included in the network model.

Additionally, a task network following the HGA structure was created for the network model. The task network defined task behavior, operator assignment, and interaction between tasks and operators. By linking together various networks or tasks, the network model attempts to replicate the behavior of a real system. The definition of the behavior of each individual task was done through the various tabs, the function list, and the goal and variable database of IPME. Candidates for automation (i.e., IAI agents) were built into the model in such a way that the systems designer could run the model with various combinations of the automation on or off. Figure 8.8 illustrates part of the task network model and how the automation could be turned on (i.e., IAI ON) or off (i.e., IAI OFF). The model was run in IP/PCT

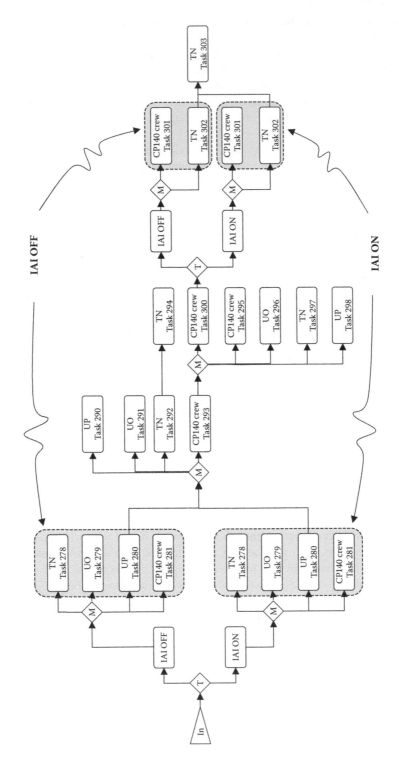

FIGURE 8.8 Part of the task network model for UAV control tasks. Automation can be turned either on or off.

mode and data were collected once with IAI agents and once without. Note that, at this stage, the IAI was a static adaptive system, as the component of operator and situation monitoring had not been implemented. However, this still presented useful opportunities to test the design concept.

IPME contains built-in data collection and reporting capabilities. The data collected were used to analyze the operator's behavior and determine the effects of using IAI agents on the operator's performance. Since the model was developed to compare missions conducted with and without IAI agents, measurements of effectiveness and performance had to provide a means of assessing the merits of incorporating these new technologies. Additionally, the measurements were required to give a clear indication of the most fruitful areas for further research and development.

The simulation output consisted of plots of two variables normalized against the mean mission timeline. This meant that the respective plot of each of the 10 runs of the model was temporally adjusted to match the mean mission time. The first variable was task conflict frequency, which was defined for each run as the percentage of time that the operator must interrupt or delay a task during a sliding 1-min window. The second variable plotted was number of ongoing tasks, which gave the average number of ongoing operator tasks with significant cognitive components. An additional output was goal completion time for critical tasks (e.g., high-level goals in HGA). It was also chosen as a means to determine not only how an IAI would affect an operator's ability to achieve an objective, but also how the IAI improved the operator's performance. These three measurements were both operationally relevant and technically feasible.

Figure 8.9 illustrates one of the main findings, which was a sample output of task conflict frequency. It indicates that operators were able to complete the mission scenario with fewer task conflicts with IAI agents (17%) than without IAI agents (38%). It also shows that operators could complete the mission scenario more efficiently with IAI agents (about 15.5 min) than without IAI agents (about 20 min).

The simulation revealed that the use of a UAV GCS with IAI agents permitted operators to achieve high-level mission goals in reduced time, even under high time pressure. This confirmed that the IAI agent candidate choice made in the design phase would be beneficial. Based on these results, the IAS design was ready to move forward to Phase II and physically mock up the UAV GCSs.

8.2.4   IAI IMPLEMENTATION

The simulation study was considered an initial estimate of the utility of IAI agents as automated decision aids; further analysis was required regarding the most beneficial IAI tasks to modify. Additionally, considerable effort was still needed to apply the IAI framework and related operator–agent interaction models to the prototype system designs. Strong empirical evidence was also required to substantiate this effort. Consequently, an experimental synthetic environment was designed and developed to conduct a human-in-the-loop experiment and validate the task network modeling method used in the simulation. Consistently with the UAV crew positions used in the performance modeling phase, the synthetic environment had three GCSs that

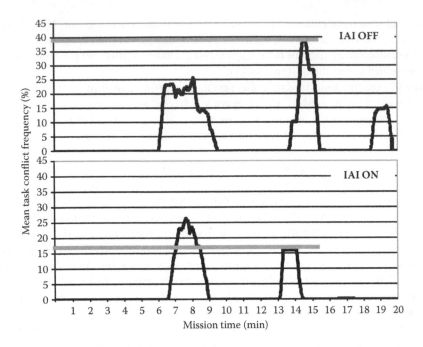

FIGURE 8.9 A sample output of task conflict frequency. Operators were able to complete the scenario more efficiently and with fewer task conflicts when using IAI agents.

replicated CP140 tactical compartment multifunction workstations. The workstations were designed to communicate with virtual UAVs through fully functional, real-world software interfaces. Each workstation had a set of appropriate displays and controls for the UAV pilot, UAV sensor operator, and tactical navigator. Figure 8.10 illustrates the positions of UAV pilot and UAV sensor operator and Figure 8.11 illustrates the primary displays of the UAV pilot and UAV sensor operator.

Figure 8.12 illustrates the experimental environment integrated video and audio data collection that enabled the empirical assessment of IAI concepts developed in the first phase of the project. The experimental environment also had multiple software components that communicated within a Microsoft Windows XP™ environment. The major software components were:

- A multiagent system embedded in the IAI
- A software package that completed data collection of operator keystrokes and vehicle motions
- A three-dimensional world through which the UAVs flew and were viewed by the participants through UAV-mounted video cameras
- A simulation and modeling software package, STRIVE, that allowed the simulation of rotary-wing and fixed-wing air vehicle flight dynamics and autonomous surface vehicle (e.g., boat) motion
- Software compliant with NATO STANAG 4586 that allowed communication between the GCSs and the simulated entities such as UAVs

FIGURE 8.10 The UAV pilot and UAV sensor operator positioned at rearward-facing workstations. (From Hou, M. and Kobierski, R. D., Intelligent adaptive interfaces: Summary report on design, development, and evaluation of intelligent adaptive interfaces for the control of multiple UAVs from an airborne platform (Report No. TR 2006-292), Defence Research and Development Canada, Toronto, Canada, p. 6, 2006.)

The IAI multiagent system was a process software component embedded in the data management system (DMS) of the UAV IAI synthetic environment. When all IAI agents were switched on (IAI ON), the various function groups coded into the interface software were activated. When all IAI agents (i.e., interface function groups) were switched off (IAI OFF), the interface became a conventional interface. The DMS was the central data processing component of the UAV IAI synthetic environment and served as a protocol gateway between the workstations and the external simulation components. Figure 8.13 illustrates the DMS architecture.

The DMS maintained a synchronized situational data repository of all relevant information related to the UAVs, CP140 ownership, and the surface and airborne tracks. The DMS processed and organized pertinent information for efficient consumption by the workstations so that the workstation operators could relay UAV control commands to the simulated UAVs and change workstation display configurations.

All entity motion data were recorded on a computer hard drive. This information included time-sequenced player positions to allow postexperiment plotting of routes for all entities and a record of all information relayed between the workstation and the STRIVE vehicle motion simulation software.

8.2.5 IAI EVALUATION

The IAI agents were functional components of the UAV control synthetic environment developed for this project. They supported the participants in accomplishing the assigned mission tasks of the experiment by (a) providing decision support to the crew, and (b) taking over certain high-workload crew tasks. The IAI agents for each crew member were tailored to suit their individual needs. The effectiveness of

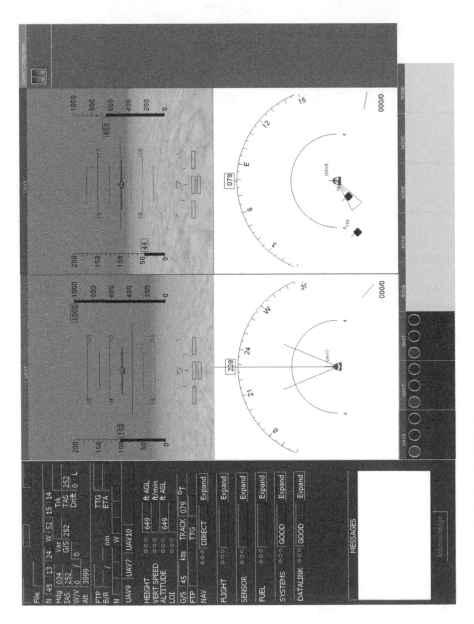

FIGURE 8.11a   The primary display for the UAV pilot.

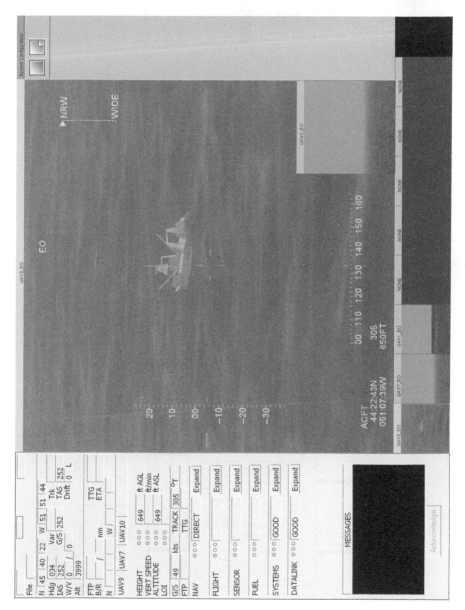

FIGURE 8.11b The primary display for the UAV sensor operator.

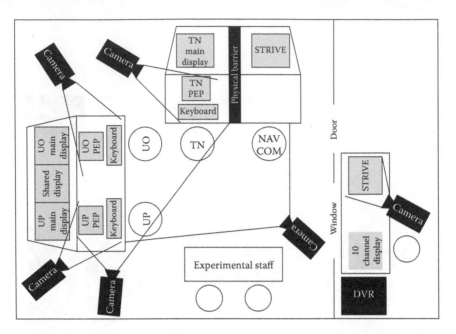

FIGURE 8.12 Layout of the UAV control experimental environment. The UAV pilot (UP), UAV sensor operator (UO), and tactical navigator (TN) are seated close to one another.

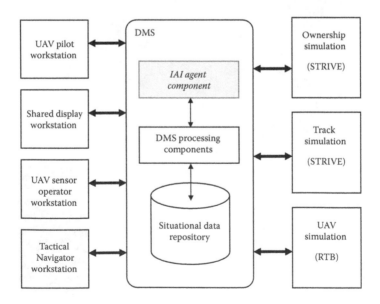

FIGURE 8.13 Data management system (DMS) architecture and interfaces. (From Hou, M. and Kobierski, R. D., Intelligent adaptive interfaces: Summary report on design, development, and evaluation of intelligent adaptive interfaces for the control of multiple UAVs from an airborne platform (Report No. TR 2006-292), Defence Research and Development Canada, Toronto, Canada, p. 14, 2006.)

the prototype UAV IAI for aiding operator performance was examined under (a) two interface conditions: IAI ON and IAI OFF; (b) three levels of operator workload: low, medium, and high; and (c) three levels of operator position complexity: UAV pilot, UAV sensor operator, and tactical navigator. Both objective and subjective measurements were used to index each operator's performance. The five objective measurements were (a) completion time for critical task sequences (CTSs); (b) percentage of CTS shedding; (c) UAV route trajectory score; (d) UAV airspace violation time; and (e) situation awareness global assessment technique (SAGAT) score. The two subjective measurements were (a) perceived workload; and (b) perceived SA. Further details regarding IAI experimentation in this context can be found in DRDC's technical report about the IAI project (Hou and Kobierski, 2006).

The main finding of the IAI project was that experiment participants performed more effectively when the IAI was ON. This was evident from both quantitative and qualitative measurements. When the IAI was ON, CTSs were shortened, fewer tasks were skipped, UAV trajectory scores were better, and no-fly areas were violated less often. Both the actual and perceived SA of operators was improved, and overall workload was reduced. Note that many of the CTSs examined in the experiment were previously modeled and simulated in the first phase of this project. They were the sequences performed by the same IAI agent groups, such as route planning and intercrew communication, used in both the simulation and experimentation phases. The consistent test results, including reduced task completion time and operator workload, of these CTSs in the experimentation phase validated the network modeling results determined in the simulation phase. Operator performance was improved through IAI use even though the operators were working in a cognitively complex situation. Figure 8.14 illustrates the results of the assessment performed by the navigator

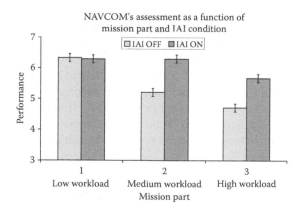

FIGURE 8.14 The navigator communicator's (NAVCOM's) determination of crew performance as a function of workload and IAI condition using a seven-point assessment scale. (Adapted from Hou, M. and Kobierski, R. D., Intelligent adaptive interfaces: Summary report on design, development, and evaluation of intelligent adaptive interfaces for the control of multiple UAVs from an airborne platform (Report No. TR 2006-292), Defence Research and Development Canada, Toronto, Canada, p. 29, 2006.)

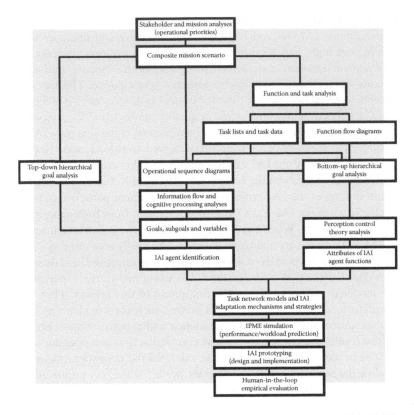

FIGURE 8.15   Processes and methods used in the design of an IAS for multiple UAV control.

communicator (NAVCOM) on the UAV crew performance over the three parts of the scenario. It depicts low (1), medium (2), and high (3) workload conditions.

Although the IAI implemented in this project only represents a small subset of a future suite of fully optimized UAV-functional agents, these findings supported the view that the control of dynamic and complex systems can be improved through the use of multiagent IASs. Perhaps a more important outcome of this research project is the experience and knowledge gained in the areas of IAI agent design and synthetic IAS prototype implementation. Many concepts were developed for designing effective IASs that show how the various design processes and methods are used to further the design. Figure 8.15 illustrates these design steps.

8.3   DESIGNING AN INTELLIGENT TUTORING SYSTEM

Case Study 2, the Intelligent Tutor for Questioning Technique (QuestionIT), was the result of an applied research project led by DRDC Toronto. The goal of this project was to enhance the distance learning capabilities of the CAF. Some of the most challenging activities for the distance education facilitators affected were responding to and facilitating student learning needs, customizing the learning experience to the student's learning style, and keeping the learner engaged. QuestionIT was conceived to support these activities within the context of instructing questioning technique within the IED Disposal (IEDD) operator course at the CAF School of Military Engineering. Prototype versions of QuestionIT were successfully tested during the summer and fall 2011 IEDD operator courses at the CAF base in Gagetown, New Brunswick, Canada.

8.3.1 THE IMPORTANCE OF QUESTIONING TECHNIQUE FOR IEDD IDENTIFICATION

Questioning technique is the cornerstone of IEDD operations; it enables the IEDD operator to determine and apply the appropriate render-safe procedures to the IED. Specifically, the IEDD operator gathers, assimilates, and analyzes important information in order to establish the device type, then plans and conducts render-safe procedures in accordance with IED disposal principles and best practices. The effective questioning of witnesses by the IEDD operator to discover key clues about the type of IED is both one of the most critical aspects of the IEDD operator's role and one of the most difficult skills to train.

Figure 8.16 illustrates, at a high level, the five-step process that is involved in the safe disposal of IEDs. The dashed line shows the aspects of the IEDD operator course that are supported by QuestionIT, including (a) arriving on scene and interviewing the on-scene commander (OSC); (b) interviewing witnesses; and (c) making an initial threat assessment (i.e., device identification). The complete five-step process for the safe disposal of IEDs involves the following steps:

1. *Arrive and interview OSC*: Before any type of questioning can occur, certain activities must be completed at the suspect IED location. Once the IEDD response team arrives at the incident site, the IEDD operator needs to coordinate with the OSC or civilian authorities to confirm the presence of medical support, fire support, and area security.
2. *Interview witnesses*: Once the IEDD operator has confirmed the appropriate cordon and evacuation procedures with the OSC, operators are free to interview witnesses. During the interview, witnesses provide clues that will help the operator to support (+) or refute (−) the likelihood of each device type (i.e., timer-based, command detonator-based or victim-initiated). In the example shown in Figure 8.16, the rightmost witness reports to the operator that he has not seen any evidence of a timer. This clue should indicate to the IEDD operator that the device is less likely to be timer-based.
3. *Make initial threat assessment*: Operators use clues acquired from interviewing witnesses to make an initial deduction about the device type prior to making a visual inspection. The suspect IED's device type will determine how the IEDD operator attempts to dispose of it.
4. *Inspect suspect device*: After making an initial threat assessment, the IEDD operator inspects the device based on the highest number of supportive clues for a particular device type. In the example shown in Figure 8.16, the IEDD operator has collected the most support for a victim-initiated device. If the visual inspection does not support this conclusion, the operator will

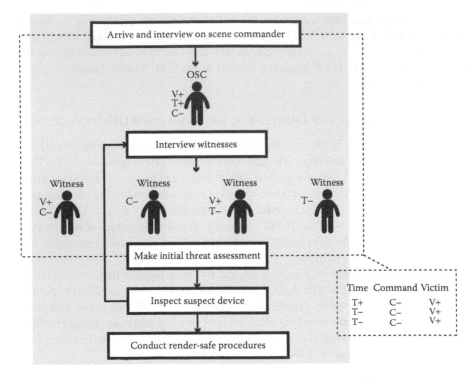

FIGURE 8.16 High-level overview of an IEDD process taught on an IEDD operator course. Principles captured by QuestionIT are represented by the dashed line.

return to interviewing witnesses to gather more evidence, as shown by the feedback loop.

5. *Conduct render-safe procedure*: Upon confirming the device type, the IEDD operator uses appropriate methods to destroy or deactivate the device.

8.3.2 IEDD Operator Course

QuestionIT was developed to support the witness-interviewing aspects of the IEDD operator course. The QuestionIT training environment allows students to practise their questioning technique in a safe, structured manner, at their own pace. Additionally, QuestionIT provides feedback on student threat-assessment skills, in terms of specific guidelines relating to clue interpretation and IED identification during the witness interviews.

QuestionIT simulates a domestic IED threat scenario from radical animal rights activists against a controversial university professor who conducts animal experiments. The scenario requires students to interview a number of witnesses to identify and reveal clues that support or refute timer-based, command detonator-based, or victim-initiated IED types. The questions posed are designed to determine the who, what, when, where, and why of the IED, and are based on specific lines of

FIGURE 8.17   An IEDD operator course student using QuestionIT. The student is receiving positive feedback from the tutor.

questioning taught in the IEDD operator course. Figure 8.17 illustrates students receiving helpful real-time feedback on their questioning technique throughout the interview process in the form of short tips highlighting instances of good and poor questioning. Students are assessed based on their ability to ask good questions and their ability to deduce the correct device type from the revealed clues.

Students begin QuestionIT by selecting a witness to interview. The student interviews the witness by selecting from a list of available questions developed by IEDD operator course instructors. Based on the type of question the student asks, the intelligent tutor provides immediate feedback, or instructional intervention, for both good and poor question choices. The instructional intervention includes the question and answer that triggered the intelligent tutor's response.

During the interview process, witnesses will occasionally provide important clues that relate directly to the classification of the IED. When a witness reveals a clue, it appears in a pop-up window and the student is asked to classify the clue as supporting either a timer-based, command detonator-based, or victim-initiated device by selecting the appropriate check box.

8.3.3   QuestionIT Design and Development Road Map

This section describes a working example of the development road map and taxonomy from the practical experience of developing QuestionIT. As with Case Study 1, the development, implementation, and evaluation of QuestionIT are discussed within the context of each step of the road map discussed in Chapter 7.

8.3.3.1   Determining Operational Priorities

The first step in determining operational priorities for QuestionIT was conducting a stakeholder analysis to investigate the requirements of stakeholders of the CAF IEDD training community (e.g., Canadian Defence Academy, CAF counter-IED task forces, CAF counter-IED training unit, and training instructors and students) for the development and evaluation of adaptive learning technologies in the course. Interviews with key course stakeholders focused on problematic aspects of the course and the practicalities of implementing intelligent tutoring technologies within an existing course.

From the stakeholder analysis, the IEDD operator course was identified as the ideal candidate for implementing and evaluating intelligent tutoring technologies. As discussed, the course teaches students how to identify, recognize, and formulate an accurate threat assessment of a suspected IED. However, the course failure rate was very high (40%). From interviews with senior course instructors, it was identified that the course failure rate was attributed to deficiencies in the threat-assessment and questioning skills of some students.

Although the course content comprised specific lists of device-related questions to use when interviewing witnesses, instructors remarked that the actual skill involved in good questioning technique could not be taught because there is no one right way to interview a witness. Furthermore, they reported that the few students who were immediately successful were naturals; the majority of students could only manage to learn the necessary skills after multiple practice scenarios. However, a significant minority of students who were not able to acquire effective questioning skills in time would, more than likely, end up failing the course.

Effective questioning is difficult because of the amount of information that must be compiled. Students are taught that information elements extracted from the witnesses and the environment count as evidence for or against each of the three device types. By the end of the interview, the device with the most evidence should indicate the highest probable threat. However, this assumes that the information acquired from witnesses is accurate. It also assumes that the accuracy of information is directly related to (a) asking the right questions; (b) reconciling the information across witnesses; and (c) making sense of it all. While an experienced operator is able to cope with this level of complexity, students can easily become overwhelmed by the amount of information available and tend to fixate on the immediate, surface-level details of physical device characteristics, while neglecting more subtle clues.

The overall goal of this project was to develop an intelligent tutoring system that addressed questioning technique, which would serve as a training environment for IEDD operator course students so that they could practise interviewing witnesses on their own in a safe environment, while receiving real-time feedback on their questioning performance from a virtual instructor. The feedback would be derived from common best practices from other domains, such as patient and clinical interviews, and tailored to fit the context of the IED-specific domain. In addition to improving the questioning technique of IEDD operator course students, the tool could also be used by existing operators to keep their questioning skills up-to-date between refresher courses (Hou et al., 2013).

8.3.3.2 Ethical and Legal Implications

During the key stakeholder meetings with the CAF IEDD training community, the scope of QuestionIT was discussed in terms of what type of scenarios, IED threats, and witnesses should be used. Feedback from the stakeholders confirmed that the project team should avoid any classified material, in terms of both the scenario type and the inclusion of material relating to render-safe procedures. In light of these constraints, the scope of QuestionIT was restricted to witness interviews and the initial threat assessment only. A domestic scenario—a bomb threat to a university professor by extreme animal rights activists—was chosen to avoid any issues relating to the use of classified material.

8.3.3.3   Selecting an Analytical Technique

The choice of analytical technique was dependent on a number of factors, including the subject matter to be analyzed, the analysis team's experiences, and the level of effort that could be deployed toward the analysis work. An additional factor was the SMEs (i.e., the course instructors), who were not in agreement on how to teach questioning technique, or even whether it could be taught. Over many years, each course instructor had developed a unique questioning style and, when asked, had great difficulty detailing how and why they had adopted a particular line of witness questioning for a specific situation.

The first step was to research how questioning is conducted in other related domains and to see whether any lessons learned about good and poor questioning technique could be integrated into IEDD witness interviews. A literature review of the use of questioning technique across the medical, professional, psychological, and police domains was conducted. These bodies of knowledge were surveyed and synthesized, and recommendations and best practices regarding how to integrate this knowledge into both the instructional material and the assessment of student questioning technique were developed. The results of the review on questioning technique demonstrated consistent themes across domains. Overlapping strategies that directly support the current teaching points of the IEDD operator course were included as good practices. Additionally, a questioning framework was derived from common themes found in the literature review of other domains that use interview-style questioning, and then tailored to fit the context of the IEDD-specific domain. The information amalgamated across these domains supported the development of instructional material within QuestionIT for topics such as (a) building witness rapport, (b) question type (e.g., open, closed, or leading), and (c) active listening strategy. In addition, a questioning performance evaluation methodology was developed for QuestionIT; course instructors were involved throughout this review process and provided critical feedback to the development team.

With a thorough understanding of questioning technique completed, an analysis to support the development of QuestionIT could be conducted, including (a) detailed training scenarios; (b) instructional content; and (c) the QuestionIT conceptual architecture. Figure 8.18 illustrates a decision tree for the development of QuestionIT. Given the cognitively complex nature of the instructional material (i.e., threat assessment and decision making), the detailed knowledge-capturing requirements to facilitate implementation of an intelligent tutoring system, and the availability of expert analysts within the project team, CTA-based techniques were used to populate various models (i.e., expert model and student model) within the conceptual architecture and develop instructional content.

CTA techniques were used to systematically decompose two training scenarios provided by the IEDD operator course instructors into their component parts; from scenario to IED elements, and then to witness knowledge, and so on. Two scenarios of similar complexity that contained different clues and device types were needed to

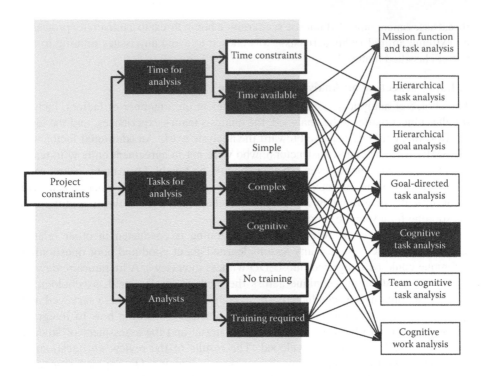

FIGURE 8.18 Decision tree highlighting the analytical techniques applicable to the development of QuestionIT. An unmarked decision tree can be found in Chapter 7.

support the experimental design of the evaluation trial. Both scenarios were related to a domestic threat from animal rights activists; in the first scenario the device was command detonator-based and in the second scenario the device was victim-initiated.

The following information was recorded in the TaskArchitect™ task analysis software tool:

1. *Scenario and witness specification*: This part of the analysis provided a description of the two QuestionIT scenarios, in which the student was required to interview a number of virtual witnesses to make a threat assessment of two suspect devices. The descriptions comprise a back story leading up to the event for each of the witnesses.
2. *Scenario IED clues specification*: This part of the analysis specified clues that must be revealed in order to correctly deduce the types of IEDs in both scenarios. Given that the scenarios detail the actions and motivations of the bomber, these clues provided the ground truth of the scenarios. This specification included the information gained from interviewing the witnesses.
3. *Witness question and answer specification*: This part of the analysis specified the questions and related answers for all the witnesses the student had to interview based on the scenario back story and the clues specified for each device.
4. *Student questioning technique evaluation criteria specification*: This part of the analysis specified how a student's questioning technique should be

assessed based on a review of questioning technique and discussions with course instructors.

5. *Instructional intervention specification*: This part of the analysis specified the instructional intervention by the intelligent tutor based on proper or improper questioning technique and threat assessment.

The analysis identified all of the possible question and answer combinations for each witness within a hierarchical tree structure. The overarching question structure was derived from CAF training material and interviews with course instructors. At this stage, each witness had a unique set of questions that were linked to specific answers. Those answers were then associated with details of the scenario scene content, such as image files and on-screen locations, that were made available to the student as they asked questions and acquired situation knowledge. At startup QuestionIT read the XML files pulled from Task Architect to control the witness dialogue and respective SA elements, or clues, associated with the answers. For each witness question and answer, information was recorded about (a) the type of device; (b) the type of question asked, which was required by the intelligent tutor to provide specific feedback; and (c) the clue revealed, which was required by the student to classify the IED. Table 8.2 shows an example of the analysis.

8.3.3.4 Determining Range of Adaptation

With the scenario in place, and the top-level priorities and constraints from the project stakeholders understood, the interaction of QuestionIT's intelligent tutor with the student was then defined and implemented within a conceptual architecture.

QuestionIT Taxonomic Analysis

As discussed in Chapter 7, the implementation of IASs can be supported by a taxonomic analysis that determines the scope of available options for the capabilities and functions of the system. This analysis also provides a road map for development, as it allows the development team to focus on specific implementations after determining all the possibilities. Figure 8.19 illustrates the four factors that were defined to create the QuestionIT taxonomy:

- *Learning objectives*: The learning objectives for each learning point in the scenario were defined and then decomposed into the knowledge, skills, and aptitudes required to be judged as competent in that learning point.
- *Instructional intervention*: The instructional intervention required to teach or evaluate student competence for each learning point was defined and then decomposed into adaptive learning, such as hinting or influencing, and intelligent tutoring, such as explicit coaching technologies. The choice of instructional intervention was determined in consultation with the course instructors.
- *Technological requirements*: The technological requirements needed to implement the instructional interventions identified for each scenario's learning points were defined, including the possible mechanisms for

TABLE 8.2

Example of the Results of the Analysis Describing the Questions and Answers Related to Each Witness Concerning Their Knowledge of the Improvised Explosive Device (IED) Threat

No.	Task	Answer	Device ID	Component Word Property	Question Type	Classification Phrase/Scenario Device Revealed
1	Questioning technique On-scene commander					
1.1	Where				Where	
1.1.1	Where is the suspect IED?	There are 2 devices we know about. One was found outside a faculty member's office on the 7th floor by a student. Another was found near the door of the parking garage on the basement level of this building by the lab tech.	Device 1 and 2	Neutral	Where	Location arrows
1.1.1.1	Could you draw me a map?	Sure, here it is.	Device 1 and 2	Visual aids	Where	Map photo
1.1.2	Is there access?	Which device are you talking about?	Device 1 and 2	Vague	Where	
1.1.2.1	Is there a safe route to the device on the 7th floor?	Yes, for now we're using only the stairs to get access; we've shut down the elevator.	Device 1 (backpack)	Neutral	Where	
1.1.2.1.1	Has anyone approached the device?	No, not since the cordon was established. We have evacuated everyone out of the building.	Device 1 (backpack)	Funneling	Where	
1.1.2.2	Has anyone been in the vicinity of the device?	Yes, several people. Faculty and staff arriving for work in the morning would have walked past the device.	Device 1 (backpack)	Neutral	Where	Device approached without incident
1.1.2.2.1	Where did they go exactly?	I don't know.	Device 1 (backpack)	Precision	Where	
1.1.2.3	Is there line of sight to the device on the 7th?	None. There are no windows in the corridor.	Device 1 (backpack)	Neutral	Where	No line of sight 7th

1.1.2.4	Is there access to the device in the garage?	Yes, through one door in the basement, or by street down the ramp. You must take the elevator or stairs down to the basement and the door is behind you.	Device 2 (garage)	Precision	Where	Choke point in garage
1.1.2.4.1	Has anyone approached the device?	Not since the cordon was established.	Device 2 (garage)	Neutral	Where	
1.1.2.4.1.1	Tell me more	About what?	Device 2 (garage)	Vague	Where	
1.1.2.4.1.2	Before the cordon, where did they go exactly?	They park their cars in a parking space, and walk to the entry door. At the door they need to swipe their ID cards to gain access to the building.	Device 2 (garage)	Precision	Where	Choke point; target selection
1.1.2.4.1.3	Are the parking spaces allocated individually to staff members?	No idea.	Device 2 (garage)	Neutral	Where	
1.1.2.4.1.4	Can I have access to security camera footage for the 7th floor for the last 12 h?	Sure, here you go.	Device 2 (garage)	Neutral	Where	Video clip; student setting up garage device
1.1.2.5	Is there line of sight to the device in the garage?	Pretty much anywhere from the parking garage.	Device 2 (garage)	Neutral	Where	Line of sight in garage
1.1.3	Are there any secondary hazards I should know about?	Yes, a gas main in the basement level of the building, but it has now been turned off.	N/A	Neutral	Where	
1.1.4	Have the emergency services been notified?	Yes, ambulance and fire trucks are on site.	N/A	Neutral	Where	

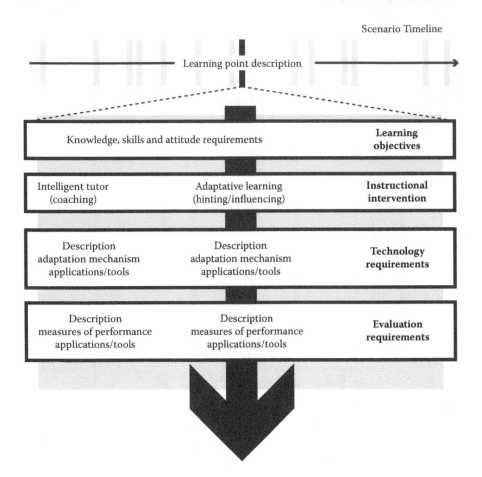

FIGURE 8.19 Taxonomic analysis framework for QuestionIT. Using the taxonomic approach facilitated the explicit identification of the instructional design for each learning point within the training.

adaptation, such as eye tracking or psychophysiological responses. The technological requirements were decomposed into adaptive learning and intelligent tutoring technologies.
- *Evaluation requirements*: The requirements for evaluating the utility of the adaptive learning and intelligent tutoring technologies for each learning point were defined, including the measurements of performance and the tools used. The evaluation requirements were decomposed into adaptive learning and intelligent tutoring technologies.

In defining these factors, the taxonomic approach allowed the instructional design to be explicitly identified for each learning point within the training through a specific adaptive learning or intelligent tutoring technology. The approach allowed both domestic IED scenarios to be developed and tailored to encompass the range of system functionality and capability identified by the taxonomy. The mission scenarios were used in both the subsequent analysis activities, as a precursor to the functional decomposition of tasks, goals, and functions, and subsequent verification activities, to help determine measures of effectiveness and performance. The taxonomic approach also allowed the systems designers to quickly understand the capability and function scope of QuestionIT in terms of priority and feasibility. This maximized the impact of QuestionIT on the IEDD operator course, while reducing the development risk within the time and budgetary constraints of the project.

QuestionIT Conceptual Architecture

Figure 8.20 illustrates a preliminary conceptual architecture for QuestionIT that was developed based on the taxonomic analysis. The preliminary conceptual architecture comprised a number of components to enable the real-time tailoring of instructional content based on the student's witness-interviewing performance during the scenario:

- *Training delivery module*: The purpose of the training delivery module is to present training content to the student. This is the component of the system with which the student interacts. The training delivery module should be capable of presenting, at minimum, text-based and multimedia content. The presentation of training content is controlled by the adaptation module. The training delivery module should report outcomes and student interaction to the evaluation module as required.
- *Evaluation module*: The purpose of the evaluation module is to update the student model based on student measurements. The evaluation module

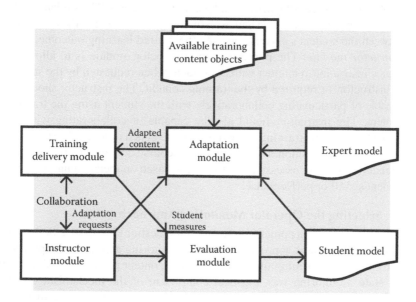

FIGURE 8.20 Preliminary conceptual architecture for QuestionIT. Ideally, the evaluation module should be capable of obtaining psychophysiological and interaction-based measurements directly from the student.

should be capable of obtaining direct measurements (e.g., eye tracking, EEG, HRV, skin conductance, etc.) from the student using psychophysiological technologies, as well as indirect interaction-based monitoring of student performance reported by the training delivery module, such as key presses, timings, and results.

- *Adaptation module*: The purpose of the adaptation module is to adapt the delivery of training based on the current state of the student model. The adaptation module should examine the student model to adapt the instructional content to the learning style of the student. The adaptation module should compare the difference between the student's grasp of the subject matter and the required understanding of the subject matter represented by the expert model. The results of this comparison are then used to adapt the delivery of instructional material to suit the learning needs of the student.
- *Expert model*: The purpose of the expert model is to represent the knowledge, skills, and behaviors that embody the desired end state, or required proficiency, of the student. The expert model helps the evaluation module and adaptation module evaluate the difference between the student's current state and the desired learning outcome. The expert model should not change during training delivery.
- *Student model*: The purpose of the student model is to represent the current knowledge, skills, and behaviors that embody the student, including their learning style. The student model should be updated during the delivery of training based on measurements of the student taken by the evaluation module. The student model may also include historical information on the student, such as learning history and previous test scores. The student model helps the evaluation module and assessment module evaluate the difference between the student's current state and the desired learning outcome.
- *Instructor module*: The purpose of the instructor module is to allow the course instructor to interact with the student when requested by the student or instructor, or required by the training content. The instructor should be capable of participating collaboratively with the student using the training content. The instructor should also be capable of collaborating with the student outside the training content in order to coach or assist the student in an unstructured fashion. Depending on the course content, instructors may be required to enter assessment information based on their evaluation of the student's skill or performance.

8.3.3.5   Selecting the Operator Monitoring Approach

Once the preliminary conceptual architecture for QuestionIT was specified, the next step was to select the most appropriate student monitoring approach (or approaches) to trigger the type and timing of the instructional content presented to the student. Operator state monitoring was a critical component of the mechanism by which QuestionIT would adapt the content and presentation of the learning material to optimize the student's learning experience. Figure 8.21 illustrates the decision tree used to identify the operator (i.e., student) state monitoring requirements and the operator state monitoring approach needed to meet those requirements.

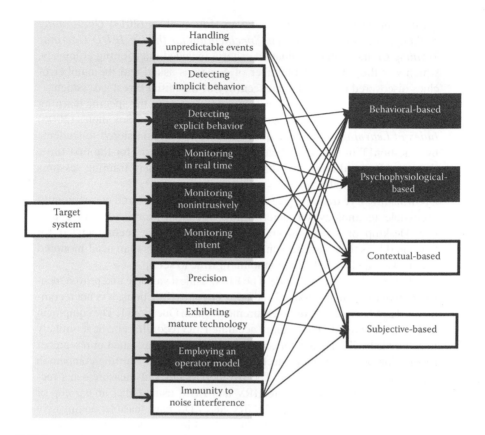

FIGURE 8.21 Decision tree highlighting the operator state monitoring approaches applicable to QuestionIT. An unmarked decision tree can be found in Chapter 7.

From the decision tree, a combination-based (behavioral-based and psychophysiological-based) approach to operator state monitoring was identified. A more in-depth review of state-of-the-art behavioral- and psychophysiological-based and behavioral-based approaches for monitoring student behavior and state for QuestionIT was then conducted (Hou et al., 2010a). Current techniques and technologies relating to eye tracking, psychophysiological indices of workload and stress, learning styles, and performance tracking that had been successfully used, or could theoretically be used, to augment an intelligent tutor were reviewed. The purpose of the review was to provide straightforward, applicable recommendations that would lead to the purchase of technologies (and questionnaires, in the case of learning styles) that could be implemented into QuestionIT. The following adaptation triggers were identified:

- *Performance tracking*: A number of evaluation criteria of the student's questioning technique and threat-assessment performance for each question were developed based on good questioning technique (e.g., precise, good rapport, or open ended) and poor questioning technique (e.g., jargon, leading, vague, closed ended, or compound). A comprehensive review of questioning technique can be found in Hou et al.' s (2013) *Questioning Technique Review and Scenario Specification for the CF IEDD Operator Training Course*. Other evaluation criteria included questioning efficiency, which was the ratio of the number of questions asked and the number of clues discovered as a result, and accuracy of the initial threat assessment.

- *Learning styles*: Adapting the instructional content to the specific learning style of each student was implemented using Felder and Solomon's (2006) *Index of Learning Styles*. The questionnaire was administered to students by QuestionIT immediately before starting the scenario for the first time. QuestionIT would reuse this information for subsequent training sessions by updating the student model.
- *Eye tracking*: Eye tracking proved to be the most mature and commercially accessible technology for determining attention, workload, and expertise. Desktop or screen-integrated eye trackers were recommended for QuestionIT, as they were determined to be less invasive than head-mounted eye trackers and took the least amount of time to set up.
- *Psychophysiological monitoring*: The EEG, while a valuable and proven measuring tool for mental workload and other cognitive functions, was not recommended as a form of adaptive measurement within QuestionIT. The equipment and setup time required to acquire accurate EEG recordings made it unlikely to be usable by IEDD operator course students without assistant or researcher supervision, and QuestionIT needed to be a standalone learning companion readily accessible by students. The psychophysiological measurements recommended for QuestionIT were HRV and EDR, using either an ear clip or a wireless Bluetooth sensor. Both HRV and EDR have feasible, noninvasive techniques for assessment, and could be used in combination to help determine whether a student reacted to specific critical clues from witness answers.

As discussed in Chapter 7, it is also important to consider what is possible from an implementation perspective for projects such as QuestionIT. Schedule constraints, budgetary constraints, and system environmental constraints had a significant impact on what IAS technologies could be implemented in this case. In particular, there were a number of constraints relating to the project schedule and the students and facilities at the IEDD School. Although four adaptation mechanisms were initially identified, after further investigation of both the maturity level of the technology and the IEDD operator course constraints, only student performance tracking was integrated as an adaptation mechanism for the current version of QuestionIT (Hou et al., 2010a,b). In the final QuestionIT implementation, performance tracking provided both real-time adaptive feedback of questioning and threat assessment, and customized training modules on completion of the QuestionIT scenario. The rationale for the exclusion of the three other possible adaptive mechanisms was summarized as follows:

- *Learning styles*: Learning styles were not implemented because a baseline study of IEDD student learning styles demonstrated that this population was a very homogeneous sample (i.e., predominantly visually based learners). No adaptation was required as all students needed visually based instructional content.

- *Eye tracking*: Eye tracking was not implemented as it was considered operationally unrealistic to allow students to visually inspect the device while interviewing witnesses. In real life, witness interviews and device inspection would be conducted at two different times, given the practical constraints of visually inspecting the IED using either a robotic device (i.e., indirectly), or wearing a protective blast suit (i.e., directly).
- *Psychophysiological tracking*: Psychophysiological monitoring was not implemented as a combination of technical difficulties and logistical challenges was associated with implementing these technologies, unsupervised, in the CAF IEDD operator course classroom.

8.3.4 QuestionIT Implementation

As discussed earlier, it is not the intent of this book to give advice pertaining to the actual design and implementation of IASs. Rather, the intent is to guide the early IAS design and development stages from an interaction-centered perspective. As a result, only a general overview of the implementation of QuestionIT is provided. For more detailed information regarding the implementation, Lepard et al.' s (2011a,b) DRDC technical reports *Intelligent Tutoring System: Architecture and Phase 1 Prototype* and *Phase II Prototype for CF IEDD Intelligent Tutoring System* are excellent resources.

QuestionIT was implemented over three build spirals in C#, WPF, XML, C++, and Prolog programming languages. Figure 8.22 illustrates the main software components of QuestionIT, including a training delivery module that contains an OMI, an evaluation module, and an adaptation module. The evaluation module compares

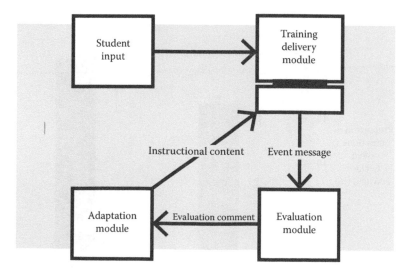

FIGURE 8.22 QuestionIT's evaluation module evaluates student performance, and then its adaptation module compares current student performance with desired student performance. The adaptation module then selects appropriate instructional content to display through the training delivery module.

student performance based on current and past question selection and updates the student model with the student's current learning state. Then, the adaptation engine compares current student performance (i.e., student model) with desired student performance (i.e., expert model), and selects appropriate instructional content to display on screen through the training delivery module as tutor feedback. The instructor module was not implemented as it was decided by IEDD operator course instructors that QuestionIT should be a completely independent learning experience for the student.

8.3.5 QuestionIT Evaluation

Two versions of QuestionIT were developed to evaluate the effectiveness of intelligent tutoring technologies for teaching questioning technique to IEDD operator course students: (a) a "tutor absent" version that did not provide adaptive feedback and instruction based on student performance; and (b) a "tutor present" version that did provide adaptive feedback and instruction based on student performance. The educational impact of presenting adaptive feedback and instruction to course students was assessed by directly comparing these two versions of QuestionIT in terms of their impact on questioning performance. Questioning performance was measured using a combination of objective measurements embedded in QuestionIT, including questioning efficiency, realism, usefulness, and ease of use.

Figure 8.23 illustrates the main finding of this study—that using the "tutor present" version of QuestionIT led to improved student questioning efficiency (i.e., proportion of questions asked to clues revealed). QuestionIT improved efficiency by teaching students how to be more selective in their questions and to avoid asking inappropriately worded questions, such as those that included technical terms a

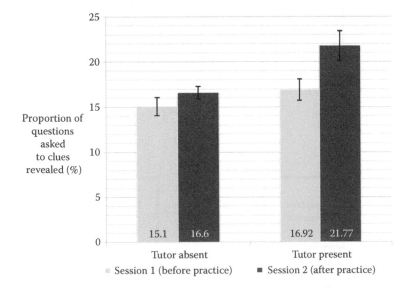

FIGURE 8.23 Student questioning efficiency results for the "tutor absent" and "tutor present" versions of QuestionIT. Efficiency improved in students who used QuestionIT.

witness would not understand. QuestionIT was found to be realistic, useful, and easy to use by students and course instructors. Most gratifying was that the failure rate of the course dropped to 6% in the year that QuestionIT was implemented.

Feedback from IEDD operator course instructors suggested that QuestionIT was easy to implement and supported their instructional methods. It also provided an effective method for course instructors to demonstrate good and poor questioning technique to students. QuestionIT also provided students with the opportunity to practise and develop questioning skills in their own time.

QuestionIT will continue to be used to train IEDD operators within the CAF. In 2012, course instructors requested the development of 12 additional unique scenarios to support future course requirements and provide refresher training for qualified IEDD operators needing recertification. QuestionIT has also been filed for patent application in both Canada and the United States. The exploitation of a QuestionIT-related intelligent tutoring system is currently being considered for other CAF courses that involve questioning, such as tactical questioning, counseling, disciplinary hearings, and medical staff training.

8.4   SUMMARY

This chapter presented two case studies detailing the IAS development process: (a) an IAI designed to support CAF personnel controlling multiple UAVs and (b) an intelligent tutoring system designed to support CAF personnel in acquiring effective questioning skills for IED identification and disposal. A number of practical constraints that shaped and influenced the individual IAS development processes were shared.

Both case studies provided worked examples of how the road map described in Chapter 7 can be used to support IAS development. Decision-making processes were discussed for each case study to illustrate how the IAS conceptual architecture, various analytical techniques, interaction-centered design methods, operator state monitoring approaches, and intelligent adaptation mechanisms were implemented. The decision trees illustrated in Chapter 7 were reused to demonstrate the development process and chosen operator state monitoring approaches for each project.

To investigate the efficacy of IAIs in a multi-UAV scenario, the UAV IAI project was divided into three phases. The first phase consisted of concept development and performance modeling, and a stakeholder analysis was conducted to understand the operational concepts and constraints of developing a composite scenario. MFTA was applied to the developed scenario, resulting in multiple function flow diagrams and OSDs. These diagrams helped systems designers understand the tasks, functions, and communications among operators conducting the multi-UAV missions. A PCT-based HGA was then conducted to generate an inventory of goals and tasks for the systems designers to work with SMEs and to identify appropriate IAI agent candidates and associated variables. The determined parameters were implemented in an IPME task network model, and operator performance was evaluated to compare interface conditions with and without IAI augmentation.

The second phase of the project involved designing and implementing a synthetic environment containing three operator workstations in order to validate the initial simulation results. Six agent groups were created, and agent adaptation mechanisms and performance measurements were built into the synthetic environment. A study conducted in the third phase provided empirical results confirming the simulation findings: the IAI improved operator SA and reduced workload while controlling multiple UAVs.

The second case study described the development of QuestionIT to improve the witness-questioning skills of IEDD operators within the CAF. A series of interviews with stakeholders of the CAF IEDD training community was conducted to determine the requirements for adaptive learning technologies within the IEDD operator training course. Once the requirements and constraints of the CAF IEDD training community were understood, the CTA technique was used to systematically decompose two training scenarios into their component parts; from scenario to IED elements, and then to witness knowledge. The next step was a taxonomic analysis of each mission scenario to identify the range of adaptation and triggering methods through which QuestionIT would be able to adapt the instructional content to the needs of the student. Once completed, the taxonomic analysis guided the development of QuestionIT's conceptual architecture and subsequent build spirals. Finally, two versions of QuestionIT were developed (i.e., tutor present and tutor absent) to evaluate the effectiveness of intelligent tutoring technologies for teaching questioning technique to IEDD operator course students. The results of the evaluation showed that QuestionIT improved questioning efficiency by teaching students how to be more selective in their questions and to avoid asking inappropriately worded questions, and was found to be realistic, useful, and easy to use by both students and course instructors.

Both case studies demonstrated that the application of IAS design processes discussed in previous chapters could be implemented in different problem domains and result in IAS designs that improve operational effectiveness. Notably, in Case Study 1 operators were able to better manage their workload when operating multiple UAVs, and in Case Study 2 all course students were able to develop questioning skills to the level of proficiency required to pass the course.

REFERENCES

Dahn, D. and Laughery, K. R. (1997). The integrated performance modeling environment—Simulating human-system performance. In *Proceedings of the 29th Conference on Winter Simulation*, pp. 1141–1145. December 7–10, Atlanta, GA: IEEE.

Felder, R. M. and Solomon, B. A. (2006). *Index of Learning Styles*. Raleigh, NC: North Carolina State University. Retrieved July 24, 2014 from http://www.engr.ncsu.edu/learningstyles/ilsweb.html.

Hou, M. and Kobierski, R. D. (2005). Performance modeling of agent-aided operator-interface interaction for the control of multiple UAVs. In *Proceedings of the 2005 IEEE International Conference on Systems, Man and Cybernetics*, vol. 3, pp. 2463–2468. October 10–12, Atlanta, GA: IEEE.

Hou, M. and Kobierski, R. D. (2006). Intelligent adaptive interfaces: Summary report on design, development, and evaluation of intelligent adaptive interfaces for the control of multiple UAVs from an airborne platform (Report No. TR 2006-292). Toronto, Canada: Defence Research and Development Canada.

Hou, M., Kobierski, R. D., and Brown, M. (2007). Intelligent adaptive interfaces for the control of multiple UAVs. *Journal of Cognitive Engineering and Decision Making*, *1*(3),
327–362.

Hou, M., Kramer, C., Banbury, S., Lepard, M., and Osgoode, K. (2013). Questioning technique
review and scenario specification for the CF IEDD operator training course (Report No.
TR 2013-061). Toronto, Canada: Defence Research and Development Canada.

Hou, M., Sobieraj, S., Kramer, C., Tryan, J. L., Banbury, S., and Osgoode, K. (2010a). Suitable
adaptation mechanisms for intelligent tutoring technologies (Report No. TR 2010-074).
Toronto, Canada: Defence Research and Development Canada.

Hou, M., Sobieraj, S., Pronovost, S., Roberts, S., and Banbury, S. (2010b). Suitable learning
styles for intelligent tutoring technologies (Report No. TR 2010-073). Toronto, Canada:
Defence Research and Development Canada.

Lepard, M., Kramer, C., and Banbury, S. (2011a). Intelligent tutoring system: architecture and
phase 1 prototype (Report No. CR 2012-026). Toronto, Canada: Defence Research and
Development Canada.

Lepard, M., Kramer, C., and Banbury, S. (2011b). Phase II prototype for CF IEDD intelligent
tutoring system (Report No. CR 2012-027). Toronto, Canada: Defence Research and
Development Canada.

Subject Index